Ernst-Christian Koch

Metal-Fluorocarbon Based
Energetic Materials

Related Titles

Mahadevan, E. G.

Ammonium Nitrate Explosives for Civil Applications

Slurries, Emulsions and Ammonium Nitrate Fuel Oils

2012
ISBN: 978-3-527-33028-7

Lackner, M., Winter, F., Agarwal, A. K. (eds.)

Handbook of Combustion

5 Volume Set

2010
ISBN: 978-3-527-32449-1

Agrawal, J. P.

High Energy Materials

Propellants, Explosives and Pyrotechnics

2010
ISBN: 978-3-527-32610-5

Meyer, R., Köhler, J., Homburg, A.

Explosives

2007
ISBN: 978-3-527-31656-4

Kubota, N.

Propellants and Explosives

Thermochemical Aspects of Combustion

2007
ISBN: 978-3-527-31424-9

Agrawal, J. P., Hodgson, R.

Organic Chemistry of Explosives

2007
ISBN: 978-0-470-02967-1

Ernst-Christian Koch

Metal-Fluorocarbon Based
Energetic Materials

**WILEY-
VCH**

WILEY-VCH Verlag GmbH & Co. KGaA

The Author

Dr. Ernst-Christian Koch
NATO Munitions Safety
Information Analysis Center (MSIAC)
Boulevard Leopold III
1110 Brussels
Belgium

Cover
The cover picture depicts the combustion
flame of Magnesium/Teflon TM/HycarTM
strand (photographed by Andrzej Koleczko,
Fraunhofer ICT, Germany) superimposed
on the assumed main combustion step
between difluorocarbene and magnesium.

Library of Congress Card No.: applied for

**British Library Cataloguing-in-Publication
Data**
A catalogue record for this book is available
from the British Library.

**Bibliographic information published by the
Deutsche Nationalbibliothek**
The Deutsche Nationalbibliothek
lists this publication in the Deutsche
Nationalbibliografie; detailed bibliographic
data are available on the Internet at
<http://dnb.d-nb.de>.

© 2012 Wiley-VCH Verlag & Co. KGaA,
Boschstr. 12, 69469 Weinheim, Germany

Cover Design Formgeber, Eppelheim
Typesetting Laserwords Private Limited,
Chennai, India
Printing and Binding Markono Print Media Pte Ltd,
Singapore

Printed in Singapore
Printed on acid-free paper

Print ISBN: 978-3-527-32920-5
ePDF ISBN: 978-3-527-64420-9
oBook ISBN: 978-3-527-64418-6
ePub ISBN: 978-3-527-64419-3
Mobi ISBN: 978-3-527-64421-6

Dedicated to my family

Contents

Foreword

We have known Dr Ernst-Christian Koch since meeting him during one of the International Pyrotechnics Seminar back in the 1990s. Even back then, we knew that he was more than just a research scientist interested in pyrolants and energetics. Dr Koch demonstrated a passion for pyrolants beyond that of a hobbyist or an employee. His enthusiasm is clearly demonstrated by the impressive number of his patents and publications. Therefore, it comes as no surprise that Dr Koch has channelled his drive for the dissemination of knowledge about pyrolants, and more specifically magnesium-Teflon-Viton (MTV) compositions into a clearly written book. Often, we will receive an email from Dr Koch that directs us to information about a new patent or publication about MTV. His ability to take a small amount of information and extrapolate beyond it is just one facet of his talent as a scientist. It is a pleasure to finally read a book that encompasses some (obviously, not all) of his knowledge in the field of pyrolants.

This book is unique in that its scope is limited to data about the MTV reaction, application of the reactions related to MTV, and metal–halogen reactions that might be substituted for the MTV reaction. The book provides the reader a single source for research results and data on all compositions related to MTV and the application thereof. The breadth of references, figures and tables demonstrate the vast and careful research Dr Koch undertook.

This book fills a void in the collection of pyrotechnic literature because it deals exclusively with research related to MTV-like compositions. Chapter 9 includes pictures that enable the reader to actually envision the combustion reaction of the different metal/fluoride reactions. Chapter 10 and 10.5 Operational Effects chapters are limited, only because of the availability and security constraints beyond the author's control. Chapters delving into previously unconsidered regions and Chapters 11 and 13 are of notable interest in the context of cyberwar and intellectual property disputes. Chapters 18 and 19 are a great compilation of the past and current practices. The history of the incidents involved with MTV manufacturing and the way processing has evolved to help mitigate explosive incidents is presented in a straightforward manner. Chapter 15 exemplifies Dr Koch's ability to look ahead. His citations in this chapter are abundant for a very limited field of research. Once again, the author illustrates his ability to take new information/ideas and to compile them in a useful and informative way.

No other known book documents MTV-like compositions in this depth. This book can be considered to be a textbook of everything associated with the MTV composition and, because of the extraordinary amount of documentation of data about MTV-like compositions, it will make an excellent reference book that all researchers of pyrolants and energetics must have.

Dr Bernard Douda and Dr Sara Pliskin
Naval Surface Warfare Center,
Crane, Indiana, USA

Preface

Metal/Fluorocarbon pyrolants, similar to black powder, are very versatile energetic materials with a great many number of applications. Over the last 50 years metal/fluorocarbon-based energetic materials have developed from secret laboratory curiosities into well-acknowledged standard payload materials for high-performance ordnance such as countermeasure flares and both strategical and tactical missile igniters. However, the long-lasting obligation to maintain secrecy over many of these compositions in most countries affected their further development and impeded personnel involved in becoming acquainted with the particular safety and sensitivity characteristics of these materials.

When I first dealt with Magnesium/Teflon™/Viton™ (MTV) in the mid 1990s I became fascinated by these materials. However, trying to learn more about them was difficult because of the above mentioned classification issues. Thus my research aimed at exploring some of the fundamentals of MTV to have good basis to start further development on. Fortunately, in the meantime the Freedom of Information Act both in the United States and United Kingdom brought significant relief to this and has enabled access to formerly classified files. Still the information is not readily retrievable as the actual content of these files is not well documented. Thus, in order to establish a reference base for MTV, I have gathered documents from the public domain over the last 15 years. The present book now is the result of an attempt to present the most relevant information in a reasonable manner.

Even though carefully compiled, I preemptively apologize for any kind of technical errors and omissions which I am afraid cannot be completely avoided. However, I would be glad to receive your critical comments in order to improve future editions of this book.

I hope you enjoy reading the book as much as I enjoyed writing it.

Brussels and Kaiserslautern, October 2011   *Ernst-Christian Koch*

Acknowledgment

A book like this is never the achievement of a single individual. Thus I wish to express my sincere appreciation for the support I received during my research on metal fluorocarbon based energetic materials and while writing this monograph.

The late Dr. rer. nat. Peter Kalisch (†2010), retired technology and scientific director of the Diehl Stiftung/Nuremberg I gratefully acknowledge for his enduring, constructive and friendly support, and facilitating financial support for research on both combustion synthesis and obscurant properties of magnesium/graphite fluoride pyrolants.

I wish to thank the following individuals (in no certain order) for their advice both oral and written and/or their experimental support on this fascinating topic which has helped me in writing this book.

Dr. Sara Pliskin (NSWC, USA); Dr. Eckhard Lissel, Harald Franzen, Jürgen Wolf (former WIWEB/Heimerzheim); Mrs. Ilonka Ringlein, Alfred Aldenhoven, Heinz Hofmann, Johann Licha, Dr. Arno Hahma (Diehl BGT Defence); Dr. Axel Dochnahl, Daniel Krämer (former Piepenbrock Pyrotechnik/Germany); Dr. Alla Pivkina, Dr. Alexander Dolgoborodov (Semenov Institute/Moscow); Jim Callaway (DSTL/Fort Halstead); Dr. Trevor T. Griffiths (Qinetiq/Fort Halstead); Prof. Steven Son PhD., Aaron B. Mason, Dr. Cole Yarrington (Purdue University/USA); Prof. Michelle Pantoya PhD. (Texas Tech. University/USA); Dr. Harold D. Ladouceur (NRL/USA); Prof. Dr. Takuo Kuwahara (NST/Japan); Rutger Webb, Chris van Driel (TNO/Netherlands); Mrs. Evelin Roth, Mrs. Angelika Raab, Mrs. Heidrun Poth, Mrs. Sara Steinert, Andreas Lity, Sebastian Knapp, Uwe Schaller, Dr. Lukas Deimling, Dr. Manfred Bohn (Fraunhofer-ICT/Germany), Prof. Dr. Thomas Klapötke (LMU Munich/Germany), Dr. Henrik Radies (Rheinmetall/Germany); Patrice Bret (A3P/France); Dr. Jim Chan (Orica/Canada); Dr. Dana Dattelbaum (LANL/USA); Prof. Dr. Jean'ne M. Shreeve (University of Idaho/USA), Prof. Dr. Anton Meller (Göttingen/Germany), Prof. Dr. Richard Kaner (University of California/USA), Prof. Dr. Edward Dreizin (New Jersey Institute of Technology/USA)), Prof. Dr. Wolfgang Kaminsky (Hamburg/Germany), Dr. Dave Dillehay (Cobridge/USA); Dr. Michael Koch (WTD 91/Germany); Dr. Martin Raftenberg (ARL/USA); Bernard Kosowski (MACH I /USA); Dr. Stany Gallier (SNPE/France); Dr. Günther Diewald (former Oerlikon Contraves/Switzerland); Olivier Azzola (Ecole Polytechnique/France); and Patrice Bret (CEHD/France).

I thank Mrs. Leslie Belfit, project editor at Wiley-VCH/Weinheim, and Mrs. Rajalakshmi Gnanakumar Laserwords/India for the good cooperation and smooth support with a difficult manuscript.

Further I am very grateful to the following people for giving valuable advice and for taking the burden to review selected chapters of this book: Prof. Dr. Stanisław Cudziło (MUT-Warzaw/Poland); Beat Berger (Armasuisse/Switzerland); Dr. Nigel Davies (University of Cranfield/UK); Neal Brune (Armtech/USA); and Volker Weiser, Dr. Stefan Kelzenberg, Dr. Norbert Eisenreich (Fraunhofer-ICT/Germany).

Finally I wish to thank Dr. Bernard E. Douda for inviting me to become actively involved with the International Pyrotechnics Society. I thank him for his help and the friendly support he provided in response to the many questions I addressed to him over the years. I greatly appreciate his reviewing of parts of this book and his efforts to write a preface together with Dr. Sara Pliskin. It is his report on the "Genesis of IRCM" that has definitely triggered me to write this book.

Thank you Bernie!

Brussels and Kaiserslautern, October 2011 *Ernst-Christian Koch*

1
Introduction to Pyrolants

Energetic materials are characterised by their ability to undergo spontaneous ($\Delta G < 0$) and highly exothermic reactions ($\Delta H < 0$). In addition, the specific amount of energy released by an energetic material is always sufficient to facilitate excitation of electronic transitions, thus causing known luminous effects such as glow, spark and flame. Energetic materials are typically classified according to their effects. Thus, they can be classified into high explosives, propellants and pyrolants (Figure 1.1). Typical energetic materials and some of the salient properties are listed in Table 1.1.

When initiated, high explosives undergo a detonation. That is a supersonic shockwave supported by exothermic chemical reactions [1–3]. In contrast, propellants and pyrolants undergo subsonic reactions and mainly yield gaseous products as in the case of propellants [4, 5] or predominantly condensed reaction products as in the case of pyrolants. The term *pyrolant* was originally coined by Kuwahara to emphasise on the difference between these materials and propellants [6]. Thus, the term aims at defining those energetic materials that upon combustion yield both hot flames and large amount of condensed products. Hence, pyrolants often find use where radiative and conductive heat transfer is necessary. Pyrolants also prominently differ from other energetic materials in that they have both very high gravimetric and volumetric enthalpy of combustion and very often densities far beyond $2.0\,\mathrm{g\,cm^{-3}}$ (see Table 1.1 for examples).

Pyrolants are typically constituted from metallic or non-metallic fuels (e.g. Al, Mg, Ti, B, Si, $C_{(gr)}$ and S_8) and inorganic (e.g. Fe_2O_3, $NaNO_3$, $KClO_4$ and $BaCrO_4$) and/or organic (e.g. C_2Cl_6 and $(C_2F_4)_n$) oxidizers or alloying partners (e.g. Ni and Pd). In contrast to propellants, they are mainly fuel rich and their combustion is influenced by afterburn reactions with atmospheric oxygen or other ambient species such as nitrogen or water vapour.

Pyrolants serve a surprisingly broad spectrum of applications such as payloads for mine-clearing torches ($Al/Ba(NO_3)_2/PVC$) [7, 8], delays ($Ti/KClO_4/BaCrO_4$) [9], heating charges ($Fe/KClO_4$) [10, 11], igniters (B/KNO_3) [12, 13], illuminants ($Mg/NaNO_3$) [14, 15], thermites (Al/Fe_2O_3) [16, 17], obscurants ($RP/Zr/KNO_3$) (RP, red phosphorus) [18], ($Al/ZnO/C_2Cl_6$) [20], tracers ($MgH_2/SrO_2/PVC$) [21], initiators (Ni/Al) [22] and many more. Recently, pyrolant combustion is increasingly used for the synthesis of new materials.

Metal-Fluorocarbon Based Energetic Materials, First Edition. Ernst-Christian Koch.
© 2012 Wiley-VCH Verlag GmbH & Co. KGaA. Published 2012 by Wiley-VCH Verlag GmbH & Co. KGaA.

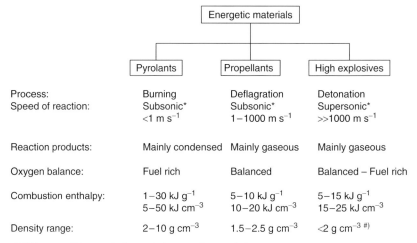

Figure 1.1 Classification of energetic materials.

Table 1.1 Performance parameters of selected energetic materials.

Class of energetic materials	Material, formula, weight ratio	ρ^a(g cm^{-3})	$\Delta_c H^c$ (kJ g^{-1})	$\Delta_c H^q$(kJ cm^{-3})	T_{ig}(°C)
High explosive	HMX, $C_4H_8N_8O_8$	1.906	9.459	18.028	287
	TNT, $C_7H_5N_3O_6$	1.654	14.979	24.775	300
	PETN, $C_5H_8N_4O_{12}$	1.778	8.136	14.465	148
	Nitroglycerine, $C_3H_5N_3O_9$	1.593	6.717	10.699	180
	Nitrocellulose[b], $C_6H_7N_3O_{11}$	1.660	9.118	15.135	200
Pyrolant	KNO$_3$/S$_8$/charcoal (75/10/15)	1.940	3.790	7.353	260–320
	Al/KClO$_4$(34/66)	2.579	9.780	25.223	446
	Fe/KClO$_4$ (20/80)	2.916	1.498	4.360	440–470
	Mg/PTFE/Viton (60/30/10)	1.889	22.560	42.616	540
	Zn/C$_2$Cl$_6$(45/55)	3.065	4.220	12.934	420
	Ta/THV-500[d] (74/26)	5.802	6.338	36.773	310

[a]At TMD = Theoretical Maximum Density.
[b]14.4 wt% N.
[c]With liquid H_2O.
[d]THV-500 is copolymer of tetrafluoroethylene (TFE), Hexafluoropropene (HFP) and Vinylidene difluoride (VF$_2$) ratio: 60/20/20, $C_{2.223}H_{0.624}F_{3.822}$, $\rho = 2.03$ g cm^{-3}. PTFE, polytetrafluoroethylene.

An important group of pyrolants are those constituted from metal powder and halocarbon compounds [19]. The high energy density of metal–halocarbon pyrolants stems from the high enthalpy of formation of the corresponding metal–halogen bond (M−X). Thus, chlorocarbon but mainly fluorocarbon compounds are used as oxidizers.

On the basis of metal fluorocarbon combinations, pyrolants show superior exothermicity compared to many of the aforementioned fluorine-free systems [22]. This advantage is due to the high enthalpy of formation of the metal–fluorine bond not outperformed by any other combination of the respective metal. Thus, the exothermic step

$$M^w + wF \longrightarrow MF_w$$

is the driving force behind the reaction (w = maximum valence).

Owing to a great number of metallic elemental fluorophiles ($\sim$70), metal fluorocarbon pyrolants (MFPs) offer a great variability in performance. In addition, many alloys and binary compositions of fluorophiles may also come into play to further tailor the performance of the pyrolant: Mg_4Al_3, MgH_2, MgB_2, Mg_3N_2, $Mg(N_3)_2$, Mg_2Si and so on [23]. Very often MFPs find use in volume-restricted applications where other materials would not satisfy the requirements – see, for example, payloads for infrared decoy flares (see Chapter 10). Within the scope of this book, the following applications are discussed:

- agent defeat payloads
- countermeasure flares
- cutting torches
- heating devices
- igniters
- incendiaries
- material synthesis
- obscurants
- propellants
- reactive fragments
- stored chemical energy propulsion systems
- tracers
- tracking flares
- underwater flares.

This book focuses only on specialised pyrotechnic applications; thus, for a more generalised introduction to pyrotechnics, the interested reader is referred to the books by Shidlovski [24], Ellern [25], McLain [26], Conkling [27, 28], Hardt [29] and Kosanke *et al.* [30].

References

1. Fickett, W. and Davis, W.C. (2000) *Detonation – Theory and Experiment*, Dover Publications Inc., Mineola, New York.

2. Zukas, J.A. and Walters, W.P. (1998) *Explosive Effects and Applications*, Springer Publishers, New York.

3. Cooper, P.W. (1996) *Explosives Engineering*, Wiley-VCH Verlag GmbH, New York.

4. Kubota, N. (2007) *Propellants and Explosives, Thermochemical Aspects of Combustion*, 2nd completely revised and extended edn, Wiley-VCH Verlag GmbH, Weinheim.

5. Assovskiy, I.G. (2005) *Physics of Combustion and Interior Ballistics*, Nauka, Moscow.

6. Kuwahara, T. and Ochiai, T. (1992) Burning rate of magnesium/TF pyrolants. *Kogyo Kayaku*, **53** (6), 301–306.

7. Kannberger, G. (2005) Test and Evaluation of Pyrotechnical Mine Neutralisation Means. ITEP Work Plan Project Nr. 6.2.4, Final Report, Bundeswehr Technical Center for Weapons and Ammunition (WTD 91), Germany.

8. N.N. (2005) *Operational Evaluation Test of Mine Neutralization Systems*, Institute for Defense Analyses, Alexandria, http://en.wikipedia.org/wiki/Political_divisions_of_the_United_States VA.

9. Wilson, M.A. and Hancox, R.J. (2001) Pyrotechnic delays and thermal sources. *J. Pyrotech.*, **13**, 9–30.

10. Callaway, J., Davies, N. and Stringer, M. (2001) Pyrotechnic heater compositions for use in thermal batteries. 28th International Pyrotechnics Seminar, Adelaide Australia, November 4–9, 2001, pp. 153–168.

11. Czajka, B. and Wachowski, L. (2005) Some thermochemical properties of high calorific mixture of Fe-KClO$_4$. *Cent. Eur. J. Energetic Mater.*, **2** (1), 55–68.

12. Klingenberg, G. (1984) Experimental study on the performance of pyrotechnic igniters. *Propellants Explos. Pyrotech.*, **9** (3), 91–107.

13. Weiser, V., Roth, E., Eisenreich, N., Berger, B. and Haas, B. (2006) Burning behaviour of different B/KNO$_3$ mixtures at pressures up to 4 MPa. 37th International Annual ICT Conference, Karlsruhe Germany, June 27–30, p. 125.

14. Beardell, A.J. and Anderson, D.A. (1972) Factors affecting the stoichiometry of the magnesium-sodium nitrate combustion reaction. 3rd International Pyrotechnics Seminar, Colorado Springs, CO, 21–25 August, pp. 445–459.

15. Singh, H., Somayajulu, M.R. and Rao, B. (1989) A study on combustion behaviour of magnesium – sodium nitrate binary mixtures. *Combust. Flame*, **76** (1), 57–61.

16. Fischer, S.H. and Grubelich, M.C. (1998) Theoretical energy release of thermites, intermetallics, and combustible metals. 24th International Pyrotechnics Seminar, Monterey CA, July 27–31, pp. 231–286.

17. Weiser, V., Roth, E., Raab, A., del Mar Juez-Lorenzo, M., Kelzenberg, S. and Eisenreich, N. (2010) Thermite type reactions of different metals with iron-oxide and the influence of pressure. *Propellants Explos. Pyrotech.*, **35** (3), 240–247.

18. Koch, E.-C. (2008) Special materials in pyrotechnics: V. Military applications of phosphorus and its compounds. *Propellants Explos. Pyrotech.*, **33** (3), 165–176.

19. Koch, E.-C. (2010) *Handbook of Combustion*, Wiley-VCH Verlag GmbH, pp. 355–402.

20. Ward, J.R. (1981) MgH$_2$ and Sr(NO$_3$)$_2$ pyrotechnic composition. US Patent 4, 302,259, USA.

21. Gash, A.E., Barbee, T. and Cervantes, O. (2006) Stab sensitivity of energetic nanolaminates. 33rd International Pyrotechnics Seminar, Fort Collins CO, July 16–21, pp. 59–70.

22. Cudzilo, S. and Trzcinski, W.A. (2001) Calorimetric studies of metal/polytetrafluoroethylene pyrolants. *Pol. J. Appl. Chem.*, **45**, 25–32.

23. Koch, E.-C., Weiser, V. and Roth, E. (2011) Combustion behaviour of binary pyrolants based on MgH$_2$, MgB$_2$, Mg$_3$N$_2$, Mg$_2$Si, and polytetrafluoroethylene. EUROPYRO 2011, Reims, France, May 16–19.

24. Shidlovski, A.A. (1965) Fundamentals of Pyrotechnics.

25. Ellern, H. (1968) *Military and Civilian Pyrotechnics*, Chemical Publishing Company, New York.

26. McLain, J.H. (1980) *Pyrotechnics from the Viewpoint of Solid State Chemistry*, The Franklin Institute Press, Philadelphia, PA.

27. Conkling, J. (1985) *Chemistry of Pyrotechnics – Basic Principles and Theory*, Marcel Dekker, Inc., Basel.

28. Conkling, J. and Mocella, C.J. (2011) *Chemistry of Pyrotechnics – Basic Principles and Theory*, CRC Press, Boca Raton, FL.

29. Hardt, A. (2001) *Pyrotechnics*, Pyrotechnica Publications, Post Falls, ID.

30. Kosanke, K., Kosanke, B., Sturman, B., Shimizu, B., Wilson, A.M., von Maltitz, I., Hancox, R.J., Kubota, N., Jennings-White, C., Chapman, D., Dillehay, D.R., Smith, T. and Podlesak, M. (2004) *Pyrotechnic Chemistry*, Pyrotechnic Reference Series, Journal of Pyrotechnics Inc., Whitewater, CO.

2
History

2.1
Organometallic Beginning

Although the unambiguous discovery of metal–fluorocarbon-based energetic materials did not occur until the mid-twentieth century, the way to these materials began nearly 100 years earlier. To begin with, in 1849, the British chemist Edward Frankland (1825–1899) (Figure 2.1), who was working with Robert Bunsen in Marburg/Germany, made attempts to isolate the ethyl radical, $\cdot C_2H_5$. Therefore, he treated iodoethane, C_2H_5I, with a surplus of zinc powder [2]. However, he did not get the radical but obtained a mixture of zinc(II) iodide and diethylzinc(0), $Zn(C_2H_5)_2$ (Eq. 2.1). This was the first ever reported reaction of an electropositive metal with a halocarbon compound:

$$2\,Zn + 2\,C_2H_5I \longrightarrow 2\,C_2H_5ZnI \longrightarrow Zn(C_2H_5)_2 + ZnI_2 \qquad (2.1)$$

In 1855, the Alsatian chemist Charles Adolphe Wurtz (1817–1884) observed the high reactivity of alkali metals with aliphatic halides (Eqs. 2.2a,b) and developed a C–C-coupling method that was later named after him [3]:

$$R - Br + 2\,Na \longrightarrow R - Na + NaBr \qquad (2.2a)$$

$$R - Na + R - Br \longrightarrow R - R + NaBr \qquad (2.2b)$$

Wilhelm Hallwachs (not to be taken for the German physicist Wilhelm Hallwachs (1859–1920)) and Adalbert Schafarik in 1859 reported that magnesium undergoes fierce reaction with iodoethane to yield magnesium(II) iodide and a product of the unknown constitution, which was probably diethylmagnesium(0) (Eq. 2.3):

$$2\,C_2H_5I + 2\,Mg \longrightarrow (C_2H_5)_2Mg + MgI_2 \qquad (2.3)$$

In the same paper, they also reported about the reaction of aluminium with iodoethane, yielding what they believed was "ethylaluminium Spontaneous combustion of this material in air yields brown-violet smoke, indicating the presence of iodine. Today, it is known that their reaction gave both diethylaluminium iodide and ethylaluminium diodide [4] (Eq. 2.4):

$$2\,Al + 3\,C_2H_5I \longrightarrow C_2H_5Al(I)_2 + (C_2H_5)_2Al - I \qquad (2.4)$$

Metal-Fluorocarbon Based Energetic Materials, First Edition. Ernst-Christian Koch.
© 2012 Wiley-VCH Verlag GmbH & Co. KGaA. Published 2012 by Wiley-VCH Verlag GmbH & Co. KGaA.

Figure 2.1 Edward Frankland [1].

Surprisingly, their report had no major impact on organomagnesium-chemistry and was not noticed in the scientific community until Barbier and Grignard's work (see below).

Although Antoine Lavoisier (1743–1794) had assumed the existence of a new element in hydrofluoric acid, it was not until 1886 when Henri Moissan (1852–1907) succeeded in synthesizing elemental fluorine. He electrolysed potassium fluoride dissolved in liquid hydrofluoric acid at $-55\,^{\circ}$C in a platinum apparatus [5], and he was awarded the Nobel Prize for the same in 1906. As Moissan used graphite electrodes, the nascent fluorine reacted with carbon and produced a blend of perfluorocarbon compounds mainly containing tetrafluoromethane, CF_4. Moissan noted that what he assumed to be pure CF_4 would fiercely react with sodium to give carbon and sodium fluoride [6].

In 1898, Philippe Barbier (1848–1922) modified Saytzeff's alcohol synthesis by replacing zinc with the more reactive magnesium [7] (Eq. (2.5)). His student Victor Grignard (1871–1935) further developed this substitution [8] and was awarded the Nobel Prize for the same in 1912:

$$R-X + Mg \xrightarrow[\Delta]{Et_2O} R-Mg-X \xrightarrow[-MgOHX]{H_2O} R-OH \qquad (2.5)$$

where $X = I$, Br (Cl).

After nearly 70 years, in 1972 Reuben Rieke showed that even kinetically inert alkyl fluorides reacts with activated magnesium – the latter is prepared by *in situ* reduction of $MgCl_2$ with potassium sand [9] (Eq. (2.6)):

$$MgCl_2 \xrightarrow[-2KCl]{K, THF} Mg_{(act.)} \xrightarrow[162\,°C,\ 1h,\ diglyme]{C_6H_5F} \{C_6H_5Mg - F\} \xrightarrow{CO_2} C_6H_5 - CO_2MgF$$

(2.6)

2.2
Explosive & Obscurant Properties

In 1907, Camille Matignon (1867–1934) who was exploring ways to make aluminium carbide, Al_4C_3 (which then was sought as a source for technical methane production), observed a very exothermic reaction between chlorocarbons and aluminium. Heating a mixture of hexachlorobenzene with aluminium in a test tube lead to ignition at 225 °C and subsequent vigorous combustion to give both aluminium carbide and clouds of aluminium chloride ($T_{subl.}$: 128 °C) [10]:

$$10\ Al + C_6Cl_6 \xrightarrow[225\,°C]{\Delta} 2\ Al_4C_3 + 2\ AlCl_3$$

(2.7)

This is the first documented record of a combustion reaction between a metal and halocarbon compound.

In 1913, the German chemist Hermann Staudinger (1881–1965) tried to synthesize illusive ethylenedione (O=C=C=O). Therefore, he aimed at abstracting the halogen atoms (X) from either oxalyl chloride and -bromide, $(COX)_2$ by a reaction with potassium or its sodium alloy (NaK). He observed that mixtures of the reactants after some initial delay at ambient temperature would become highly sensitive to impact and would fiercely detonate if struck. Thus, he speculated about the formation of an unknown highly reactive intermediate [11].

Ernest E. F. Berger (1876–1934) (Figure 2.2) was appointed as a staff chemist at the chemical laboratory of the French artillery to work on explosives in 1908 [12]. In 1916, he proposed to use mixtures from metals and chlorocarbons to generate smoke. Being aware of the underlying chemical concept, he investigated a great variety of chlorine sources including also inorganic chlorine sources such as chlorides of antimony, lead, and iron (Table 2.1). However, the most famous ones are based on tetrachloromethane and zinc, which are widely known as *Berger's mixture*.

The following is an example of mixture taken from [13].

Berger mixture

---

25 wt% Zinc
50 wt% Tetrachloromethane
20 wt% Zinc oxide
 5 wt% Diatomaceous earth

Figure 2.2 Photograph of Captain Ernest Edouard Frédéric Berger (1897). (Reproduced with kind permission by Bibliotheque de l'Ecole Polytechnique Palaiseau.)

Table 2.1 Claimed components of Berger's smoke mixtures [15, 16].

Chlorine sources		Fuels	Auxiliary components	
Name	Formula		Name	Formula
Tetrachloromethane	CCl_4	Al	Sodium chlorate	$NaClO_3$
Hexachloroethane	C_2Cl_6	$CaSi_2$	Sodium nitrate	$NaNO_3$
Tetrachloroethane	$C_2Cl_4H_2$	Ca	Naphthalene	$C_{10}H_8$
Trichloroethylene	C_2Cl_3H	Fe	Zinc oxide	ZnO
Pentachlorobenzene	C_6Cl_5H	Mg	–	–
Hexachlorobenzene	C_6Cl_6	Mn	–	–
Lead(II) chloride	$PbCl_2$	Na	–	–
Iron(III) chloride	$FeCl_3$	$P_{(red)}$	–	–
Antimony(III) chloride	$SbCl_3$	Zn	–	–

The French forces applied these obscurants in the First World War with both navy and army in the so-called smoke generators. Later in the war, the Berger mixture saw widespread application with all belligerent countries. Berger was honored with the "Grand Prix de la Marine" in 1918 for his contribution to French warfare [14]. After the war, Berger reported that he had been in contact with Victor Grignard by early 1916 on these mixtures. Grignard had also proposed to him to use hexachlorobenzene as an alternative source of chlorine. In the same report, Berger also refers to Matignon's work [14]. Thus, it is very likely that he was inspired to apply these highly exothermic reactions in pyrotechnic obscurant formulations [15].

In 1919, the US First World War veteran Richard Clyde Gowdy (1886–1946), a citizen of Cincinatti, invented a signal smoke mixture based on magnesium, hexachloroethane, and anthracene [17]. Although Berger had already proposed to use magnesium as a fuel, it was noted by him that these mixtures would burn almost too vigorously. Gowdy, a mechanical engineer, modified Berger's mixture in that he applied anthracene both to cool down the combustion temperature and to generate soot that would turn the generated smoke black. It can be assumed that Gowdy learned about the Berger mixture in his military deployment to Europe.

Further refinements and modifications of Berger's smoke mixture were successively undertaken by Metivier (1926) [18, 19] and Brandt (1937) [20] both in France.

In 1922, Staudinger reported about explosive reactions of alkali and alkaline earth metals with partially and perhalogenated solvents such as CH_2Cl_2 and CCl_4. He intuitively assumed the formation of a very instable species on contact that would be very sensitive to mechanical impact and thus trigger an explosive reaction. However, he was unable to identify the actual species [21] but proposed to exploit these fierce reactions in detonating charges for ammunition [22, 23]. Staudinger was a visionary and he even tried to apply the explosive reaction between sodium and tetrachloromethane to make diamond [24]. He thought that both the high temperature and pressure from the explosion would enable to force the formed carbon soot to undergo phase transition to diamond. Although he was not successful with his experiment, about 60 years later nanodiamonds were isolated from TNT/RDX detonation soot and proved Staudinger's basic idea that a detonation would furnish the necessary physical conditions for the formation of diamond [25]. Still 10 years after that (in 1998), a Chinese group reported about the successful nanodiamond synthesis based on the original Staudinger set up with sodium and tetrachloromethane in an autoclave [26].

2.3
Rise of Fluorocarbons

In the course of the electrolytic preparation of beryllium from molten KHF_2/BeF_2, in 1926, Paul Lebeau (1858–1959) and his collaborator Damien were able to purify and isolate tetrafluoromethane, CF_4, by fractional distillation with a liquid air-cooled condenser. They observed vigorous combustion reactions of CF_4 with both sodium and calcium [27].

In 1930, Otto Ruff (1871–1939) and his coworker Otto Bretschneider synthesized CF_4 by a reaction of F_2 with charcoal and reported about violent combustion reactions of tetrafluoromethane with both calcium and magnesium at elevated temperatures [28]. Ruff and Brettschneider were also the first to obtain the important monomer tetrafluoroethylene, C_2F_4 (TFE), from tetrafluoromethane by arc discharge between graphite electrodes (Scheme 2.1) in 1933. They described its fierce combustion reaction with sodium [29].

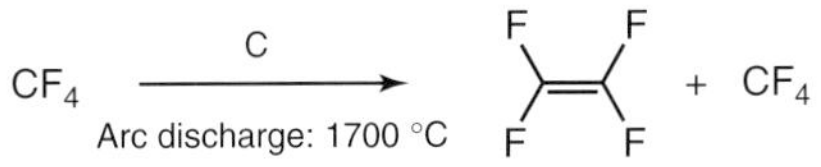

Scheme 2.1   TFE synthesis via arc discharge between carbon electrodes.

In 1934, Ruff and Brettschneider synthesized the first *all*-fluorinated polymer, graphite fluoride, $(CF_x)_n$, by fluorination of norite (Scheme 2.2) and investigated its thermal stability and decomposition mechanisms [30].

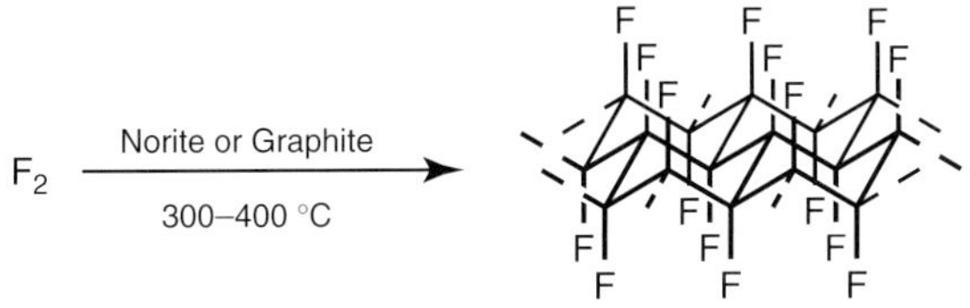

Scheme 2.2   Synthesis of graphite fluoride by fluorination of norite or graphite.

The first ever reported fluorinated flexible polymers based on fluorinated ethylenes were developed in 1934 by Fritz Schloffer (1901–1978) and Otto Scherer (1903–1987), both were from IG Farben/Frankfurt Germany. They described polymerization of chlorotrifluoroethylene to give polychlorotrifluoroethylene, $(CF_2ClF)_n$, according to Scheme 2.3 [31].

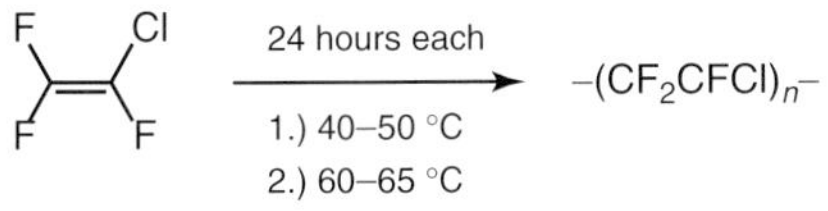

Scheme 2.3   Polymerization of chlorotrifluoroethylene (CTFE) according to Ref. [31].

They also described the synthesis of polychlorodifluoroethylene, $(CClHCF_2)_n$, polybromotrifluoroethylene $(CBrFCF_2)_n$, and mixed polyvinylchloride-*co*-polychlorotrifluoroethylene $(CClFCF_2)_m(CH_2CHCl)_n$, where all these occur under similar reaction conditions. Mass production of these started shortly after that. Later these materials were adopted in the United States by Kellog Company and marketed under the brand name Kel-F$^®$. Former Hoechst scientist Walter Wetzel (born 1925) in his paper on the discovery of polytetrafluoroethylene (PTFE) has revealed, from the company archives of Hoechst, that Schloffer and Scherer had investigated the polymerization of TFE as well but the resulting polymer, PTFE,

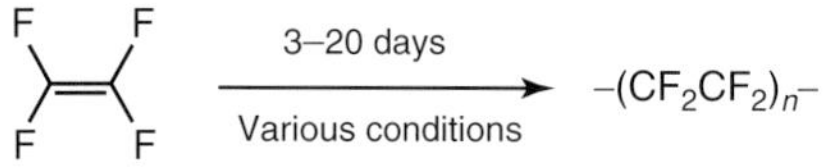

$$\text{(structure: } F_2C=CF_2 \text{)} \xrightarrow[\text{Various conditions}]{\text{3–20 days}} -(CF_2CF_2)_n-$$

Scheme 2.4 Polymerization of TFE according to Ref. [33].

owing to its chemical inertness and insolubility was not considered a useful material [32]. The application department at Hoechst rejected the material with the rather rhetoric question: "Was sollen wir mit diesem klitschigen Ding?" literally translated: "What do you expect us to do with this repellent material?" In view of the problems to access technical reasonable quantities of TFE to run into mass production – a technical problem that should remain for another 15 years – Hoechst unfortunately decided to disregard this material and did not include PTFE in the upcoming disclosure [31].

In 1939, Roy Plunkett (1911–1994) at Dupont discovered the polymerization of TFE to give PTFE the chemically and thermally most resistant fluoropolymer ever made (Scheme 2.4) [33]. However, similar to Schloffer's and Scherer's experience with the same material, Plunkett would not obtain any royalties from Dupont as the company would not work on PTFE for another four years for the same reasons [32]. It was only in 1943 when the Manhattan project created a demand for corrosion-resistant liners and gasket for reactors and valves to handle highly corrosive UF_6 [34]. Then, people at Dupont remembered the highly hydrophobic and chemically inert material. This is when PTFE came into play again and its small-scale production started. At the end of the 1940s, PTFE was produced on a small scale for the civilian market under the brand name Teflon®.

From a BIOS report, it is known that successive chlorination and fluorination of "Cerisin," a hydrocarbon wax frequently used in Germany, was applied before the war to obtain chemically highly resistive waxes with 57 wt% chlorine and 20 wt% fluorine for chemical engineering purposes [35].

In addition, in 1939, Hugo Stoltzenberg (1883–1974) invented a new obscurant based on magnesium and a mixture of liquid and solid halocarbon compounds [36]. His company was one of the few private firms beside big IG Farben to be involved with the complete span of chemical warfare development in Germany. He developed all kinds of warfare supplies for German Wehrmacht such as pyrotechnics (illuminants, obscurants, and incendiaries), chemical warfare agents, and the corresponding protective equipment. Thus, he had the most advanced materials such as the newly invented fluoropolymers at his disposal. Hence, it is likely that he had experimented with them as oxidizers as well. At the time of writing this book, any of the FIAT/CIOS/BIOS[1] interrogation files of Hugo Stoltzenberg were still not released by either UK or US government (*sic*).

---

1) Combined Intelligence Objectives Subcommittee (CIOS), British Intelligence Objectives Subcommittee (BIOS), Field Information Agency Technical (FIAT) are allied post-war reports on German Science and Industry.

2.4
Rockets Fired Against Aircraft

By the end of World War II, German air force and air force armament were desperate at fighting the bombing squads that raided the country day and night and feverishly they sought for technical solutions to fight enemy bombers and their accompanying fighters to possibly avert the final collapse or at least delay. Despite the general agony, innumerous discoveries and developments were made then [37]. The most noticeable achievements in this context include the development of the first ever unguided 50 mm air-to-air missile, the R4M, nicknamed "Orkan," which did not came into service until late March 1945 but produced an excellent track record. Another weapon system that did not see operational use at all was the radio guided antiaircraft missile "Enzian." For the Enzian missile, the first ever designed fuel-air warhead was in development that was supposed to break off the wings of enemy air planes by enhanced blast effects [38]. In addition, the Enzian missile was to be equipped with a self-steering seeker based on a noncooled PbS-detector, codenamed "Madrid," that was supposed to home on the hot exhaust pipes of the enemy air planes [39, 40] (Figure 2.3). Work published after the war reveals that the development of a homing head with a cooled PbS detector and an improved reticle was underway [41].

After Germany's surrender, its technology and scientific results were exploited by the allies leading to the development of the first operational infrared guided air-to-air missile, the Aerial Intercept Missile 9 (AIM-9), nicknamed Sidewinder that was ingeniously designed by Dr Burdette McLean at Naval Ordnance Test Station (NOTS) China Lake in 1948 [42] (Figure 2.4).

For a cutaway sketch of the missile and a seeker head, see Figure 2.5.

With the advent of infrared-guided missiles, training targets were needed that would exhibit sufficient radiant intensity in the PbS-infrared band (1.8–2.5 µm)

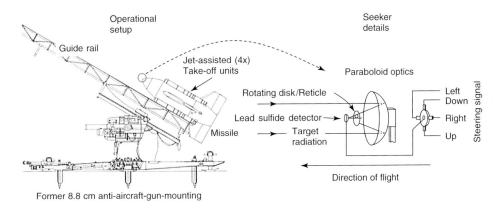

Figure 2.3 Technical sketch of Enzian E1 Flab missile with details of Madrid infrared seeker unit. (After Ref. [39].)

Figure 2.4 Photograph of Dr William Burdette McLean [43].

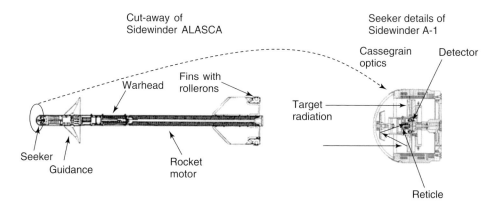

Figure 2.5 Cutaway sketch of Sidewinder missile with details of infrared seeker unit.

to facilitate homing. The first compositions employed for this purpose, however, had been designed for maximum visible signature and were classic metal/nitrate combinations such as aluminium/strontium nitrate/ammonium perchlorate [44] and magnesium/sodium nitrate such as US RITA flare [45] or British illuminating flare formulation SR 580 [46]. Thus, it is not surprising that these compositions did not meet the requirements.

A composition tested in the United Kingdom parallel to SR 580 was SR 107 [46]. The spectral efficiency of both compositions in two spectral bands corresponding to the PbS ($\lambda = 1.9-2.6\,\mu m$) and PbTe ($\lambda = 3.2-5.1\,\mu m$) spectral detector range is given below.

SR 107 [46]	SR 580 [46]
Composition	
35 wt% Magnesium	60 wt% Magnesium
65 wt% Ferric oxide	36 wt% Sodium nitrate
	4 wt% Acaroid resin
Spectral efficiency	
E_{PbS}: 56 J g^{-1} sr^{-1}	E_{PbS}: 21 J g^{-1} sr^{-1}
E_{PbTe}: 31 J g^{-1} sr^{-1}	E_{PbTe}: 10 J g^{-1} sr^{-1}
Spectral radiance	
L_{PbS}: 33 W sr^{-1} cm^{-2}	L_{PbS}: 27 W sr^{-1} cm^{-2}
L_{PbS}: 18 W sr^{-1} cm^{-2}	L_{PbS}: 18 W sr^{-1} cm^{-2}

2.5
Metal/Fluorocarbon Pyrolants

The first unambiguously documented use of fluorocarbon as oxidizers in pyrotechnics was 1956 in a patent that was not published until 1964. The chemist Edgar A. Cadwallader (1918–2006) disclosed the first pyrotechnic material to include a fluoropolymer, Kel-F, polychlorotrifluoroethylene, and a metal such as magnesium or aluminium for a visual flare composition (Table 2.2) [47].

Cadwallader had worked on organometallic reactions previously [48], obviously a prerequisite for many researchers involved with metal–halocarbon reactions. The first reported use of a metal/fluorocarbon material in infrared tracking flares then was made in Spring 1956. At NOTS, the type 702A target augmentation flare was designed, which used the below given composition based on Mg and PTFE [49]. The flare material had a heat of reaction of 9.2 kJ g^{-1} and a specific energy in the PbS band that would outperform both SR 107 and 580 by two or six times.

Table 2.2 Table with ingredient proportions and performance of a 0.5 in diameter flare candle. (Taken from Cadwallader's disclosure [47].)

Mg (%)	Kel-F (%)	Candle power
40	60	18
50	50	19
60	40	25
70	30	40

Name of flare	Composition	Watts/ steradian per Sq. In. Burning Surface (0.8 – 3.5 μm)	
		Ambient	65 000 Feet
1. BuOrd Mk 21 Mk O	54% Mg 34% NaNo$_3$ 12% Laminac	677	500
2. Applicants' flare	54% Mg 23% Teflon 23% Kel-F	2283	1070
3. Army "Rita" flare	66.7% Mg 28.5% NaNO$_3$ 4.8% Binder	1000	—
4. Optimum aluminum-Teflon	48% Al 52% Teflon	1700	—
5. Optimum boron-Teflon	56% B 44% Teflon	445	—
6. Optimum zirconium-Teflon	54% ZrH$_2$ 46% Teflon	428	—

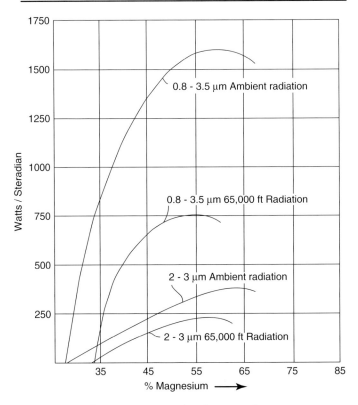

Figure 2.6 Ingredient proportions and performance of an infrared flare candle. (Taken from Hahn's disclosure [45].)

NOTS 702 [49]
54 wt% Magnesium
30 wt% Polytetrafluoroethylene
16 wt% Kel-F wax
Performance
$E_{1.8-2.7\,\mu m}$: 125 J g^{-1} sr^{-1}
Burn rate: 3.4 mm s^{-1}
$I_{1.8-2.7\,\mu m}$: 300 W sr^{-1}

A similar composition was finally filed by Hahn, Rivette, and Weldon in a patent application in 1958, which was not disclosed until nearly 39 years later in 1997 [45].

The spectral performance in both 2–3 μm band is given in Figure 2.6.

A report on magnesium/Teflon®/Kel-F-based infrared flare compositions and their performance was issued in 1959 and declassified from secret to unclassified in 1980 [50].

In 1962, Hugo Stoltzenberg and his coworker Martin Leuschner (1913–1982) filed a patent on obscurants based on titanium and fluorocarbons and/or chlorocarbons that was published in 1965 [51]. In 1963, the IIT Research Institute located in Illinois Chicago issued a report under Air-Force contract on thermochemical properties of a large number of metal/oxidizer systems. Their considerations also included both CF_4 and C_2F_6 as oxidizers [52].

In 1965, the Russian chemist Alexander Alexandrovich Shidlovskii (1911–1985) in his book on pyrotechnics refers to the possibility to use PTFE as an oxidizer in pyrotechnic mixtures [53]. Finally, Herbert Ellern (1902–1987) in his epochal monograph on pyrotechnics in 1968 for the first time in the open and accessible literature refers to the use of magnesium/PTFE mixtures in infrared decoy flares and gives details on caloric data and ignition sensitivity [54]. From then on, magnesium/PTFE has made its way into numerous applications as will be discussed in the following chapters.

References

1. *http://de.wikipedia.org/w/index.php?title= Datei:Edward_Frankland.jpg&filetimestamp =20050608204915* (accessed September 1 2011).

2. Frankland, E. (1848) Ueber die Isolirung der organischen Radicale. *Ann. Chem.,* **71**, 171–213.

3. Wurtz, C.A. (1855) Sur une nouvelle classe de radicaux organiques. *Ann., Chim., Phys.,* **44**, 275–312.

4. Hallwachs, W. and Schafarik, A. (1859) Ueber die Verbindungen der Erdmetalle mit organischen Radicalen. *Liebigs Ann. Chem.,* **109**, 206–109.

5. Moissan, H. (1886) Nouvelles experiences sur la decomposition de l'acide fluorhydrique par un courant electrique. *Comptes rendus hebdomadaires des séances de l'Académie des sciences,* **103**, 256–258.

6. Moissan, H. (1890) Sur la preparation et les proprietes du tetrafluorure de carbone. *Comptes rendus hebdomadaires des séances de l'Académie des sciences,* **110**, 951–954.

7. Barbier, P. (1899) Synthese du demethyl-heptenol. *Comptes rendus hebdomadaires des séances de l'Académie des sciences,* **128**, 110–111.

8. Grignard, V. (1900) Sur quelques nou-vells combinaisons organometalliques du magnesium et leur application a des syntheses d'alcools et d'hydrocarbures. *Comptes rendus hebdomadaires des séances de l'Académie des sciences*, **130**, 1322–1324.

9. Rieke, R.D. and Hudnall, P.M. (1972) Activated metals. I. Preparation of highly reactive magnesium metal. *J. Am. Chem. Soc.*, **94**, 7178–7179.

10. Matignon, C. (1907) Formation et prepa-ration du carbure d'aluminium. *Comptes rendus hebdomadaires des séances de l'Académie des sciences*, **145**, 676–679.

11. Staudinger, H. and Anthes, E. (1913) Oxalylchlorid. V.: Über Oxalylbro-mid und Versuche zur Darstellung von Di-kohlenoxid. *Chem. Ber.* **46**, 1426–1437.

12. Bret, P. (2006) Les Laboratoires Français et l'étude des Munitions et Matériels Allemands Pendant la Grande Guerre, Cahier du CEHD, No. 33, Les relations Franco-Allemande.

13. Hanslian, R. (1937) *Der Chemische Krieg, I. Militärischer Teil*, Verlag von E.S. Mittler & Sohn, Berlin, p. 630.

14. Berger, E. (1920) Production de chlorures par reactions amorcees. *Comptes rendus hebdomadaires des séances de l'Académie des sciences*, **171**, 29–32.

15. Berger, E. (1919) Noveau procede, D'obtention de fumees par combustion des melanges. FR Patent 501,836, filed Nov. 1, 1916, France.

16. Berger, E.-E.-F. (1920) Nouveau procede d'obtention de fumes par combustion de mélanges. FR Patent 501,836 1ere addition, France.

17. Gowdy, R.C. (1919) Smoke-making compound for signal rocket. US Patent 1,318,074, filed April 24, 1919, USA.

18. Metivier, M. (1926) Engins fumigenes a l'ethane hexachlore et aux derives chlores du naphthalene. FR Patent 613,884.

19. Metivier, M.-A.-E. (1928) Engins funi-genes a l'ethane hexachlore au propane octochlore et aux derives chlores du naphthalenes: engins ordinaries, engins a flotteur. FR Patent 649,853.

20. Brandt, E.W. (1937) Artifice fumigene. FR Patent 807,502, France.

21. Staudinger, H. (1922) Erfahrungen über einige Explosionen, *Angew. Chem.*, **35**, 657–659.

22. Staudinger, H. (1922) Verfahren zur Initialzündung von Sprengstoffen. DE 391,346, Deutschland.

23. Staudinger, H. (1922) Verfahren zur Herstellung von Sprengmitteln. DE Patent 396,209, Deutschland.

24. Staudinger, H. (1961) *Arbeitserinnerun-gen*, Hüthig Verlag, Heidelberg.

25. Greiner, R.N., Phillips, D.S., Johnson, J.D. and Volk, F. (1988) Diamonds in detonation soot. *Nature*, **333**, 440–442.

26. Li, Y., Qian, Y., Liao, H., Ding, Y., Yang, L., Xu, C., Li, F. and Zhou, G. (1998) A reduction-pyrolysis-catalysis synthesis of diamond. *Science*, **281**, 246–247.

27. Lebeau, P. and Damiens, A. (1926) Sur le tetrafluorure de carbone, *C. R. Acad. Sci.* **182**, 1340–1342.

28. Ruff, O. and Keim, R. (1930) Die Reaktionsprodukte der verschiede-nen Kohlearten mit Fluor. I. Das Kohlenstoff-4-fluorid (Tetrafluormethan). *Z. Anorg. Allg. Chem.*, **192**, 249–256.

29. Ruff, O. and Bretschneider, O. (1933) Die Bildung von Hexafluorethan und Tetrafluoräthylen aus Tetrafluorkohlen-stoff. *Z. Anorg. Allg. Chem.*, **210**, 173–183.

30. Ruff, O. and Bretschneider, O. (1934) Die Reaktionsprodukte der verschiede-nen Kohlearten mit Fluor. II (Das Kohlenstoff-monofluorid). *Z. Anorg. Allg. Chem.*, **217**, 1–18.

31. Schloffer, F. and Scherer, O. (1939) Verfahren zur Darstellung von Polymeri-sationsprodukten. DE Patent 677,071, Filed Oct. 7, 1934, Germany.

32. Wetzel, W. (2005) Entdeckungs-geschichte der Polyfluorethylene–Zufall oder Ergebnis gezielter Forschung? *NTM Z. Gesch. Wissen. Tech. Med.*, **13**, 79–91.

33. Plunkett, R. (1939) Tetrafluoroethylene polymers. US Patent 2,230,654, USA.

34. Kirsch, P. (2004) *Modern Fluoroorganic Chemistry: Synthesis, Reactivity, Appli-cations*, Wiley-VCH Verlag GmbH, Weinheim.

35. British Intelligence Objectives Sub-Committee (BIOS) (1946) German Fluorine and Fluoride Industry, Final Report 1595, Item No. 22.

36. Stoltzenberg, H. (1942) Verfahren zur Herstellung von Magnesiumhalogenidnebeln. DE Patent 724,807, Deutschland.

37. Lusar, R. (1964) *Die deutschen Waffen und Geheimwaffen des 2. Weltkrieges und ihre Weiterentwicklung*, 5th edn, J. F. Lehmann, München.

38. Zippermayr, M. (1945) Testimony Given Towards CIOS Mission 214 "Salzburg Area".

39. Simon, L.E. (1971) *German Research in World War II*, WE Inc. Publishers, Old Greenwich, p. 11, 23, 126.

40. Simon, L.E. (1947) *German Scientific Research Establishments*, Mapleton House, New York, p. 162.

41. Pick, H. (1948) Über lichtelektrische Leitung am Bleisulfid. *Ann. Phys.*, **3**, 255–269.

42. Blanchard, D.G. (1996) A brief history of air-intercept missile 9 (Sidewinder). 32nd Joint Propulsion Conference, Lake Buena Vista, FL, 1–3 July, AIAA-1996-3154.

43. *http://en.wikipedia.org/wiki/File:2008-12_mclean_ship_name01.jpg*, (accessed September 1 2011).

44. Loedding, A.C. (1958) Tracking flares. US Patent 2,829,596, filed 1954.

45. Hahn, G.T., Rivette, P.G. and Weldon, R.G. (1997) Infra-red tracking flare. US Patent 5,679,921, filed 1958, USA.

46. Moss, T.S., Brown, D.R. and Hawkins, T.D.F. (1957) Infra-Red Decoys, Technical Note No. RAD.702 Royal Aircraft Establishment, September 1957.

47. Cadwallader, E.A. (1964) Flare composition. US Patent 3,152,935, filed 1956, USA.

48. Cadwallader, E.A., Fookson, A., Mears, T.W. and Howard, F.L. (1948) Aliphatic halide-carbonyl condensations by means of sodium. *J. Res.*, **41**, 111–118.

49. Tyroler, J.F. and Knapp, C.A. (1957) An Evaluation of the NOTS Type 702A Infrared, Picatinny Arsenal, PL-R-TN No. 6, 25 October 1957.

50. Knapp, C.A. (1959) *New Infrared Flare and High Altitude Igniter Compositions*, Feltman Research and Engineering Laboratories, Picatinny Arsenal, Dover, NJ.

51. Stoltzenberg, H. and Leuschner, M. (1965) Nebelsatz zur Herstellung von beständig gefärbten anorganischen Nebeln, DE Patent 1,188,490, filed 1962, Deutschland.

52. Raisen, E., Katz, S. and Franson, K.D. (1963) *Survey of the Thermochemistry of High-Energy Reactions*, IIT Research Institute, Chicago, IL, Prepared under Contract No. AF 33(616)-7835.

53. Shidlovskii, A.A. (1965) Fundamentals of Pyrotechnics, Technical Memorandum 1615, Translated by U.S. Joint Publication Research Service from a Russian Textbook. Osnovy Pirotekhniki issued in 1964.

54. Ellern, H. (1968) *Military and Civilian Pyrotechnics*, Chemical Publishing Company, New York, p. 412.

Further Reading

Douda, B.E. (2009) *Genesis of Infrared Decoy Flares*, Naval Surface Warfare Center, Crane, IN.

Henne, A.L. and Renoll, M.W. (1936) Fluoroethanes and fluoroethylenes. *V. J. Am. Chem. Soc.*, **58**, 889–890.

Julian, E.C., Crescenzo, F.G. and Meyers, R.C. (1973) Igniter composition. US Patent 3,753,811, filed 1957, USA.

3
Properties of Fluorocarbons

3.1
Polytetrafluoroethylene (PTFE)

Polytetrafluoroethylene (PTFE) is a crystalline translucent solid polymer with a high molecular weight ranging between 10^6 and 10^7 g mol^{-1}. It is prepared from the monomer C_2F_4 by polymerization in aqueous medium and is obtained as fine powder. A number of trade names exist for PTFE. Thus it is often referred to as *Teflon*® *(DuPont), Hostaflon*® *(Dyneon), Fluon*® *(ICI), Halon*® *(Allied Chemical)* or *Fluoroplast*® *(Ftoroplastoviye Tekhnologii JSC)* [1, 2].

PTFE undergoes a series of first- and second-order transitions at 19 °C (first), 30 °C (second) and 90 °C (first), and finally at 130 °C (second), it expands to about 1.2 times its volume. At 340 °C, pristine PTFE changes into a transparent amorphous gel and expands to about 1.3 times its solid state volume [1]. However, this melt process is irreversible, and remelting of the recrystallized polymer will occur at 327 °C [1]. At temperatures above 350–400 °C, pyrolysis of the polymer into gaseous products occurs. This occurs without significant charring, as the C–C bond is significantly weaker (348 kJ mol^{-1}) than the C–F bond (507 kJ mol^{-1}), thus favouring C–C scission over C–F bond breaking [3, 4]. The main decomposition products of PTFE are tetrafluoroethylene (TFE) and CF_2, which depending on the reaction conditions undergo successive reactions and yield complex product mixtures [4]. Hydrogen and chlorine atmosphere inhibit PTFE decomposition, whereas steam, oxygen and sulfur dioxide accelerate its decomposition [4]. The thermal decomposition in moist air yields carbonyl difluoride, CO_2F and the monomer [5, 6]. The monomer itself will also undergo further reaction to both carbonyl difluoride and difluorocarbene. In the presence of moisture, both carbonyl difluoride and difluorocarbene give hydrofluoric acid and carbon dioxide or monoxide [5].

$$(C_2F_4)_n \xrightarrow[\text{Air}]{\Delta} n\,COF_2 + \frac{n}{2}\,C_2F_4 \tag{3.1}$$

$$C_2F_4 \xrightarrow[\text{Air}]{\Delta} COF_2 + CF_2 \tag{3.2}$$

$$COF_2 + H_2O \longrightarrow 2\,HF + CO_2 \tag{3.3}$$

$$CF_2 + H_2O \longrightarrow CO + 2\,HF \tag{3.4}$$

Metal-Fluorocarbon Based Energetic Materials, First Edition. Ernst-Christian Koch.
© 2012 Wiley-VCH Verlag GmbH & Co. KGaA. Published 2012 by Wiley-VCH Verlag GmbH & Co. KGaA.

Table 3.1 Main products of pyrolysis of PTFE.

Products	Formula	Unit	Temperature (°C)		
			605	650	700
Tetrafluoroethylene	C_2F_4	weight percentage	78.5	75.0	60.2
Hexafluoropropene	C_3F_6	weight percentage	4.6	4.7	6.0
cyclo-Octafluorobutane	C_4F_8	weight percentage	3.7	8.7	16.1
Soot	C_n	weight percentage	0.8	0.5	0.4

After Refs. [4, 6].

The main products of PTFE pyrolysis as a function of temperature obtained in a fluidized-bed reactor are given in Table 3.1 [4]. In the absence of oxygen, pyrolysis of PTFE at temperatures below 600 °C yields perfluoroisobutene (PFIB) [6], which is about 10 times more toxic than phosgene ($LC_{50} < 1$ ppm). The target organs of PFIB are the lungs and the liver [7]. However, PFIB formation has not been observed on pyrolysis of PTFE in air or oxygen [4, 6].

Ignition of PTFE in air occurs at specific irradiance levels of $E = 4.3$ W cm^2 or at an irradiance of $H = 1$ kJ cm^{-2}, equalling a temperature of 630 °C [8]. PTFE burns under oxygen pressure (Limiting Oxygen Index (LOI: 96%)). Figure 3.1 depicts the burn rate as a function of oxygen pressure after Ref. [9]. Burning PTFE rods attain a conical shape with some soot at the apex. At pressures below 200 kPa, the liquid layer formed at the burning surface shows bubbles indicative of gas formation. The unsaturated fluorocarbon compounds formed on pyrolysis of PTFE (C_2F_4, C_3F_6, C_4F_8) are combustible at ambient pressure, whereas the saturated fluorocarbons formed (CF_4, C_2F_6) will not burn at standard temperature and pressure (STP) [10].

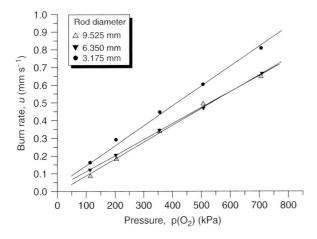

Figure 3.1 Burn rate of PTFE rods under oxygen as a function of pressure and rod diameter [9].

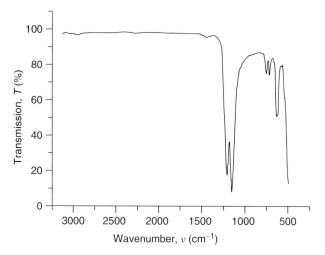

Figure 3.2 FTIR spectrum of PTFE.

Table 3.2 Characteristic IR vibrations of PTFE [11].

Mode	Wave number (cm^{-1})	Intensity
CF$_2$ asymmetrical stretching	1230	vs
CF$_2$ symmetrical stretching	1149	vs
	774	m
CF$_2$ scissoring	729	m
CF$_2$ wagging	635	s
CF$_2$ wagging	553	s
CF$_2$ rocking	520	s

The FTIR spectrum of PTFE is depicted in Figure 3.2. The characteristic vibrations are given in Table 3.2 [11].

3.2
Polychlorotrifluoroethylene (PCTFE)

Polychlorotrifluoroethylene (PCTFE) is a crystalline translucent solid polymer with a high molecular weight ranging between 10 000 and 50 000 g mol^{-1}, thus available as either viscous oil or hard plastic. It is made by polymerization of the bulk monomer or in solution, emulsion or dispersion by using free-radical starters, UV and γ-radiation. The presence of the chlorine atom improves the attractive forces between the polymer strands, thus leading to the greatest hardness and tensile strength encountered within the group of fluoropolymers [2]. Commercial brand names for PCTFE are *Kel-F*® *(3M), Plaskon*® *(Allied Chemical Corporation), Hostaflon C*® *(Dyneon), Neoflon*® *(Daikin Industries)* and *Aclar*® *(Honeywell)*. Solid PCTFE has its glass transition between 71 and 99 °C and melts at 211–216 °C. It is thermally stable up to 250 °C but starts to decompose when subjected to $T > 300$ °C. However, rapid decomposition does not occur until $T > 400$ °C. The decomposition of PCTFE in air produces the monomer as the main product followed by equal amounts of CO_2, $C_2F_2Cl_2$, C_3F_5Cl and $C_2F_3Cl_3$ [12, 13]. The IR spectrum of low-molecular-weight PCTFE is given below together with the assignment of the most important vibrational bands. High-molecular weight, solid PCTFE does not exhibit the 902 cm^{-1} vibrational mode (Figure 3.3 and Table 3.3) [11, 14].

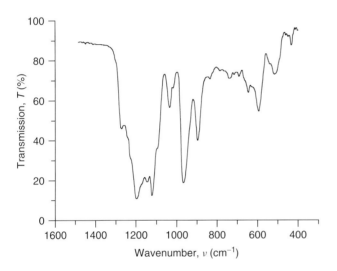

Figure 3.3 FTIR spectrum of low molecular weight PCTFE [14].

Table 3.3 Characteristic IR vibrations of PCTFE [11, 14].

Mode	Wave number (cm^{-1})	Intensity
CF stretching	1279	–
CF$_2$ stretching	1200	–
CF$_2$ stretching	1150	–
CF$_2$ stretching	1126	–
	1041	–
CCl stretching	970	–
	902	–
CF$_2$ wagging	599	–
CF$_2$ bending	520	–

3.3
Polyvinylidene Fluoride (PVDF)

Polyvinylidene fluoride (PVDF) (PVF$_2$) is made from 1,1-difluoroethylene by poly-merization in bulk, solution or dispersion with starters such as peroxides or γ-radiation. Commercial PVDF products are *Kynar*® *(Pennwalt Corporation), Solef*® *(Solvay)* and *Vidar*® *(AWK Trostberg)* [2].

Because of its alternating CF$_2$ and CH$_2$ groups, PVDF has a dipole moment (2.1 Debye), which makes it soluble in highly polar solvents such as DMF, THF, acetone and esters. PVDF is the only known polymer to occur in four different polymorphs [2]. These phases are present in the polymer at varying contents depending on the manufacture and both thermal and mechanical history of the sample. The orthorhombic phase (β-polymorph) is obtained from crystallization of PVDF from solvents. Figure 3.4 shows FTIR spectra of both orthorhombic and monoclinic (α-polymorph) phases of PVDF (Table 3.4) [15].

The density of the α-polymorph is 1.98 g cm^{-3}; amorphous PVDF has a density of 1.68 g cm^{-3}. Thus, commercial samples with a density of 1.75–1.78 g cm^{-3} have ~45% crystallinity. The α-polymorph melts at 170 °C; however, the processed polymer, because of its polymorphism, displays no sharp melting point but melts between 150 and 190 °C. The thermal decomposition becomes significant at $T > 300$ °C. Pyrolysis of PVDF yields hydrogen fluoride, the monomer C$_2$H$_2$F$_2$ and C$_4$F$_3$H$_3$ [12]. Up to 600 °C, pyrolysis also yields polyaromatic structures by cyclization of polyenic intermediates formed through HF elimination [16]. This is a particular advantage over PTFE, which is less likely to yield carbonaceous products. Thus in obscurant applications, PVDF is preferred over PTFE as a fluorine source (see Chapter 11).

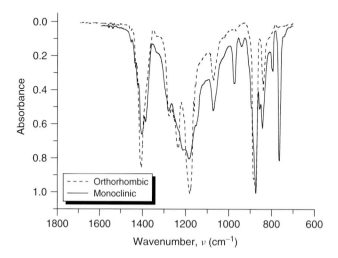

Figure 3.4 FTIR spectrum of both orthorhombic and monoclinic PVDF [15].

Table 3.4 Characteristic IR vibrations of PVDF [15].

Mode	Wavenumber (cm^{-1})	Intensity
CCH, CC, CF	1408	vs
	1385	s
	1281	s
	1240	s
CC, CCH, CCF	1180	vs
CCH	976	m
	868	vs
CCH	798	m
CCF, CF, CCC	764	vs

3.4
Polycarbon Monofluoride (PMF)

Depending on its fluorine content, polycarbon monofluoride (PMF), also known as *graphite fluoride*, is a white cream to dark grey, highly hydrophobic microcrystalline powder. It is obtained by fluorination of graphite or norite at temperatures between 400 and 700 °C [17–19]. Fluorination of less-ordered carbon materials also yields graphite fluoride, although at the expense of greater amounts of CF$_4$. The FTIR spectrum of a highly fluorinated sample (61% by weight F) is shown in Figure 3.5. The vibrational assignment is given in Table 3.5.

In contrast to any of the other fluorinated polymers, graphite fluoride does not melt or soften before its decomposition, which occurs slowly at temperatures above 300 °C (see DSC diagram in Figure 3.6). However, the onset of thermal

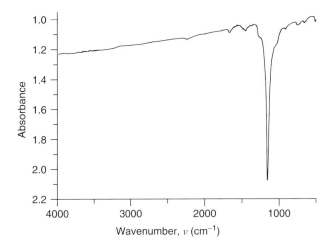

Figure 3.5 FTIR spectrum of PMF.

Table 3.5 Characteristic IR vibrations of PMF [17].

Mode	Wave number (cm^{-1})	Intensity
CF$_2$ symmetric stretching^a	1350	sh
CF stretching	1219	vs
CF$_2$ asymmetric stretching^a	1076	sh

^{a}Owing to terminal groups at the edge of graphite sheet.

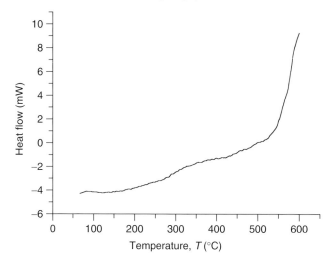

Figure 3.6 Differential Scanning Calorimetry (DSC) plot of $(CF_{0.989})_n$ at 10 K min^{-1} under 50 ml min^{-1} Ar.

decomposition with a steep slope occurs at ~500 °C for a material with $x = 0.989$. The decomposition of graphite fluoride has been investigated by Watanabe *et al.* [20–22]. They found that decomposition in both vacuum and oxygen proceeds via a similar mechanism, provided the temperature is above 588 °C. With a material containing 59.7% by weight fluorine, approximate composition $(CF_{0.937})_n$, it was found that the rate constant changes at 588 °C, indicative of change of the mode of decomposition. Below that temperature, the decomposition in oxygen proceeds very slowly and yields fluorine-containing residues, whereas at temperatures above 588 °C, no residue is obtained. Above 588 °C, the fraction of decomposed material, α, is given by the following expression known as the *Avrami–Erofeyev* equation, with k = rate constant and $n = 2$:

$$-\ln(1 - \alpha) = (kt)^n \tag{3.5}$$

The decomposition of graphite fluoride under oxygen yields mainly difluorocarbene, TFE and exfoliated graphite according to the following reaction scheme:

$$\left(\frac{2x+3y}{n}\right)(CF)_n \xrightarrow[\text{Air}]{\Delta} (x+2\,y)C_{(exp.gr)} + x\,CF_2 + y\,C_2F_4 \qquad (3.6)$$

with $n = (x + 2\,y)$.

3.5
Vinylidene Fluoride–Hexafluoropropene Copolymer

Vinylidene fluoride, VDF, –hexafluoropropene HFP, copolymer is a rubbery translucent milk-white polymer. Figure 3.7 shows a typical slab of the material. The monomer ratio is about 78 : 22. It is commercially available as *Viton*® *(Dupont), Fluorel FC-2175 (3M), Tecnoflon*® *(Monteflos)* or *Dai-el*® *(Daikin)*. However, these types differ slightly in composition, as is manifested in different solution behaviour in supercritical solvents [23] (see Chapter 18.3.1). In general, a VDF-HFP copolymer is soluble in ketone-type solvents. Typical molecular weights range between 1.5×10^5 and 10^6. Adiabatic compression heating in pressurized oxygen yields COF_2 and CF_4 beside CO_2 and CO, which constitute the main combustion products [13]. The thermal decomposition kinetics has been investigated in Ref. [24].

Figure 3.7 FC-2175 slab. (Reproduced with kind permission of MACH I Inc.)

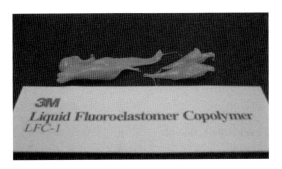

Figure 3.8 LFC-1 viscous chunks. (Reproduced with kind permission of MACH I Inc.)

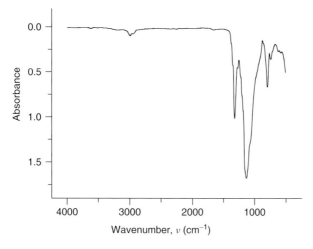

Figure 3.9 FTIR spectrum of LFC-1-polymer [25].

Table 3.6 Characteristic IR vibrations of LFC-1 [25].

Mode	Wave number (cm^{-1})	Intensity
CCH	745	w
CCH	798	m
CF	1136	vs
CF	1323	s
CH	2948	w
CH	2994	w

3.5.1
LFC-1

A special low-molecular-weight copolymer of VDF and HFP in about the same ratio as FC-2175 yields a viscous material (Figure 3.8) that enables processing energetic materials without using solvents. It is commercially available as *LFC-1*® *(3M)* and has a tan to brown colour. Figure 3.9 depicts the FTIR spectrum of LFC-1. Table 3.6 gives the vibrational assignment. It decomposes above 400 °C (Figure 3.10). With increasing temperature, its viscosity drops as depicted in Figure 3.11 [25].

The low shear viscosity of LFC-1 is shown in Figure 3.11.

3.6
Vinylidene Fluoride–Chlorotrifluoroethylene Copolymer

VDF–CTFE (chlorotrifluoroethylene) copolymer is a colourless granular material (Figure 3.12) that is commercially available as Kel-F 800® or FK 800® (3M) in

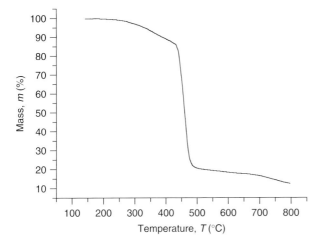

Figure 3.10 Thermogravimetry (TG) diagram of LFC-1 in air [25].

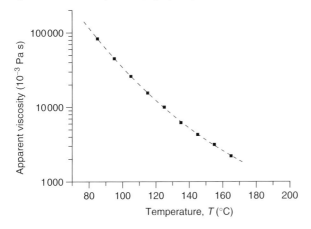

Figure 3.11 Low shear viscosity of LFC-1 [25].

Figure 3.12 FK-800 powder. (Reproduced with kind permission of MACH I Inc.)

a monomer ratio VDF : CTFE = 1 : 3. It is soluble in ketone- and acetate-type solvents, and at 10–30% by weight content, it forms clear, air-drying lacquers that are applicable by brush, dip or spray techniques. In addition, it is soluble in supercritical solvent such as CO_2 [23]. Using VDF–CTFE, compositions can be made in which binder amounts to less than 5% of the total ingredients [26]. The thermal degradation kinetics has been investigated in Ref. [24]. The thermal decomposition of a polymer with a different molar ratio of VDF : CTFE) = 4 : 1 has been investigated in Ref. [12]. The main decomposition products comprise HF, HCl, CTFE, VDF, the dimer of VDF and C_3F_5Cl.

3.7
Copolymer of TFE and VDF

TFE and VDF can be polymerized in any proportion, thus giving rise to a broad variety of products. The copolymers are obtained in either emulsion or suspension process. An approximate composition of TFE : PVDF of 20 : 80 is the eutectic point in the system, with a melting point of 120 °C. It is widely used as a technical polymer and is available under the brand names Kynar® 7200 and Kynar® SL (Pennwalt Corporation). A composition with the an approximate composition of TFE : PVDF 29 : 71) is available as Fluoroplast® 42 (Russia). Both copolymers are soluble in ketones and esters but are insoluble in alcohols and chlorinated hydrocarbons and are mainly processed via melt extrusion at temperatures between 190 and 260 °C [27, 28]. The low-wave-number FTIR spectrum of the copolymer is depicted in Figure 3.13.

Owing to the higher TFE content, Fluoroplast®-42 melts at temperatures slightly higher than the melting point (135 °C) of Kynar® 7200 (Table 3.7) and decomposes above 400 °C as is depicted in Figure 3.14 showing both DSC and TG-signal [30].

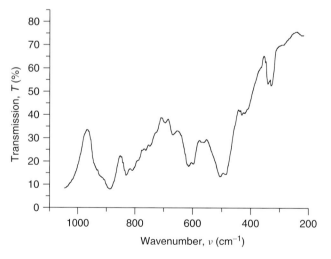

Figure 3.13 FTIR spectrum of Kynar 7200 [29].

Table 3.7 Characteristic properties of various VDF-TFE copolymer Kynar® 7200 [30, 31].

Property	Kynar® 7200	Fluoroplast®-42
Ratio of TFE : PVDF	20 : 80	29 : 71
Density (g cm^{-3})	1.88	–
Melting point (°C)	122–126	135
Heat of fusion (J g^{-1})	13–21	–
Thermal decomposition (°C)	>400	>400
Soluble in solvents	++	?
Fluorine content (wt%)	64.0	65.8

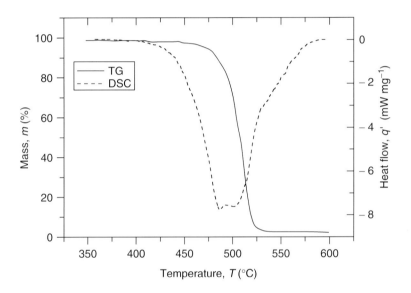

Figure 3.14 Thermoanalysis of Fluoroplast®-42 [30].

3.8
Terpolymers of TFE, HFP and VDF

Terpolymers of TFE, HFP and VDF are melt-processable fluoropolymers. They are polymerized in aqueous emulsion. Depending on the ratio of the monomers, a large number of different terpolymers are available. Figure 3.15 depicts the elastomeric region in the ternary system [28]. These materials are commercially available as *THV*® *(Dyneon)* [32].

The density of THV varies with composition between 1.95 and 1.98 g cm^{-3}.

Table 3.8 gives information on the principal properties of a series of commercially available Terpolymers, THV.

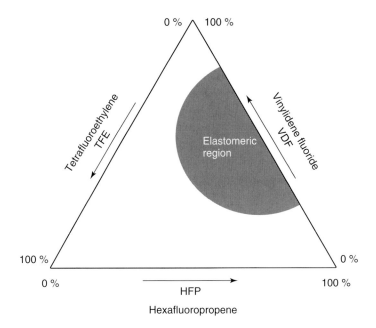

Figure 3.15 Elastomeric region in the ternary system TFE, HFP, VDF. (After Ref. [27].)

Table 3.8 Characteristic properties of various THV® grades [32–34].

Property	THV-200	THV-221	THV-X310	THV-400	THV-500	THV-610	THV-815
Density (g cm⁻³)	1.95	1.93	n.a.	1.97	2.00	2.04	2.06
Melting point (°C)	115–125	115	140	150–160	165	185	225
Thermal decomposition (°C)	420	370[a]	n.a.	430	440	370[a]	370[a]
Soluble in solvents	+	+	−	−	−	−	−
Hardness shore, D	44	44	n.a.	53	54	56	n.a.
Glass transition temperature (°C)	−	5	n.a.	−	26	34	36
Fluorine content (weight percentage)	n.a.	70.5	71–72	71–71	72.4	n.a.	n.a.

[a]Onset of thermal decomposition in presence of metal powders.

3.9
Summary of chemical and physical properties of common fluoropolymers

The following tables summarize the chemical and physical properties of common fluoropolymers.

Table 3.9 Characteristic properties of various fluorinated polymers [18, 31, 35].

Name	Formula	CAS-no	m_r (g mol^{-1})	F-content (wt%)	Tg (°C)	Mp (°C)	Dp, air (°C)
PTFE	$(C_2F_4)_n$	9002-84-0	100.016	76.0	–	328, 340	>400
PCTFE	$(C_2ClF_3)_n$	9002-83-9	116.470	48.9	71–99	211	>400
PVDF	$(C_2H_2F_2)_n$	24937-79-9	64.038	59.4	–40	154–184	350
Viton A	$(C_2H_2F_2)_n$ $(C_3F_6)_m$ $n = 3.5, m = 1$	9011-17-0	374.145	66.0	–27	–	>400
FK-800	$(C_2H_2F_2)_n$ $(C_2ClF_3)_m$ $n = 1, m = 3$	9010-75-7	413.445	50.6	28–38	105	–
PMF	$(CF_x)_n$	11113-63-6	31.01	61.3	n.a.	n.a.	610

Name	ρ (g cm^{-3})	melting enthalpy, $\Delta_m H$ (kJ mol^{-1})	enthalpy of formation, $\Delta_f H$ (kJ mol^{-1})	specific heat, Cp (J g^{-1} K^{-1})	thermal conductivity, τ (W k^{-1} m^{-1})	Limiting Oxygen Index LOI (% O$_2$)
PTFE	2.20	3.6	–809	1.020	0.024	99.5
PCTFE	2.10–2.13	4.659	–	0.835	–	99.5
PVDF	1.75–1.80	5.966	–	1.26–1.42	0.1–0.13	43.6
Viton A	1.82–1.85	0.3–0.7	–2784	–	0.226	31.5
FK-800	2.02	2.5	–2418	1.3	0.053	–
PMF	2.65	n.a.	–196	0.89	–	–

Abbreviations: m_r, molecular weight; Tg, glass-transition temperature; Mp, melting point; Dp, decomposition temperature.

References

1. Gangal, S. (2000) *Fluorine-Containing Polymers, Polytetrafluoroethylene,* Kirk-Othmer Encyclopedia of Chemical Technology, Wiley-VCH Verlag GmbH.
2. Carlson, D.P. and Schmiegel, W. (2000) *Fluoropolymers, Organic, Ullmann's Encyclopedia of Industrial Chemistry,* Wiley-VCH Verlag GmbH.
3. Arai, N. (1979) Transient ablation of Teflon in intense radiative and convective environments. *AIAA J.,* **17,** 634–640.
4. Simon, C.M. and Kaminsky, W. (1998) Chemical recycling of polytetrafluoroethylene by pyrolysis. *Polym. Degrad. Stab.,* **62,** 1–7.

5. Baker, B.B. and Kasprzak, D.J. (1993) Thermal degradation of commercial fluoropolymers in air. *Polym. Degrad. Stab.*, **42**, 181–188.

6. Arito, H. and Soda, R. (1977) Pyrolysis products of polytetrafluoroethylene and polyfluoroethylenepropylene with reference to inhalation toxicity. *Ann. Occup. Hyg.*, **20**, 247–255.

7. Patocka, J. and Bajgar, J. (1998) Toxicology of perfluoroisobutene, *ASA Newsl.*, **5**.

8. Lyon, R.E. and Quintiere, J.G. (2007) Criteria for piloted ignition of combustible solids. *Combust. Flame*, **151**, 551–559.

9. Ladouceur, H.D. (2005) An overview of the known chemical kinetics and transport effects relevant to Mg/PTFE combustion. 2nd International Workshop on Pyrotechnic Combustion Mechanisms, Pfinztal Germany, June 27.

10. Matula, R.A., Orloff, D.I. and Agnew, J.T. (1970) Burning velocities of fluorocarbon-oxygen mixtures. *Combust. Flame*, **14**, 97–102.

11. Liang, C.Y. and Krimm, S. (1956) Infrared spectra of high polymers. III polytetrafluoroethylene and polychlorotrifluoroethylene. *J. Chem. Phys.*, **25**, 563–571.

12. Zulfiqar, S., Zulfiqar, M., Rizvi, M., Munir, A. and McNeill, C. (1994) Study of the thermal degradation of polychlorotrifluoroethylene, poly(vinylidene fluoride) and copolymers of chlorotrifluoroethylene and vinylidene fluoride. *Polym. Degrad. Stab.*, **43**, 423–430.

13. Hsieh, F.-Y. and Beeson, H.D. (2004) A preliminary study on the toxic combustion products testing of polymers used in high-pressure oxygen systems. http://ntrs.nasa.gov/archive/nasa/casi.ntrs.nasa.gov/20100041342_201000 44104.pdf. (accessed 1 September 2011).

14. Kawano, Y. and de Araujo, S.C. (1996) Raman and infrared spectra of polychlorotrifluoroethylene. *J. Braz. Chem.*, **7**, 491–496.

15. Roth, E.P., Nagasubramanian, G., Tallant, D.R. and Garcia, M. (1999) Instability of Polyvinylidene Fluoride-Based Polymeric Binder in Lithium-Ion Cells: Final Report. Sandia Report SAND99-1164, Sandia National Laboratories, Albuquerque, New Mexico.

16. O'Shea, M.L., Morterra, C. and Low, M.J.D. (1990) Spectroscopic studies of carbons. XVII. Pyrolysis of polyvinylidene fluoride. *Mater. Chem. Phys.*, **26**, 193–205.

17. Watanabe, N., Nakajima, T. and Touhara, H. (1988) *Graphite Fluorides, Studies in Inorganic Chemistry*, Vol. **8**, Elsevier, Amsterdam.

18. Nakajima, T. and Watanabe, N. (1990) *Graphite Fluorides and Carbon-Fluorine Compounds*, CRC Press, Boca Raton, FL.

19. Kamarchik, P. and Margrave, J.L. (1976) Poly(carbon monofluoride): a solid layered fluorocarbon. *Acc. Chem. Res.*, **11**, 296–300.

20. Watanabe, N., Koyama, S. and Imoto, H. (1980) Thermal decomposition of graphite fluoride. I. Decomposition products of graphite fluoride, $(CF)_n$ in a vacuum. *Bull. Chem. Soc. Jpn.*, **53**, 2731–2734.

21. Watanabe, N. and Koyama, S. (1980) Thermal decomposition of graphite fluoride. II. Kinetics of thermal decomposition of $(CF)_n$ in a vacuum. *Bull. Chem. Soc. Jpn.*, **53**, 3093–3099.

22. Watanabe, N., Kawamura, T. and Koyama, S. (1980) Thermal decomposition of graphite fluoride. III. Thermal decomposition of $(CF)_n$ in oxygen atmosphere. *Bull. Chem. Soc. Jpn.*, **53**, 3100–3103.

23. CPIA Bulletin (1996) CPIA presents: technical area task on solubility of PEP binders in supercritical carbon dioxide.

24. Burnham, A.K. and Weese, R.K. (2005) Kinetics of thermal degradation of explosive binders Viton A, Estane and Kel-F. *Thermochim. Acta*, **426**, 85–92.

25. Hunter, R.W. (1997) A new binder ingredient for explosives. JANNAF Propellant Development and Characterization Subcommittee and Safety and Environmental Protection Subcommittee Joint Meeting, USA, pp. 113–120.

26. Mach I Inc. (2000) Product Bulletin for FK-800.

27. Arcella, V. and Ferro, R. (1997) in *Modern Fluoropolymers* (ed. J. Scheirs), John Wiley & Sons, Inc., pp. 71–90.
28. Hull, D.E., Johnson, B.V., Rodricks, I.P. and Staley, J.B. (1997) in *Modern Fluoropolymers* (ed. J. Scheirs), John Wiley & Sons, Inc., pp. 257–270.
29. Latour, M. (1977) Infra-red analysis of poly(vinylidene fluoride) thermoelectrets. *Polymer*, **18**, 278–280.
30. Tarasov, A.V., Alikhanian, A.S., Kirakosyan, G.A. and Arkhangel'skii, I.V. (2010) Chemical interaction of metallic titanium with tetrafluoroethylene-vinylidene fluoride copolymer. *Inorg. Mater.*, **46**, 1308–1312.
31. Hirsch, D.B. and Beeson, H.D. (2001) Improved test method to determine flammability of aerospace materials.

Halon Options Technical Working Conference, USA April 24–26, 2001.
32. Dyneon® Fluorothermoplastic Product Comparison Guide (2009) Dyneon, Oakdale, MN.
33. Tournut, C. (1997) in *Modern Fluoropolymers* (ed. J. Scheirs), John Wiley & Sons, Inc., pp. 257–270.
34. Nielson, D.B., Truitt, R.M. and Rasmussen, N. (2005) Low temperature, extrudable, high density reactive materials. US Patent 6,962,634, USA.
35. Yeager, J.D., Dattelbaum, A.M., Orler, E.B., Bahr, D.F. and Dattelbaum, D.M. (2010) Adhesive properties of some fluoropolymer binders with the insensitive explosive 1,3,5-triamino-2,4,6-trinitrobenzene (TATB). *J. Colloid Interface Sci.*, **352**, 535–541.

4
Thermochemical and Physical Properties of Metals and their Fluorides

The essential thermochemical properties of metals and their fluorides are given in the following tables [1–4]. In this context the following abbreviations apply.

- m_r: atomic and molecular weight (g mol^{-1})
- $\rho_{20\,°C}$: density (g cm^{-3})
- M_p: melting point (°C)
- B_p: boiling point (°C)
- C_p: specific heat (J g^{-1} K^{-1})
- $\Delta_m H$: enthalpy of fusion (J g^{-1})
- $\Delta_v H$: enthalpy of vaporization (J g^{-1})
- τ thermal conductivity (W m^{-1} K^{-1})
- decomposition (dp).

Metal-Fluorocarbon Based Energetic Materials, First Edition. Ernst-Christian Koch.
© 2012 Wiley-VCH Verlag GmbH & Co. KGaA. Published 2012 by Wiley-VCH Verlag GmbH & Co. KGaA.

Metal	Symbol	m_r	$\rho_{20°C}$	M_p	B_p	C_p	$\Delta_m H$	$\Delta_v H$	τ	Symbol	m_r	$\rho_{20°C}$	$\Delta_f H°$	$\Delta_f H°$	$\Delta_f H°$	$\Delta_m H$	$\Delta_v H$	M_p	B_p	C_p
		g mol⁻¹	g cm⁻³	°C	°C	J g⁻¹ K⁻¹	J g⁻¹	kJ g⁻¹	W m⁻¹ K⁻¹		g mol⁻¹	g cm⁻³	kJ mol⁻¹	kJ g⁻¹	kJ cm⁻³	J g⁻¹	kJ g⁻¹	°C	°C	J g⁻¹ K⁻¹
Lithium	Li	6.941	0.534	180	1347	3.547	432	21.28	84.7	LiF	25.939	2.635	−616	−88.9	−47.49	1044	8.22	845	1717	1.604
Natrium	Na	22.99	0.97	97.8	883	1.224	113	4.32	141	NaF	41.988	2.558	−576	−13.7	−35.09	794	4.98	993	1695	1.117
Potassium	K	39.098	0.862	63.4	764	0.749	60	2.02	102.4	KF	58.097	2.49	−569	−9.6	−23.97	468	2.97	857	1502	0.842
Rubidium	Rb	85.468	1.532	39.7	688	0.363	25	0.89	58.2	RbF	104.47	3.557	−558	−5.3	−19.00	245	1.58	795	1410	0.483
Caesium	Cs	132.905	1.873	28.5	675	0.242	16	0.51	35.9	CsF	151.90	4.115	−554	−3.7	−15.01	143	0.95	703	1229	0.342
Beryllium	Be	9.012	1.848	1287	2468	1.825	877	32.4	200	BeF₂	47.01	1.986	−1027	−21.8	−43.39	101	4.24	552	1167	1.102
Magnesium	Mg	24.305	1.738	649	1088	1.024	368	5.24	171	MgF₂	62.31	3.13	−1124	−18.0	−56.46	942	4.40	1263	2262	0.988
Calcium	Ca	40.078	1.55	839	1482	0.632	213	3.83	200	CaF₂	78.08	3.18	−1229	−15.7	−50.10	381	3.95	1417	2484	0.879
Strontium	Sr	87.62	2.54	777	1412	0.305	85	1.57	35.3	SrF₂	125.62	4.24	−1217	−9.7	−41.08	236	2.55	1477	2486	0.557
Barium	Ba	137.327	3.594	729	1894	0.205	56	1.03	18.4	BaF₂	175.34	4.893	−1208	−6.9	−33.70	133	1.63	1368	2270	0.412
Scandium	Sc	44.956	2.99	1541	2836	0.567	313	7.40	15.8	ScF₃	101.95	2.53	−1612	−15.8	−40.00	626	–	1552	1791	0.830
Yttrium	Y	88.906	4.472	1526	3336	0.298	128	4.11	17.2	YF₃	145.9	4.01	−1718	−11.8	−47.22	192	–	1155	2466	0.653
Lanthanum	La	138.906	6.145	920	3453	0.195	47	2.98	13.5	LaF₃	195.90	5.9	−1700	8.7	51.4	255	–	1493¹	2296	0.461
Cerium	Ce	140.115	6.773	798	3422	0.192	39	2.96	11.4	CeF₃	197.12	6.16	−1689	8.6	52.8	396	–	1430	2279	0.474
Praseodymium	Pr	140.908	6.773	931	3507	0.195	49	2.11	12.5	PrF₃	197.90	6.18	−1689	8.5	52.7	389	–	1397	2296	0.468
Nedoymium	Nd	144.24	7.007	1016	3064	0.190	49	1.89	16.5	NdF₃	201.24	6.506	−1679	8.3	54.3	273	–	1376	2304	0.459
Samarium	Sm	150.36	7.520	1072	1788	0.196	57	1.11	13.3	SmF₃	207.35	6.928	−1700	8.2	56.8	251	–	1298	2331	0.442
										SmF₂	188.35	6.16	−1180	6.3	38.6	–	–	1417	2400	–
Europium	Eu	151.965	5.243	817	1525	0.182	61	0.95	13.9	EuF₃	208.96	6.78	−1612	7.7	52.3	201	–	1276	2389	0.469
										EuF₂	189.96	6.495	–	–	–	–	–	1380	2400	–

Element										Fluoride									
Gadolinium	Gd	157.25	7.900	1312	3263	0.237	64	2.29	10.6	GdF$_3$	214.25	7	−1699	7.9	55.5	243	–	1228	2391	0.413
Terbium	Tb	158.925	8.229	1357	3219	0.182	68	2.08	11.1	TbF$_3$	215.92	7.2	−1696	7.9	56.6	241	–	1173	2428	0.419
Dysprosium	Dy	162.5	8.550	1409	2558	0.173	68	1.42	10.7	DyF$_3$	219.5	7.45	−1692	7.7	57.4	223	–	1153	2466	0.405
Holmium	Ho	164.930	8.795	1470	2691	0.165	74	1.47	16.2	HoF$_3$	221.93	7.644	−1698	7.7	58.5	252	–	1143	2269	0.399
Erbium	Er	167.26	9.066	1522	2859	0.168	119	1.56	14.3	ErF$_3$	224.28	7.806	−1694	7.6	59.0	125	–	1140	2427	0.402
Thulium	Tm	168.934	9.321	1545	1944	0.160	100	1.13	16.8	TmF$_3$	225.93	7.9	−1694	7.5	61.5	230	–	1158	2312	0.402
Ytterbium	Yb	173.04	6.965	824	1192	0.155	44	0.75	38.5	YbF$_3$	230.04	8.2	−1655	7.2	59.0	147	–	1162	2307	0.389
										YbF$_2$	211.04	7.985	−1184	5.6	44.8	–	–	1407	2380	–
Lutetium	Lu	174.967	9.840	1663	3391	0.153	107	2.03	16.4	LuF$_3$	231.96	8.29	−1680	7.2	60.0	134	–	1182	2230	0.375
Vanadium	V	50.942	6.110	1902	3406	0.489	411	8.87	30.7	VF$_3$	107.94	3.363	−1297	12.0	40.4	–	–	1406	–	0.833
										VF$_4$	126.94	3.15	−1403	11.1	34.8	–	–	325	–	0.843
										VF$_5$	145.93	2.177	−1481	10.1	22.1	–	–	19.5	48.3	–
Niobium	Nb	92.906	8.57	2472	4799	0.266	323	7.43	53.7	NbF$_5$	187.898	3.293	−1814	−9.7	31.8	64	–	78	238	0.795
Tantalum	Ta	180.948	16.65	3000	5534	0.773	174	4.19	57	TaF$_5$	275.95	4.74	−1902	6.89	32.67	–	–	96	228	–
Titanium	Ti	47.88	4.54	1666	3358	0.523	296	8.56	21.9	TiF$_3$	104.9	3.4	−1436	13.69	46.54	2.1	–	–	950	0.877
										TiF$_4$	123.89	2.798	−1649	13.31	37.24	0.7	–	283	–	0.920
Zirconium	Zr	91.224	6.506	1852	4430	0.276	229	6.15	22.7	ZrF$_3$	148.219	–	−1442	9.73	–	–	–	1190	dp	–
										ZrF$_4$	167.21	4.43	−1911	11.43	50.63	–	~1.2	–	903	0.616
Hafnium	Hf	178.49	13.310	2227	4598	0.144	135	3.22	23.0	HfF$_4$	254.48	7.13	−1931	7.59	54.10	–	–	–	–	–
Chromium	Cr	51.996	7.190	1857	2669	0.449	326	6.62	93.7	CrF$_4$	127.99	–	−1248	9.75	–	–	–	277	–	0.784
										CrF$_3$	108.99	3.8	−1173	10.76	40.90	–	–	1407	–	0.723
										CrF$_2$	89.99	4.11	−778	8.65	35.53	–	–	894	1300	0.720
Molybdenum	Mo	95.94	10.22	2624	4705	0.249	408	6.07	142	MoF$_6$	209.93	2.543	−1586	7.55	19.21	44	0.13	17.5	35	0.810
Tungsten	W	183.84	19.30	3407	5658	0.132	193	4.38	178	WF$_{6(g)}$	297.84	0.013	−1721	5.75	0.07	14	0.09	2.3	17.1	0.400
Manganese	Mn	54.938	7.44	1244	2059	0.479	220	4.13	29.7	MnF$_2$	92.93	3.891	−849	9.14	35.56	315	–	900	–	0.730
										MnF$_3$	111.93	3.54	−1071	9.57	33.87	–	2.54	–	–	–

Metal	Symbol	m_r	$\rho_{20°C}$	M_p	B_p	C_p	$\Delta_m H$	$\Delta_v H$	τ	Symbol	m_r	$\rho_{20°C}$	$\Delta_f H°$	$\Delta_f H°$	$\Delta_f H°$	$\Delta_m H$	$\Delta_v H$	M_p	B_p	C_p
		g mol⁻¹	g cm⁻³	°C	°C	J g⁻¹ K⁻¹	J g⁻¹	kJ g⁻¹	W m⁻¹ K⁻¹		g mol⁻¹	g cm⁻³	kJ mol⁻¹	kJ g⁻¹	kJ cm⁻³	J g⁻¹	kJ g⁻¹	°C	°C	J g⁻¹ K⁻¹
Rhenium	Re	186.207	21.02	3180	5591	0.136	178	3.84	71.2	ReF_6	300.2	3.616	−1150	3.83	13.85	70	0.1	18.8	33.7	–
Iron	Fe	55.847	7.874	1536	2859	0.447	247	6.26	80.2	FeF_3	112.84	3.52	−1039	9.21	32.41	–	1.79	–	926_{sb}	0.806
										FeF_2	93.84	4.09	−706	7.52	30.77	485	–	940	1800	0.726
Ruthenium	Ru	101.07	12.41	2250	4146	0.238	240	5.89	117	RuF_3	158.0	–	$-314_{(g)}$	$1.99_{(g)}$	–	–	–	650	–	–
										RuF_5	196.06	2.963	−893	4.55	13.50	405	0.28	85	230	0.832
Osmium	Os	190.23	22.570	3027	5008	0.130	167	3.92	87.6	OsF_8	342.19	3.87	–	–	–	21	0.08	34.4	47.3	–
Cobalt	Co	58.933	8.90	1495	2925	0.421	275	6.39	100	CoF_3	115.92	3.89	−790	6.82	26.51	–	–	927	–	0.792
										CoF_2	96.93	4.46	−672	6.94	30.94	–	–	1127	1747	0.710
Rhodium	Rh	102.906	12.41	1960	3694	0.242	209	4.79	150	RhF_3	159.90	5.38	–	–	–	–	–	–	600_{sb}	–
										RhF_5	197.90	3.95	–	–	–	–	–	95.5	–	–
										RhF_6	216.90	3.13	–	–	–	–	–	70	–	–
Iridium	Ir	192.22	22.420	2443	4424	0.130	136	3.14	147	IrF_6	306.19	6.00	−544	1.78	10.66	27	0.02	43.8	53.6	0.395
Nickel	Ni	58.693	8.902	1453	2911	0.444	298	6.29	83	NiF_2	96.71	4.63	−658	6.80	31.50	–	–	1450	1441_{sb}	0.662
Palladium	Pd	106.42	12.02	1552	2961	0.244	165	3.36	71.8	PdF_2	144.4	5.8	−469	3.25	18.83	–	–	–	350_{dec}	0.456
Platinum	Pt	195.08	21.450	1772	3823	0.133	101	2.61	71.6	PtF_4	271.08	6.12	–	–	–	–	–	–	–	–
Copper	Cu	63.546	8.96	1085	2570	0.415	223	5.10	401	CuF	82.54	7.07	−280	3.39	23.98	–	–	–	1100_{sb}	0.605
										CuF_2	101.54	4.23	−539	5.31	22.45	544	0.80	836	1670	0.646

Metal / **Fluoride**

Silver	Ag	107.868	10.500	960	2160	0.236	105	2.32	429	Ag$_2$F	234.74	8.75	−211	0.90	7.87	–	–	100$_{dec}$	–	–
										AgF	126.87	7.076	−205	1.62	11.43	–	–	435	1160	0.409
										AgF$_2$	145.87	5.942	−354	2.43	14.42	–	–	690	700$_{dec}$	–
Gold	Au	196.967	19.320	1065	2854	0.129	64	1.70	317	AuF$_3$	253.96	6.75	−363	1.43	9.65	–	–	–	–	–
Zinc	Zn	65.39	7.133	419	906	0.388	112	1.76	121	ZnF$_2$	103.37	4.95	−764	7.39	36.59	386	1.79	872	1500	0.635
Cadmium	Cd	112.411	8.650	321	766	0.231	55	0.87	96.8	CdF$_2$	150.40	6.64	−700	4.65	30.90	150	1.34	1072	1751	0.445
Mercury	Hg	200.59	13.546	−38.7	356	0.139	11	0.30	8.34	Hg$_2$F$_2$	439.18	8.73	−485	1.10	9.60	–	–	570	676$_{dec}$	–
										HgF$_2$	238.59	8.59	−423	1.77	15.22	96	0.39	645	646	0.313
Boron	B	10.811	2.340	2077	3866	1.047	4662	44.45	27	BF$_3$	67.81	0.003	−1137	16.77	0.05	63	0.28	−129	−100	0.737
Aluminium	Al	26.981	2.698	660	2517	0.900	396	10.90	237	AlF$_3$	83.98	3.197	−1510	17.98	57.48	–	3.25	–	1275	0.894
Gallium	Ga	69.723	5.907	29.8	2202	0.375	80	3.71	33.5	GaF$_3$	126.72	4.47	−1175	9.27	41.45	–	–	–	1077	0.702
Indium	In	114.828	7.310	157	2070	0.233	28	2.02	81.6	InF$_3$	171.82	4.39	−1189	6.92	30.37	372	–	1172	1904	–
Thallium	Tl	204.383	11.850	304	1471	0.129	20	0.80	46.1	TlF	223.37	8.36	−324.7	1.45	12.15	62	–	327	826	0.239
										TlF$_3$	261.37	8.7	−650	2.49	21.64	–	–	550	–	–
Silicon	Si	28.0855	2.329	1412	3217	0.711	1788	13.65	83.7	SiF$_4$	104.08	0.0047	−1615	15.51	0.07	–	–	−90.3	−86	0.707
Germanium	Ge	72.61	5.323	937	2831	0.322	509	4.56	58.6	GeF$_4$	148.58	0.007	−1190	8.01	0.05	–	0.22	−15	−40	0.549
Tin	Sn	118.710	7.310	232	2606	0.227	59	2.49	66.6	SnF$_2$	156.69	4.85	−649	4.14	20.09	67	–	213	850	0.462
										SnF$_4$	194.68	4.78	−1188	6.10	29.17	–	–	701	–	–
Lead	Pb	207.2	11.35	328	1746	0.130	23	0.86	35.2	PbF$_4$	283.18	6.7	−942	3.33	22.29	–	–	600	–	0.321
										PbF$_2$	245.19	8.37	−677	2.76	23.11	60	0.65	830	1304	0.295
Arsenic	As	74.922	5.780	817	1134	0.329	370	0.46	50	AsF$_5$	169.91	0.008	−1236	7.24	0.06	67	0.12	−79	−52	0.574
										AsF$_3$	131.92	3.01	−821	6.22	18.73	79	0.25	−6	58	0.961
Antimony	Sb	121.757	6.691	631	1587	0.207	181	1.36	25.9	SbF$_5$	216.74	2.99	−1328	6.13	18.32	–	0.04	8.3	141	0.505
										SbF$_3$	178.75	4.379	−915	5.12	22.42	128	–	292	376	–
Bismuth	Bi	208.980	9.747	272	1564	0.122	54	0.83	7.87	BiF$_5$	303.97	5.4	–	–	–	–	–	154	230	–
										BiF$_3$	265.98	8.25	−909	3.42	28.19	71	–	649	904	0.323

References

1. Binnewies, M. and Milke, E. (2002) *Thermochemical Data of Elements and Compounds*, 2nd Revised and Extended edn, Wiley-VCH Verlag GmbH, Weinheim.

2. Blachnik, R., D'Ans, J. and Lax, E. (1998) *Taschenbuch für Chemiker und Physiker*, Elemente, Anorganische Verbindungen und Materialien, Minerale, Band **3**, 4th Auflage, Springer-Verlag, Heidelberg.

3. Konings, R.J.M. and Kovacs, A. (2003) in *Handbook on the Physics and Chemistry of Rare Earths Reviews*, vol. **33** (eds K.A. Gschneider, J.-C.G. Bünzli and V.K. Pecharsky), Elsevier Science, pp. 147–247.

4. Wiberg, N. (2007) *Holleman-Wiberg, Lehrbuch der Anorganische Chemie*, 102th Auflage, Walter de Gruyter, Berlin.

5
Reactivity and Thermochemistry of Selected Metal/Fluorocarbon Systems

5.1
Lithium

Lithium–halocarbon reactions are amongst the basic pathways to lithiated carbon compounds [1, 2]. Hot lithium vapour (800–1000 °C) reacts with fluorobenzenes, C_6H_5F and C_6F_6 to yield both lithium fluoride and a series of Li_xC_y compounds given in Table 5.1 indicative of fragmentation of the benzene ring due to transient formation of highly reactive benzyne species [3–5] (Scheme 5.1).

$$12\,Li + C_6F_6 \xrightarrow{800\,°C} 2\,LiF + [C_6F_4]* + 10\,Li \longrightarrow 4\,LiF + Li_2C{=}C{=}CLi_2 + LiC{\equiv}CLi \qquad (1)$$

Scheme 5.1 Assumed *'direct lithiation'* reaction of perfluorobenzene.

The Differential Scanning Calorimetry (DSC) analysis of a binary mixture of lithium/polyvinylidene fluoride (PVDF), $(CH_2F_2)_n$, in air reveals two endotherms at both 175 and 180 °C indicative of fusion of both PVDF and Li and the onset of an extended exothermic reaction at 355 °C [6] with significant subsequent heat release,

Table 5.1 Relative distribution of fragment products from the reaction of Li-vapour with various fluorocarbons after hydrolysis with D_2O.

Li-species formed	C_6H_5F, slow (%)	C_6H_5F, fast (%)	C_6F_6 (%)
C_2D_6	30	14	12
C_2D_4	4	2	Traces
C_2D_2	5	24	10
C_3D_8	19	10	28
C_3D_6	34	14	13
C_3D_4	8	36	37
C_4D_{10}	Traces	Traces	Traces

After Ref. [4].

Metal-Fluorocarbon Based Energetic Materials, First Edition. Ernst-Christian Koch.
© 2012 Wiley-VCH Verlag GmbH & Co. KGaA. Published 2012 by Wiley-VCH Verlag GmbH & Co. KGaA.

not seen with either of the single components (Figure 5.1). A similar vivid reaction is seen between Li_xC and PVDF with rising exothermicity as x increases. The reaction product reveals reductive elimination of HF and formation brown soluble carbon compounds showing sp^2-CF units according to the following reaction

$$n \, Li + (H_2C-CF_2)_n \longrightarrow n \, LiF + \frac{n}{2} H_2 + (HC{=}CF)_n \tag{5.1}$$

The reaction rate, k, of liquid lithium with a number of gaseous oxidizers including sulfur hexafluoride, SF_6 and several halocarbons has been investigated in Ref. [7]. For perfluorocyclobutane, C_4F_8 k is sensitive to both temperature and pressure and is depicted below in Figure 5.2. The onset of the reaction is at $\sim$360 °C which

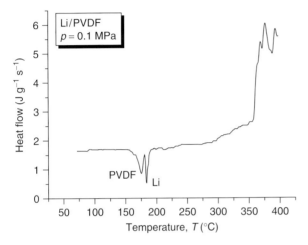

Figure 5.1 DSC of Li/PVDF sample at ambient pressure at a heating rate of 10 K min under air [6].

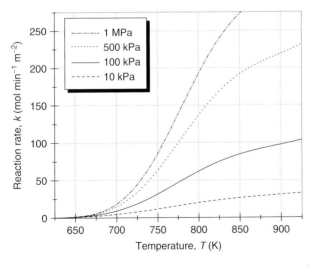

Figure 5.2 Molar reaction rate of molten lithium with perfluorocyclobutane. (After Ref. [7].)

shows similarity with the reaction behaviour of liquid lithium with liquid PVDF which commences at 355 °C.

The adiabatic combustion temperature of Li/PTFE has been calculated with NASA CEA at 0.1 MPa [8] and is depicted in Figure 5.3. The temperature profile is quite unusual as it exhibits a narrow sharp peak at 21.8 wt% Li with 3045 K. This coincides with a maximum content in the dimeric species Li_2F_2 as is depicted in Figure 5.4.

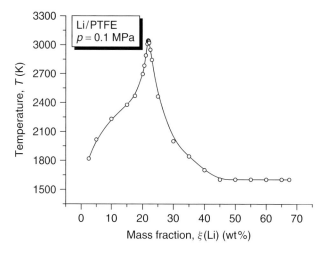

Figure 5.3 Adiabatic combustion temperature for Li/PTFE at ambient pressure [8].

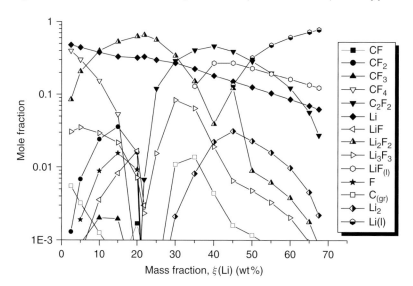

Figure 5.4 Adiabatic combustion equilibrium composition for Li/PTFE combustion at ambient pressure [8].

5.2
Magnesium

Classified work by Ladouceur on the reaction kinetics of Mg/PTFE has been summarized by Douda in 1991 [9]. Ladouceur concluded that CF_2 is the main oxidising species in Mg/PTFE flames formed by fast dissociation of tetrafluoroethylene (TFE) (Table 5.2) [10]. He also found that neither F nor F_2 play a significant role in the combustion process as there is no efficient and fast path providing either species. Table 5.2 gives three alternative thermokinetic coefficients for decomposition reactions of TFE and coefficients for reaction of either TFE or CF_2 with oxygen.

De Yong looked into kinetic modelling of Magnesium/Teflon/Viton (MTV) combustion and due to lack of experimental or theoretical data assumed rate constants for the important elementary reaction step given in Table 5.3 [11].

De Yong found that the kinetic data of the combustion step even though varying by orders or magnitude do not yield significant differences with respect to attained temperature but rather influence the time necessary to reach the maximum temperature. Figure 5.5 shows the time necessary to reach maximum temperature for MT with three different rate constants designated K1–K3 listed in Table 5.2.

A similar effect is observed as function of stoichiometry. Thus for a fuel rich composition it was found that a fast temperature rise would occur in contrary to slight fuel rich and even stoichiometric ratio (Figure 5.6). The rate of temperature increase would increase until the systems reached about 1600 K. Then within 7 μs the system temperature would increase to 3600 K. Likewise the maximum turnover of the ingredients was observed at temperatures above 1600 K. Transient species

Table 5.2 Important elementary reaction steps in anaerobic and aerobic decomposition of tetrafluoroethylene.

Step	$\Delta_R H$ (kJ mol^{-1})	A	E_a (J mol^{-1})	References
$C_2F_4 \rightarrow 2\,CF_2$	268.6	8.49×10^{17}	$-197\,004$	[10] K3
$C_2F_4 + Ar \rightarrow Ar + 2\,CF_2$	–	7.82×10^{15}	$-232\,894$	[12] K2
$C_2F_4 + M \rightarrow Ar + 2\,CF_2$	–	6.78×10^{16}	$-228\,865$	[13] K1
$C_2F_4 + O \rightarrow COF_2 + CF_2$	-395.4	8.13×10^{12}	-5271	[10]
$CF_2 + O_2 \rightarrow COF_2 + O$	–	4.00×10^{11}	$-72\,379$	[10]

Table 5.3 Important elementary reaction steps in anaerobic and aerobic combustion of Mg/PTFE [11, 14].

Step	$\Delta_R H$ (kJ mol^{-1})	A	E_a (J mol^{-1})
$Mg + F_2 \rightarrow F + MgF$	268.6	3.65×10^{10}	$-56\,500$
$Mg + CF_2 \rightarrow C + MgF_2$	–	4.00×10^{14}	$-80\,000$

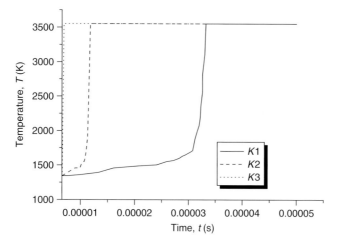

Figure 5.5 Time to maximum temperature as function of kinetic coefficient for TFE-decomposition reaction [11].

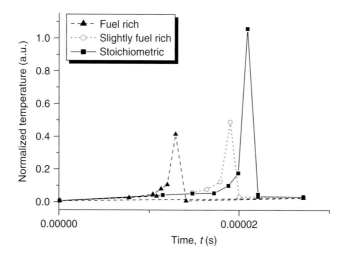

Figure 5.6 Time to maximum temperature as function of stoichiometry [11].

such as MgF, CF and CF_2 peak in this zone and decline thereafter at the benefit of MgF_2 and C.

Based on the same coefficients [11], Christo confirmed de Yong's work [15]. The thermochemistry of MTV has been dealt with in great detail in a report by De Yong and Smit [16] and later again by Christo [15].

For the ternary system Mg/PTFE/Viton® the adiabatic combustion temperature is depicted in Figure 5.7. The maximum temperature coincides with the baseline of the binary Mg/PTFE system at 32 wt% Mg. Subsequent substitution of PTFE with Viton successively lowers the combustion temperature, by about 1000 K at 75 wt% Viton.

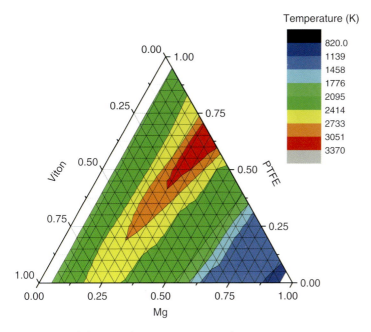

Figure 5.7 Adiabatic combustion temperature of Mg/PTFE/Viton at 0.1 MPa calculated with NASA CEA [8].

The adiabatic combustion temperature for the combustion of Mg/PTFE/air is shown in Figure 5.8. Even though neglecting actual combustion kinetics the picture shows the beneficial influence of atmospheric oxygen on the exothermicity of the overall reaction. It is also seen that the stoichiometric range of high combustion temperatures with oxygen is significantly broader than with fluorine.

In a comparative study five different thermochemical codes were used to calculate the equilibrium composition of five pyrotechnics including a particular MTV formulation described down below in Table 5.4 [17]. The results are given in the following tables (Tables 5.5–5.8). As is evident from the comparative temperature plot in Figure 5.9 all codes but the TANAKA code agree quite well and show only moderate deviation with respect to preferred reaction products. The predicted adiabatic combustion temperature is in good accord with experimental determination reported data (Chapter 9.3) (Figures 5.10–5.13).

5.3
Titanium

Both differential scanning calorimetry (DSC) and differential thermal analysis (DTA) of binary Ti/PTFE pyrolants with different particle sizes and stoichiometries has been investigated by a number of researchers listed below in Table 5.9.

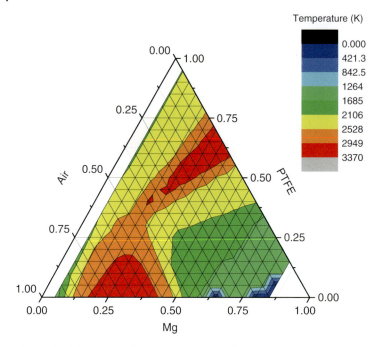

Figure 5.8 Adiabatic combustion temperature of Mg/PTFE/air at 0.1 MPa calculated with NASA CEA [8].

Table 5.4 Composition and defined thermochemical properties of MTV constituents investigated in comparative study [17].

Weight percentage	Component	Formula	Enthalpy of formation (kJ mol^{-1})
60	Magnesium	Mg	0
25	Polytetrafluoroethylene	$(C_2F_4)_n$	-809
15	Hexfluoropropene-vinylidene fluoride copolymer	$(C_{10}H_7F_{13})_n$	-2784
Pressure (MPa)		0.01, 0.1, 1, 10	

The DSC spectrum from Kuwahara reveals the PTFE-melt endotherm at 330 °C, the onset of an exothermal shoulder at 520 °C and the onset of a sharp exothermal reaction at 560 °C with peak at 570 °C (Figure 5.14). Reed *et al.* also analysed a stoichiometric mixture of Ti and Kel-F® and found the exothermal onset to occur at 470 °C [19].

Table 5.5 Species at 0.01 MPa (moles/mole).

Species	CERV	CHEETAH	EKVI	ICT	NASA	REAL	TANAKA
Mg	0.47	0.47	0.39	0.47	0.48	0.40	0.48
C(s)	0.26	0.26	0.32	0.26	0.26	0.11	0.30
$MgF_2(l,s)$	0.17	0.14	0.22	0.19	0.17	0.31	0.00
MgF_2	0.03	0.05	0.03	0.03	0.04	0.13	0.01
MgF	0.03	0.04	0.01	0.03	0.02	0.05	0.00
H_2	0.03	0.04	0.03	0.04	0.04	0.00	0.04
HF	0.00	0.00	0.00	0.00	0.00	0.00	0.00
MgH	0.00	0.00	0.00	0.00	0.00	0.00	0.01

Table 5.6 Species at 0.1 MPa (moles/mole).

Species	CERV	CHEETAH	EKVI	ICT	NASA	REAL	TANAKA
Mg	0.47	0.46	0.38	0.47	0.48	0.40	0.53
C(s)	0.26	0.26	0.32	0.26	0.26	0.11	0.28
$MgF_2(l,s)$	0.19	0.17	0.26	0.19	0.19	0.36	0.00
MgF_2	0.02	0.03	0.01	0.02	0.02	0.07	0.04
MgF	0.03	0.04	0.01	0.03	0.02	0.05	0.00
H_2	0.04	0.04	0.03	0.04	0.04	0.00	0.04
HF	0.00	0.00	0.00	0.00	0.00	0.00	0.00
MgH	0.00	0.00	0.00	0.00	0.00	0.00	0.00

Table 5.7 Species at 1 MPa (moles/mole).

Species	CERV	CHEETAH	EKVI	ICT	NASA	REAL	TANAKA
Mg	0.47	0.46	0.36	0.47	0.48	0.40	0.53
C(s)	0.26	0.26	0.33	0.26	0.26	0.10	0.29
$MgF_2(l,s)$	0.20	0.19	0.26	0.20	0.20	0.41	0.00
MgF_2	0.01	0.01	0.00	0.00	0.01	0.03	0.01
MgF	0.02	0.03	0.01	0.02	0.02	0.04	0.00
H_2	0.04	0.04	0.03	0.04	0.04	0.00	0.04
HF	0.00	0.00	0.00	0.00	0.00	0.00	0.00
MgH	0.00	0,01	0,01	0.01	0.00	0.00	0.00

Table 5.8 Species at 10 MPa (moles/mole).

Species	CERV	CHEETAH	EKVI	ICT	NASA	REAL	TANAKA
Mg	0.38	0.44	0.29	0.47	0.44	0.40	0.48
C(s)	0.26	0.26	0.32	0.26	0.26	0.11	0.30
$MgF_2(l,s)$	0.21	0.21	0.27	0.21	0.21	0.45	0.00
Mg(l)	0.06	0.00	0.06	0.00	0.04	0.00	0.00
MgF_2	0.00	0.00	0.00	0.00	0.00	0.00	0.01
MgF	0.01	0.02	0.00	0.01	0.01	0.00	0.00
H_2	0.03	0.03	0.02	0.03	0.04	0.00	0.04
HF	0.00	0.00	0.00	0.00	0.00	0.00	0.00
MgH	0.02	0.02	0.01	0.02	0.00	0.00	0.01

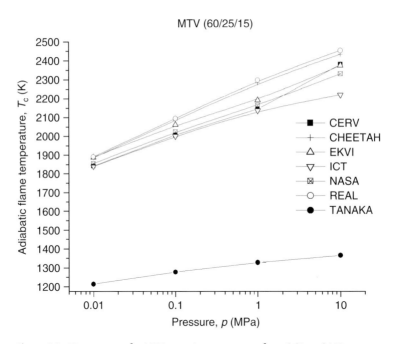

Figure 5.9 Temperature for MTV at various pressures from 0.01 to 0 MPa.

The adiabatic combustion temperature of Ti/PTFE is depicted in Figure 5.15.

Tarasov has investigated the reactivity of a special titanium alloy (87% Ti rest aluminium and transition metals: Cr, Fe, Cd, Mo and Ta) towards Fluoroplast®-42 [21]. The DSC shows a distinct onset of an exothermal reaction at 345 °C (Figure 5.16).

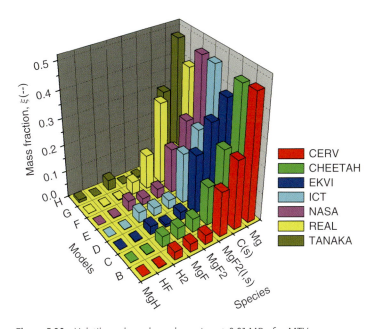

Figure 5.10 Volatile and condensed species at 0.01 MPa for MTV.

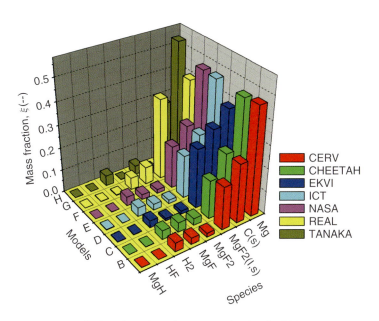

Figure 5.11 Volatile and condensed species at 0.1 MPa for MTV.

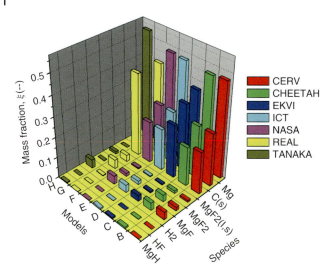

Figure 5.12 Volatile and condensed species at 1 MPa for MTV.

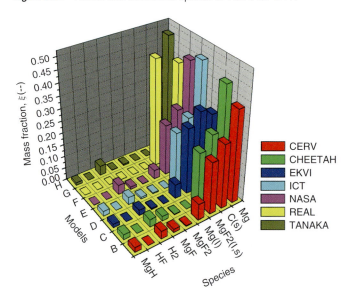

Figure 5.13 Volatile and condensed species at 10 MPa for MTV.

5.4
Zirconium

Cudziło has investigated Zr/PTFE (62.7/37.3), (20 μm Zr) under nitrogen and found the onset of a first exothermal event at 510 °C [20] (Figure 5.17).

The adiabatic combustion temperature of Zr/PTFE calculated with TIGER Code is depicted below in Figure 5.18.

Table 5.9 Properties of Ti/PTFE pyrolants investigated.

Properties	Kuwahara [18]	Reed [19]	Cudziło [20]
Stoichiometry (wt% fuel/oxidizer)	40/60	40/60	47.4/52.6
ϕ/Ti(μm)	22	20	150
ϕ/PTFE(μm)	5	35	<40
Exothermic onset ($^\circ$C)	527	500	477
Method	DSC	DTA	DTA
Heating rate (K min^{-1})	10	10	10
Atmosphere	He	Ar (80 ml min^{-1})	N$_2$

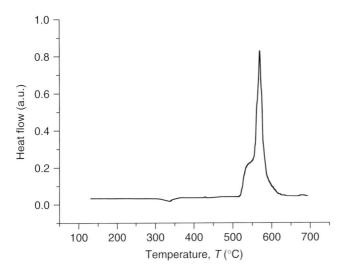

Figure 5.14 DSC of Ti/PTFE. (After Kuwahara *et al.* [18].)

5.5
Hafnium

Mixtures of hafnium/THV 220 (70/30) show slight but non-sustaining exothermal reactions in an ARC experiment at 450 $^\circ$C (Figures 5.19 and 5.20) [22].

5.6
Niob

Niobium powder reacts exothermically with Fluoroplast-42 in a stoichiometric mixture and displays onset at 317 $^\circ$C [23].

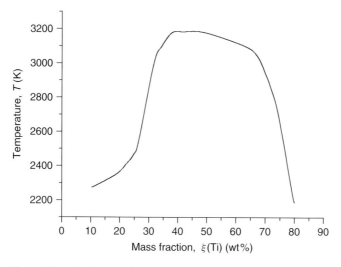

Figure 5.15 Adiabatic combustion temperature of Ti/PTFE calculated with TIGER Code [20].

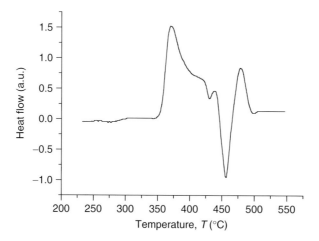

Figure 5.16 DSC of Ti-alloy/Fluoroplast-42 according to Ref. [21].

5.7
Tantalum

Stoichiometric mixtures of Tantalum/PTFE (61.7/38.3) undergo exothermic reaction commencing at 275 °C in DSC experiment [22]. Autoignition for this material has been reported to occur at 307. Likewise stoichiometric Ta/THV 220 explodes at about the same temperature (310 °C) in an ARC experiment (Figure 5.20). Stoichiometric Ta/Fluoroplast-42 shows onset of exotherm in DSC at 275 °C [23] combined with weight loss of ~18%.

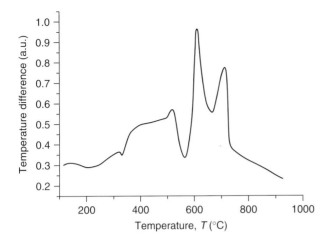

Figure 5.17 DTA plot of Zr/PTFE. (After Ref. [20].)

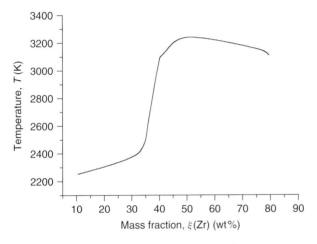

Figure 5.18 Adiabatic combustion temperature of Zr/PTFE calculated with TIGER code. (After Ref. [20].)

5.8
Zinc

DSC analysis of Zinc/PTFE mixtures reveals a pre-ignition reaction commencing at 270 °C that is followed by both fusion endotherm signal of PTFE and Zinc (Figure 5.21). At the fusion point of Zinc the actual combustion reaction starts vigorously. In nitrogen atmosphere the combustion reaction is delayed and does not start at the fusion of Zn [24].

A combined DTA/TG analysis from Cudziło [25, 26] is depicted in Figure 5.22. It shows that the combustion reaction is followed by mass loss in the order of 23 wt%. This may be associated with the volatilisation of fluorocarbon species.

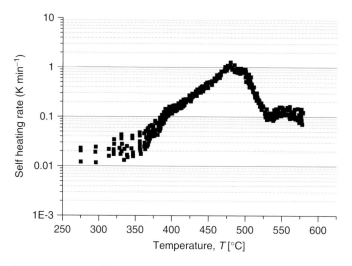

Figure 5.19 ARC Self-heating rate for Hf/THV 220 (70/30). (After Ref. [22].)

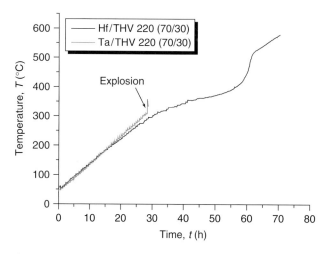

Figure 5.20 ARC temperature/time for Hf/THV 220 (70/30) and Ta/THV 220 (70/30). (After Ref. [22].)

5.9
Cadmium

Cadmium (particle diameter $= 10\,\mu m$)/PTFE (Fluon 169B) (79/21 wt%) undergoes an exothermic reaction with PTFE at $410\,°C$ in a DSC. Samples heated to $500\,°C$ comprise a black residue composed from CdF_2 and carbon and partly unreacted PTFE [27].

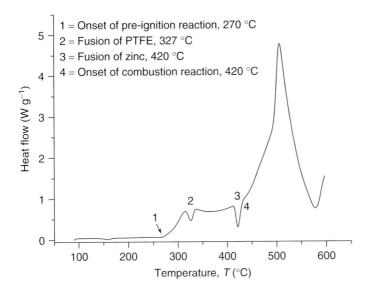

Figure 5.21 DSC of PTFE/Zn at 10 K min in air.

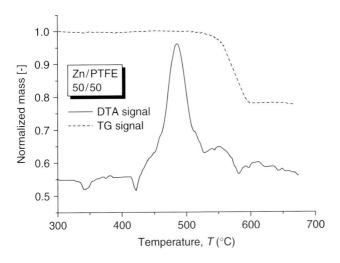

Figure 5.22 Combined DTA/TG of PTFE/Zn at 10 K min in air [25].

5.10
Boron

Bhingarkar *et al.* have investigated ignition sensitivity and interior ballistic performance of MTV igniter modified with boron [28]. In this context they performed the TG on BTV (25/65/10) (Figure 5.23).

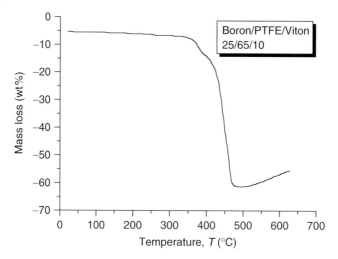

Figure 5.23 TG of BTV at 10 K under nitrogen. (After Ref. [28].)

Cudziło investigated the B/PTFE reaction recorded both DTA and TG signal (Figure 5.24) [20]. At 557 °C an exothermic reaction peaks followed by two endothermic reactions with local minima at 575 and 596 °C. These events are followed by two exothermic signals at 614 and 639 °C. These processes coincide with an overall mass loss of 45%. Thus perfectly matching the amount of BF_3 formed at the given stoichiometry.

$$4.625 \, nB + 0.5(C_2F_4)_n \longrightarrow 0.667nBF_3 \uparrow + \sim nB_4C \qquad (5.2)$$

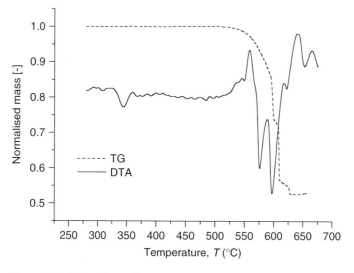

Figure 5.24 DTA and TG of B/PTFE (50/50) at 10 K under air. (After Ref. [20].)

5.11
Aluminium

The reaction of aluminium with PTFE has been studied by numerous researchers [18–20, 29]. Both a variety of different stoichiometries and particle sizes were investigated (Table 5.10). However the onset of the first exothermic reaction is obtained in a range between 500 and 550 °C.

Figure 5.25 depicts the DSC plots obtained from Turetsky *et al.* [29] superimposed from the DSC plots for PTFE in both Ar and air. Whereas micrometric aluminium

Table 5.10 Properties of Al/PTFE pyrolants investigated.

Properties	Kuwahara [18]	Reed [19]	Reed [19]	Cudziło [20]	Turetsky [29]
Stoichiometry	26.5/73.5	26.5/73.5	26.5/73.5	41.5/58.5	79.5/20.5
ϕ/Al(μm)	–	20	0.2	50	74
ϕ/PTFE(μm)	5	35	35	<40	35
Exothermic onset (°C)	550	530	520	502	510
Method	DSC	DTA	DTA	DTA	DSC
Heating rate (K min^{-1})	10	10	10	10	20
Atmosphere	He	Ar (80 ml min^{-1})	Ar	N$_2$	Ar (18 ml min^{-1})

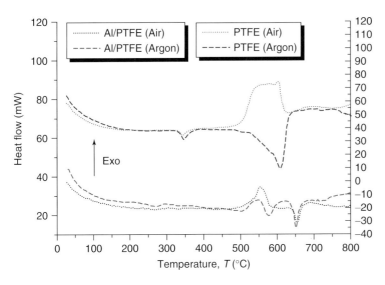

Figure 5.25 DSC of PTFE and Al/PTFE (79.5/20.5) at 10 K under air and Ar. (After Ref. [29].)

reacts only after decomposition of PTFE around 520 °C mixtures from nanometric aluminium and PTFE shows a distinct exotherm with its onset at 400 °C [30]. Another reaction has its onset at 510 °C. Pantoya *et al.* have assigned these signals to the reaction between alumina and PTFE and a subsequent exothermal phase transition of amorphous AlF_3 to α-AlF_3 as described in Eqs. (5.3) and (5.4) [30].

$$Al_2O_3 + 1.5\ (C_2F_4)_n \longrightarrow 2AlF_3 + 3CO + 230\ kJ\ mol^{-1} \tag{5.3}$$

$$AlF_3 \longrightarrow \alpha\text{-}AlF_3 + heat \tag{5.4}$$

The superimposed DSC plots for both Al/PTFE (70/30) and Al_2O_3/PTFE (66/34) are shown in Figure 5.26.

Chaudhuri *et al.* studied the reaction of atomic aluminium with PTFE decomposition products using *ab initio* calculations and obtained thermokinetic data by coupled cluster theory [31]. They considered the following anaerobic (Eqs. (5.5–5.7)) and aerobic reactions (Eqs. (5.8) and (5.9)):

$$Al + CF \longrightarrow AlF + C \tag{5.5}$$

$$Al + CF\ (Al{-}C{-}F) \longrightarrow C{-}Al{-}F \tag{5.5a}$$

$$C{-}Al{-}F \longrightarrow AlF + C \tag{5.5b}$$

$$Al + CF_2 \longrightarrow AlF + CF \tag{5.6}$$

$$Al + CF_3 \longrightarrow AlF + CF_2 \tag{5.7}$$

$$Al + COF \longrightarrow AlF + CO \tag{5.8}$$

$$Al + COF_2 \longrightarrow AlF + COF \tag{5.9}$$

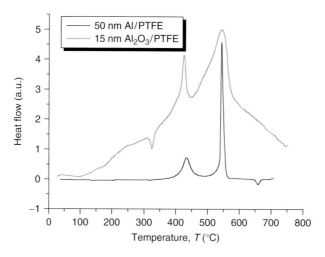

Figure 5.26 DSC of Al (15 nm)/PTFE (70/30) and Al_2O_3 (15 nm)/PTFE (66/34) at 10 K under Ar. (After Ref. [30].)

The reaction enthalpies and thermokinetic data for two temperature ranges are given below in Table 5.11

As aluminium does not undergo vapour phase diffusion flame with PTFE it is more reasonable to investigate the reaction of solid aluminium particles with PTFE decomposition products. As the topology of the $[Al_{13}]^-$-cluster (Figure 5.27) is very much like the closed packed structure of bulk aluminium it has been investigated as a model compound to study the reaction between solid aluminium and gaseous oxidizers such as hydrogen chloride [32] and chlorine [33]. Chaudhuri has adopted

Table 5.11 Gas phase reaction enthalpies and thermokinetic data for elementary reaction steps between atomic Al and PTFE decomposition products at CCSD(T)/aug-cc-pVTZ level of theory in two different temperature ranges [31].

Reaction	$\Delta_R H$ (kJ mol^{-1})	500–900 K		1000–2000 K	
		E_a (J mol^{-1})	A	E_a (J mol^{-1})	A
CF	−123.4	–	–	–	–
A	−278.6	−116 315	1.112×10^{13}	−117 696	1.345×10^{13}
B	+158.6	−153 595	1.0983×10^{14}	−154 306	1.2018×10^{14}
CF$_2$	−145.9	−50 836	1.068×10^{13}	−53 974	1.247×10^{13}
CF$_3$	−306.1	−16 777	1.359×10^{13}	−16 820	1.442×10^{13}
COF	−518.0	−7740	5.58×10^{12}	−7824	5.63×10^{12}
COF$_2$	−150.4	−35 647	1.50×10^{12}	−36 610	1.71×10^{12}

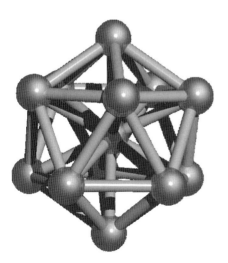

Figure 5.27 $[Al_{13}]^-$-cluster with 12-coordinated central aluminium atom.

Table 5.12 Gas phase reaction thermokinetic data for elementary reaction steps between $[Al_{13}]^-$-clusters and PTFE decomposition products at DFT/PBE0/6-311G(d) level [35].

Reaction	500–900 K		1000–2000 K	
	E_a (J mol^{-1})	A	E_a (J mol^{-1})	A
CF$_2$	−86 190	1.55×10^{12}	−87 948	1.91×10^{12}
CF$_3$	−16 777	1.359×10^{13}	−16 820	1.442×10^{13}
COF	−40 375	2.63×10^{12}	−41 254	2.95×10^{12}
COF$_2$	−18 410	9.69×10^{11}	−18 870	1.05×10^{12}

that approach and studied the thermokinetics of reactions between the $[Al_{13}]^-$ cluster and PTFE decomposition products [34, 35]. The results of this are given in Table 5.12. The reactivity of clusters is slightly different to atomic Al. Thus CF$_2$ and COF$_2$ display higher rates, whereas CF$_3$ and COF are far less reactive with $[Al_{13}]^-$.

However, based on known PTFE decomposition kinetics CF$_3$ (Table 5.2) is known to be formed at very low level only. Thus more abundant CF$_2$ will play a more significant role under anaerobic conditions. This is similar to MTV where CF$_2$ has been identified as the major oxidizing species. The adiabatic combustion temperature of Al/PTFE is depicted in Figure 5.28 [8]. The equilbrium composition for Al/PTFE at ambient pressure is depicted in Figure 5.29.

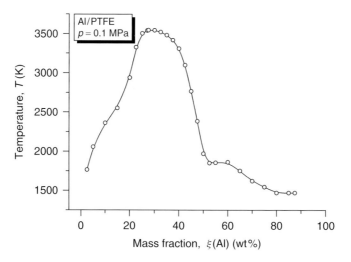

Figure 5.28 Adiabatic combustion temperature of Al/PTFE at 0.1 MPa [8].

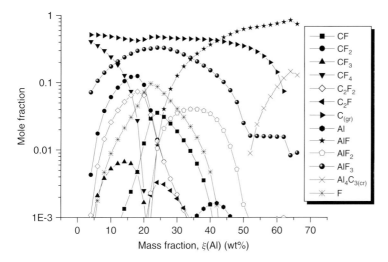

Figure 5.29 Adiabatic combustion equilibrium composition for Al/PTFE at ambient pressure [8].

5.12
Silicon

The reactivity of crystalline silicon with tetrafluoromethane, CF_4, has been investigated by Holcombe *et al.* [36]. They found a mild exothermic reaction with onset at about $500\,°C$ (peak at $530\,°C$) with no significant mass change and the onset of a strongly exothermic reaction at $990\,°C$ (peak at $1020\,°C$) with a significant mass loss indicative of volatile SiF_x formation. Samples held under CF_4 atmosphere for 75 min at $1030\,°C$ exhibited both β-SiC and metastable carbon formation on the surface of Si. The exotherm at $500\,°C$ is believed to occur from highly reactive amorphous silicon present on the surface of crystalline Si. The formation of metastable carbon, particularly diamond phase on both Si and SiC substrates by thermal treatment in fluorocarbon atmosphere has been the subject of a disclosure by Holcombe *et al.* [37].

Lee and Choi [38] observed the formation of SiF_4 upon passing CF_4 through a column filled with coarse crystalline Si – particles heated to $930\,°C$. It is observed that presence of NaF increases the conversion and reduces the onset – temperatures down to $<900\,°C$ possibly via facilitating fluoride ion migration in the condensed phase (Figure 5.30).

Wavers for electronics composed from layered Si, SiO_2, SiC and/or Si_3N_4 structures are typically etched at ambient temperature in RF-coupled fluorocarbon plasmas (CF_4/O_2). Although the reaction conditions are not quite like in a combustion of an energetic material – the Si-based substrate is not heated to high level temperatures but stays rather at ambient or moderate $T \sim 200\,°C$ – the trends in reactivity, but particularly etch-rate versus burn rate may be indicative of the trend of similar processes occurring in combustion of Si-based fluorocarbon pyrolants.

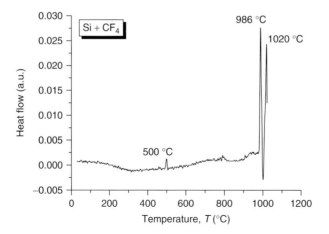

Figure 5.30 First derivative DTA signal Si waver exposed to CF_4. (After data from Ref. [36].)

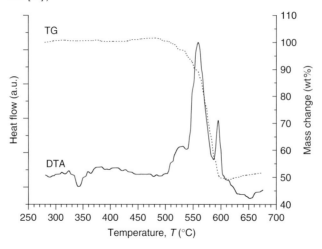

Figure 5.31 DTA and TG plot for micrometric amorphous Si/PTFE at ambient pressure. (After Ref. [25].)

Binary amorphous silicon/PTFE pyrolants have been investigated by Cudziło *et al.* [25, 39]. A combined TG/DTA trace for (Si/PTFE) (50/50) is displayed in Figure 5.31. Following the PTFE melting endotherm at ∼330 °C, the exothermic reaction commencing at ∼500 °C is due to formation of volatile SiF_4 causing the observed mass loss of ∼50%.

5.13
Calcium Silicide

The DSC/DTA TG analysis of a binary pyrolant ($CaSi_2$/PTFE:50/50) reveals onset of a first exothermic reaction at 450 °C with peak at ∼490 °C and a consecutive

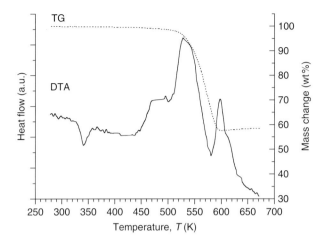

Figure 5.32 Thermal analysis of CaSi$_2$/PTFE (50/50) under air. (From Ref. [25].)

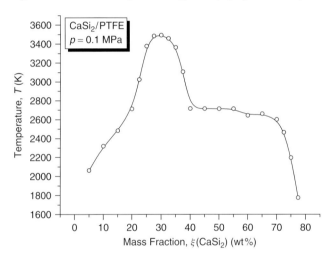

Figure 5.33 Adiabatic combustion temperature of CaSi$_2$/PTFE [8].

exothermic reaction with peak at 526 °C, the latter being accompanied with a mass loss of about ~40% that could corresponds to either release of SiF$_4$ or C$_2$F$_4$. However FTIR-spectroscopy of combustion gases of same pyrolant indicates presence of SiF$_4$ [40]. Finally another exothermic reaction is seen at 600 °C (Figures 5.32 and 5.33).

5.14
Tin

Sn(10 µm)/PTFE (Fluon L169 B) (72/28) undergoes a broad exothermal reaction at temperatures between 450 and 500 °C [41].

References

1. Haiduc, I. and Zuckerman, J.J. (1985) *Basic Organometallic Chemistry*, Walter de Gruyter, Berlin, p. 46.
2. Elschenbroich, C. (2008) *Organometallchemie*, Teubner, Wiesbaden, p. 36.
3. Sneddon, L.G. and Lagow, R.J. (1975) Vapour synthesis of polylithium compounds: the effect of halogens in activating the carbon-hydrogen bond. *J. Chem. Soc., Chem. Commun.*, 302–303.
4. Shimp, L.A., Chung, C. and Lagow, R.J. (1978) The reaction of lithium vapour with benzene and halobenzenes. *Inorg. Chim. Acta*, **29**, 77–81.
5. Landro, F.J., Gurak, J.A., Chinn, J.W. and Lagow, R.J. (1983) Synthesis of lithiumethanes ($CH_{4-n}Li_n$) and characterization of the vapour species of lithiomethanes by flash vaporization mass spectroscopy. *J. Organomet. Chem.*, **249**, 1–9.
6. Roth, E.P., Nagasubramanian, G., Tallant, D.R. and Garcia, M. (1999) Instability of Polyvinylidene Fluoride-Based Polymeric Binder in Lithium-Ion Cells: Final report. SAND Report 99-1164, Sandia National Laboratories, Albuquerque, NM.
7. Little, T.E. (1973) Reactivity of nitrogen, oxygen and halogenated gases with molten lithium metal. PhD thesis. Pennsylvania State University.
8. Gordon, S. and McBride, B.J. (1994) Computer Program for Calculation of Complex Chemical Equilibrium Compositions and Applications. NASA Reference Publications 1311, Lewis Research Center, Cleveland, OH.
9. Douda, B.E. (1991) Survey of military pyrotechnics. 16th International Pyrotechnics Seminar, Jönköping, Sweden, June 24–28, p. 1.
10. Douglass, C.H., Ladouceur, H.D., Shamamian, V.A. and McDonald, J.R. (1995) Combustion chemistry in premixed C_2F_4-O_2 flames. *Combust. Flame*, **100**, 529–540.
11. de Yong, L.V. and Griffiths, T.T. (1994) The use of equilibrium and kinetic computer programs to study the combustion of MTV formulations. 19th International Pyrotechnics Symposium, Christchurch, New Zealand, February 20–25, p. 556.
12. Modica, A.P. and LaGraff, J.E. (1965) Decomposition and oxidation of C_2F_4 behind shock waves. *J. Chem. Phys.*, **43**, 3383–3392.
13. Keating, E.L. and Matula, R.A. (1977) The high temperature oxidation of tetrafluoroethylene. *J. Chem. Phys.*, **66**, 1237–1244.
14. NIST Chemical Kinetics Database (2011) Standard Reference Database 17, Version 7.0 (Web Version), Release 1.5.3 Data Version 2011.05, *http://kinetics.nist.gov/kinetics* (accessed 1 September 2011).
15. Christo, F.C. (1999) Thermochemistry and Kinetics Models for Magnesium/Teflon/Viton Pyrotechnic Compositions. DSTO-TR-0938, Weapons Systems Division Aeronautical and Maritime Research Laboratory, Australia.
16. DeYong, L.V. and Smit, K.J. (1991) A Theoretical Study of the Combustion of Magnesium/Teflon/Viton Pyrotechnic Compositions. MRL-TR-91-25, Materials Research Laboratory, Maribyrnong, Australia.
17. Koch, E.-C., Webb, R. and Weiser, V. (2011) Review on Thermochemical Codes. O-138, NATO-Munitions Safety Information Analysis Center, Brussels, Belgium.
18. Kuwahara, T., Matsuo, S. and Shinozaki, N. (1997) Combustion and sensitivity characteristics of Mg/TF pyrolants. *Propellants Explos. Pyrotech.*, **22**, 198–202.
19. Lee, I., Reed, R.R., Brady, V.L. and Finnegan, S.A. (1997) Energy release in the reaction of metal powders with fluorine containing polymers. *J. Therm. Anal.*, **49**, 1699–1705.
20. Cudziło, S. and Trzciński, W.A. (2001) Calorimetric studies of metal/polytetrafluoroethylene pyrolants. *Pol. J. Appl. Chem.*, **45**, 25–32.
21. Tarasov, A.V., Alikhanian, A.S., Kirakosyan, G.A. and Arkhangel'skii, I.V. (2009) Chemical interaction of metallic titanium with tetrafluoroethylene – vinylidene fluoride copolymer. *Inorg. Mater.*, **46**, 1308–1312.

22. Nielson, D.B., Truitt, R.M. and Rasmussen, N. (2005) Low temperature, extrudable, high density reactive materials. US Patent 6,962,634, USA.

23. Tarasov, A.V., Alikhanian, A.S. and Arkhangel'skii, I.V. (2009) Chemical interaction of fluoropolymers with transition metals. *Inorg. Mater.*, **45**, 809–813.

24. Koch, E.-C. (2004) Pre-ignition reactions in metal/organohalogen pyrolants. 35th International Annual Conference of ICT, Karlsruhe, June 29–2 July, p. 126.

25. Cudziło, S. and Trzciński, W.A. (1999) Calorimetry studies of metal/TF pyrolants. 7^e Congres International de Pyrotechnie, Brest, France, Juin 7–11, p. 440.

26. Panas, A.J. and Cudziło, S. (2004) Complementary DSC and dilatometric investigation of M-PTFE pyrotechnic compositions. *J. Therm. Anal. Calorim.*, **77**, 329–340.

27. Cadman, P. and Gossedge, G.M. (1979) The chemical interaction of metals with polytetrafluoroethylene. *J. Mater. Sci.*, **14**, 2672–2678.

28. Bhingarkar, V.S., Sabnis, S.K., Phawade, P.A., Deshmukh, P.M. and Singh, H. (2000) Sensitivity and closed vessel evaluation of MTV igniter compositions containing boron and excess of magnesium. 27th International Pyrotechnics Seminar, Grand Junction, CO, July 16–12, p. 485.

29. Turetsky, A.L., Block, F. and Young, G. (1992) High temperature thermal analysis of a Pyronol composition. *Thermochim. Acta*, **212**, 197–207.

30. Pantoya, M.L. and Dean, S.W. (2009) The influence of alumina passivation on nano-Al/Teflon reactions. *Thermochim. Acta*, **493**, 109–110.

31. Losada, M. and Chaudhuri, S. (2009) Theoretical study of elementary steps in the reaction between aluminum and teflon fragments under combustive environments. *J. Phys. Chem. A*, **113**, 5933–5941.

32. Bergeron, D.E., Castleman, A.W., Morisato, T. and Khanna, S.N. (2004) Formation and properties of halogenated aluminum clusters. *J. Chem. Phys.*, **121**, 10456.

33. Burgert, R., Schnöckel, H., Olzmann, M. and Bowen, K.H. Jr (2006) The chlorination of the $[Al_{13}]^-$ cluster and the stepwise formation of its intermediate products, $[Al^{11}]^-$, $[Al^9]^-$ and $[Al^7]^-$: a model reaction for the oxidation of metals? Angewandte Chemie International Edition, **45**, 1476–1479.

34. Losada, M.E. and Chaudhuri, S. (2010) Aluminum-teflon reactions under extreme conditions: a multiscale approach. 14th International Symposium on Detonation, Cœur d'Alene, ID, April 11–16.

35. Losada, M. and Chaudhuri, S. (2010) Finite size effects on aluminum/Teflon reaction channels under combustive environment: a Rice-Ramsperger-Kassel-Marcus and transition state study of fluorination. *J. Chem. Phys.*, **133**, 134305.

36. Holcombe, C.E., Condon, J.B. and Johnson, D.H. (1978) Metastable carbon phases from CF^4 reactions. Part I. Reactions with SiC and Si. *High Temp. Sci.*, **10**, 183–195.

37. Holcombe, C.E., Condon, J.B. and Johnson, D.H. (1980) Process for producing diamond-like carbon. US Patent 4,228,142, USA.

38. Lee, M.C. and Choi, W. (2002) Efficient destruction of CF^4 through in situ generation of alkali metals from heated alkali halide reducing mixtures. *Environ. Sci. Technol.*, **36**, 1367–1371.

39. Cudziło, S., Huczko, A., Lange, H., Panas, A.J. and Trzciński, W.A. (2003) Self-propagating synthesis of ceramics in B/PTFE and Si/PTFE compositions. EURPYRO 2003, St Malo, France, June, p. 547.

40. Trzciński, W.A., Cudziło, S., Szala, M. and Gut, Z. (2007) Investigation of the combustion of calcium silicide/polytetrafluoroethylene mixtures. *Arch. Combust.*, **27**, 69–79.

41. Pocock, G. and Cadman, P. (1976) The application of differential scanning calorimetry and electron spectroscopy to PTFE-metal reactions of interest in dry bearing technology. *Wear*, **37**, 129–141.

6
Ignition and Combustion Mechanism of MTV

6.1
Ignition and Pre-Ignition of Metal/Fluorocarbon Pyrolants

Before ignition and steady-state combustion of a pyrolant, in the condensed phase, very often an energy release step takes place that influences both ignition and burn rate. This initial exothermal reaction is called pre-ignition reaction (PIR) [1]. PIRs have been observed with a number of different pyrolants such as $Mg/NaNO_3$ [2], Mg/BaO_2 [3], Zn/C_6Cl_6 [4], Al_3Si_2/C_2Cl_6 [5], $B/KClO_4$ [6] and finally metal/fluorocarbon systems.

It is generally assumed that the PIR yields a meta-stable species, $M \cdots AX$, constituted from both metallic fuel, M, and oxidant, AX, according to the general Eqs. (6.1) and (6.2):

$$M + AX \xrightarrow{k_1} M \cdots AX \qquad (6.1)$$

$$M \cdots AX \xrightarrow{k_2} MX + A \qquad (6.2)$$

where M is the metal and AX any oxidizing entity, with X being an electronegative atom or atomic group.

The onset temperature of the pre-ignition reaction, T_{PIR}, is usually independent of the melting or decomposition temperature of the oxidizer but related to the thermodynamic melting temperature, T_{melt} of the metal:

$$\frac{T_{PIR}}{T_{melt}} = \alpha \qquad (6.3)$$

For low melting metals such as Al, Mg and Zn, we find that $\alpha \sim 0.75 \pm 0.05$; however, with refractory fuels such as Ti and Zr, $\alpha \sim 0.45 \pm 0.05$. Table 6.1 lists T_{PIR} and onset temperature for steady-state combustion of a number of M/PTFE (polytetrafluoroethylene) pyrolants.

6.2
Magnesium–Grignard Hypothesis

With AX being a fluorocarbon, the corresponding intermediate $M \cdots AX$ can be considered a Grignard-type compound with a general structure shown in Scheme 6.1.

Metal-Fluorocarbon Based Energetic Materials, First Edition. Ernst-Christian Koch.
© 2012 Wiley-VCH Verlag GmbH & Co. KGaA. Published 2012 by Wiley-VCH Verlag GmbH & Co. KGaA.

Table 6.1 PIR-onset and steady-state combustion onset temperatures for selected metal/fluorocarbon pyrolants.

System	PIR ($^\circ$C)	Combustion ($^\circ$C)	References
Mg/PTFE	477	589	[7]
Mg/PMF	520	600	[8]
Al/PTFE	550	580	[7]
Zn/PTFE	270	420	[9, 10]
Ti/PTFE	564	580	[7]
Zr/PTFE	510	570	[11, 12]

$R_3C-Mg-F$

Scheme 6.1 Fluorocarbon Grignard(I).

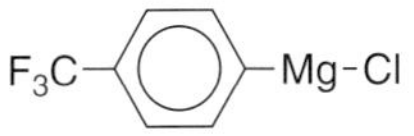

Scheme 6.2 Trifluoromethylphenylmagnesium chloride(II).

F_3C-⟨○⟩$-Mg-Cl$

The driving force of this particular PIR reaction is the exothermic metal–fluorine bond formation. On *co*-condensation of Mg atoms and alkyl halides (R–F) at cryogenic temperatures (several 10 K), the formation of the corresponding Grignard species has been observed [13–15]. Although the formation of fluorinated Grignard species is known to be hampered for kinetic reasons, both activated magnesium [16] and magnesium anthracene (Mg·$C_{14}H_{10}$) [17] give fluoro-Grignards in good yields. In view of this the formation of fluoro-Grignards, in thermal reactions appears a similar step. Both fluorinated and fluoro-Grignards are highly energetic species as they are prone to eliminate MgF_2 in an exothermal reaction [18]. Trifluoromethylphenylmagnesium chloride (**1**) (assessed enthalpy of formation, $\Delta_f H^\circ = -956$ kJ mol^{-1} [19]; Scheme 6.2) has been even observed to explode fiercely [20, 21] according to Eq. (6.4):

$$(CF_3)C_6H_4MgCl \longrightarrow MgF_2(s) + HCl + HF + \{C_7\} + H_2 + 533 \text{ kJ mol}^{-1} \text{ (6.4)}$$

The formation of **1** by Knochel reaction [22] from trifluoromethylphenylbromide with isopropylmagnesium(II) chloride in solutions >1 M can lead to a rapid increase in temperature and pressure. Hence, this reaction must be diluted down to <0.6 M to avoid catastrophic outcome [23].

The detonative potential of compositions based on Mg, benzotrifluoride and other fluorocarbons had been investigated in the 1950s; the corresponding work is still classified today (B.E. Douda, personal communication). Regardless of the known dangers of these combinations, mixtures of metal powders and benzotrifluoride have been proposed as liquid monopropellants.

At the beginning of the 1990s, it has been speculated that on combustion of Magnesium/Teflon/Viton (MTV) a similar Grignard formation step could take place [24]. For this purpose, Davis had investigated the gas-phase reaction of Mg with CH_3X, C_2H_3X (X = Cl, F) and C_2F_4 at high level of theory. He showed that all

Table 6.2 Activation energy (E_a) (kJ mol^{-1}) for Mg insertion into C–F bond of CH$_3$F, C$_2$H$_3$F and C$_2$F$_4$ and reaction enthalpy for Grignard formation step.

Fluorocarbon	Activation energy		Reaction enthalpy	
	SCF^a	MP2^b	SCF^a	MP2^c
CH$_3$F	49.5	28.5	179.1	–
C$_2$H$_3$F	56.5	25.4	168.6	226.7
C$_2$F$_4$	58.2	19.0	222.1	264.4

^{a}SCF/6-31G**. Self consistent Field
^{b}MP2/6-31G**//SCF/6-31G**. Møller-Plesset, MP
^{c}MP2/6-31G**//MP2/6-31G**.
After Refs. [25–27].

Table 6.3 Calculated harmonic vibrational frequencies for C$_2$F$_3$MgF [27, 28].

Assignment	Wavenumber, v (cm^{-1})	Intensity (km mol^{-1})
C–Mg–F	154	75
	178	79
	267	21
C–Mg	338	6
	374	8
	549	4
	655	12
	688	1
Mg–F	839	151
C–F	1098	50
C–C–F	1237	205
C–C–F	1439	196
C=C	1928	339

these reactions would exhibit sufficiently low activation energies and considerable exothermicity to allow under combustion conditions (Table 6.2) [25–27]. Table 6.3 lists the calculated vibrational frequencies for the calculated specie C$_2$F$_3$MgF [27], the structure of which is shown in Figure 6.1.

Davis also compared different possible reaction products formed in the reaction between difluorocarbene – a major high-temperature pyrolysis product of PTFE – and both Mg(^{1}S) and Mg(^{3}P). From his observations, a similar Grignard species appears energetically favoured together with a carbene-type species [24].

Thermal ignition of MTV is understood to start in the condensed phase with heat release by either Grignard-type reaction of Mg with molten PTFE (Eqs. (6.5)

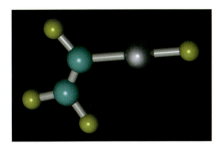

Figure 6.1 Calculated structure of C_2F_3MgF at SCF 6-31G(d) level [28] C = green, F = yellow, Mg = silver.

and (6.6)) or fluoridation of Mg with any reactive fluorocarbon specie formed by the decomposition of PTFE or Viton (Eq. (6.7)) [4]:

$$m\,Mg + (C_2F_4)_n \longrightarrow (CF_2-C(F)-Mg-F)_n + heat \qquad (6.5)$$

$$(CF_2-C(F)-Mg-F)_n \longrightarrow (CF=CF)_m-(C_2F_4)_n + m\,MgF_2(s) + heat \qquad (6.6)$$

$$Mg + 2\,R^FC-F \longrightarrow MgF_2(s) + 2\,R^FC + heat \qquad (6.7)$$

There is good reason to assume that fluoro-Grignards form as part of the PIR in the condensed phase with Mg/PTFE and Mg/PMF (polycarbon monofluoride) [9, 29]. Samples of both Mg/PTFE and Mg/PMF heated just above their observed PIR-onset temperature (500 and 520 °C) show signals in the FTIR spectrum (Figures 6.2 and 6.3), which can be assigned to a C–Mg–F units. After further exposure of the samples at $T > 700$ °C, these structures disappear and characteristic vibrations for MgF_2 are seen in both samples.

The heat released in these steps further supports decomposition of the fluorocarbon and melting the Mg. Once sufficient gaseous fluorocarbons are released,

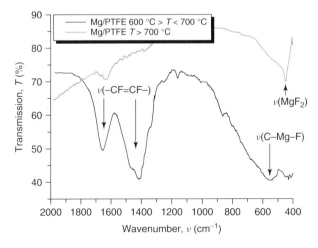

Figure 6.2 FTIR spectra of Mg/PTFE pyrolant and residues at 600 and 700 °C each [29].

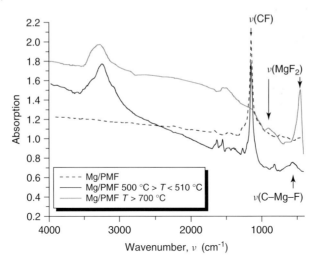

Figure 6.3 FTIR spectra of Mg/PMF pyrolant and residues at 510 and 700 °C each [9].

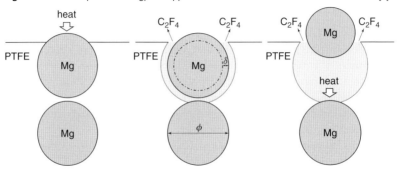

Figure 6.4 Schematic processes in the condensed phase after [35].

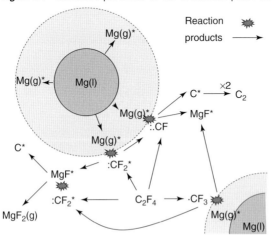

Figure 6.5 Schematic processes in the anaerobic gas phase.

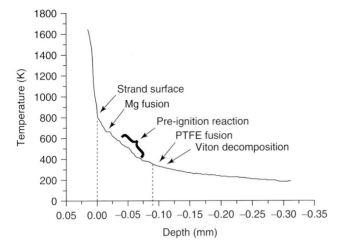

Figure 6.6 Mean-temperature profile for MTV pyrolant (30/70/3) with 22 μm Mg particles and 25 μm PTFE particle size, at 1 MPa pressure with data. (From Refs. [30–32].)

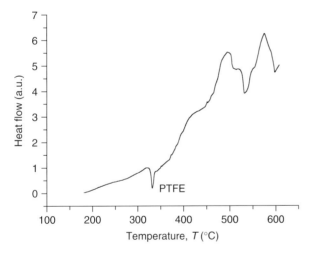

Figure 6.7 DSC for Mg/PTFE pyrolant (30/70) with 45 μm Mg particles and 5 μm PTFE particle size, at 0.1 MPa pressure.

embedded Mg particles are ejected (Figure 6.4) and react adjacent to the surface in the gas phase (Figure 6.5) as is evident from both high-speed photography and UV–VIS spectroscopy (Chapter 9). These reactions yield a heat feedback to the burning surface that, in turn, will accelerate the decomposition of PTFE and fusion of Mg until steady-state conditions are reached.

A typical temperature profile of a burning MTV strand [30–32] at 1 MPa pressure is shown in Figure 6.6. Below the surface, the strand temperature decreases

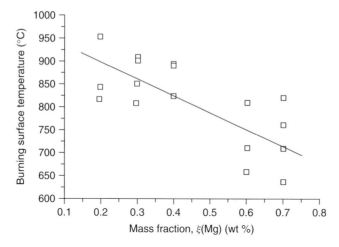

Figure 6.8 Effect of stoichiometry on surface temperature at 1 MPa. (After Refs. [30–32].)

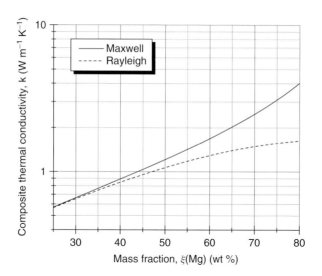

Figure 6.9 Thermal conductivity of Mg/PTFE based on either the Maywell or Rayleigh model [33].

following an exponential law. A few discontinuities are related to the fusion of Mg and heat release by PIR. Below ∼300 °C, inert heating is assumed, thus not leading to any enthalpy effects. The steep rise of temperature above 800 °C is indicative of the hot gas phase above the surface. The surface temperature profile of the combustion flame has been studied as a function of both stoichiometry and pressure. The PIR of Mg/PTFE (30/70) is obvious from the DSC plot shown in Figure 6.7.

The grain surface temperature shows influence by both pressure and temperature.

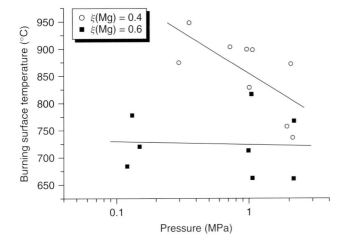

Figure 6.10 Influence of pressure on surface temperature of two different MTV pyrolant grains. (After Refs. [30–32].)

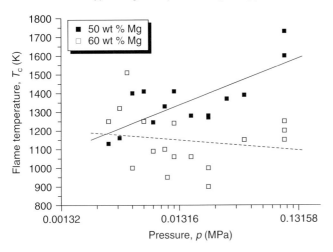

Figure 6.11 Influence of pressure on gas-phase temperature of Mg/PTFE above surface. (After Ref. [34].)

The effect of stoichiometry on the strand surface temperature is shown in Figure 6.8. It shows a temperature decrease with an increase in Mg content. This can be related to increased composite thermal conductivity (Figure 6.9), which effects faster dissipation of heat. At stoichiometries about 60 wt% Mg and above, the surface temperature scatters around the fusion temperature of magnesium (660 °C).

The influence of pressure on the temperature is less distinct. At ξ(Mg) = 60 wt%, the temperature is not affected largely by pressure changes between 0.1 and 2 MPa and scatters about 650–800 °C. In contrast, the strand surface temperature of a grain with lower Mg content (40 wt%) displays decreasing temperature with increasing pressure (Figure 6.10). This could be indicative of a Le Chatelier-type

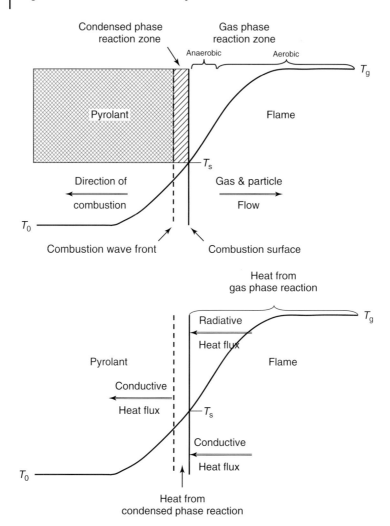

Figure 6.12 Modified qualitative structure of a pyrolant combustion wave. (After Cudziło [36].)

effect on the decomposition reactions of PTFE according to formal Eq. (6.8);

$$(C_2F_4)_n \longrightarrow n\,C_2F_4 \uparrow, \quad \Delta S < 0 \tag{6.8}$$

Figure 6.11 depicts the influence of both pressure and Mg content on the temperature of the gas phase just above the burning surface determined with W/Re thermocouples [34]. With 50 wt% Mg, a decrease in temperature is seen with decreasing pressure. However, at 60 wt% Mg, the gas-phase temperature that scatters about 1200 K seems unaffected by pressure in the range between 0.003 and 0.1 MPa [1–3]. Even though the absolute temperatures are underestimated with this method (Chapter 9), the general trend should hold valid.

Figures 6.10 and 6.11 show that the thermal equilibrium of fuel-rich pyrolants is less affected by pressure than that of fuel-lean grains, which explains the lower pressure exponents of fuel-rich grains as we will discuss later.

In summary, the combustion of MTV pyrolant is determined by processes in both gas and condensed phase. Cudziło [36] has proposed a combustion wave structure for MTV. In the pyrolant, an inert temperature increase occurs that is caused by heat conduction from the adjacent condensed-phase reaction zone. In this zone, the decomposition of the fluorocarbon occurs and initial reaction with magnesium takes place (Figure 6.12), giving rise to a further increase in the temperature. Next to the condensed-phase combustion zone is the gas-phase combustion zone that is divided into an anaerobic and aerobic part. In the former, the initial decomposition products of the pyrolant react with greater homogeneity, and in the latter, the primary combustion products mix with the atmospheric oxygen and undergo after-burn reactions, allowing for a further increase in the final temperature. The heat balance of these zones is determined by the thermal conductivity of both the condensed pyrolant, and its primary combustion products and the temperature-dependent emissivity of the pyrolant surface and the primary combustion products.

References

1. McLain, J.H. (1980) *Pyrotechnics from the Viewpoint of Solid State Chemistry*, The Franklin Institute Press, Philadelphia, PA.

2. Hogan, V.D. and Gordon, S. (1959) Pre-ignition and ignition reactions of the propagatively reacting system magnesium-sodium nitrate-laminac. *Combust. Flame*, **3**, 3.

3. Hogan, V.D. and Gordon, S. (1957) Pre-ignition and ignition reactions of the system barium peroxide-magnesium-calcium resinate. *J. Phys. Chem.*, **61** (10), 1401.

4. Gordon, S. and Campbell, C. (1955) Pre-ignition and ignition reactions of the pyrotechnic system Zn-C6Cl6-KClO4. 5th International Symposium on Combustion, The Combustion Institute, Pittsburgh, PA, p. 277.

5. Hartley, F.R., Murray, S.G. and Williams, M.R. (1984) Smoke generators. IV: the ignition of loose powder mixtures of pyrotechnic white smoke compositions containing hexachloroethane and silumin in sealed tubes. *Propellants Explos. Pyrotech.*, **9**, 108.

6. Berger, B., Brammer, A.J., Charsley, E.L., Rooney, J.J. and Wirrington, S.B. (1997) Thermal analysis studies on the boron-potassium perchlorate-nitrocellulose pyrotechnic system. *J. Therm. Anal.*, **49**, 1327.

7. Kuwahara, T., Matsuo, S. and Shinozaki, N. (1997) Combustion and sensitivity characteristics of Mg/Tf pyrolants. *Propellants Explos. Pyrotech.*, **22**, 198.

8. Koch, E.-C. (2005) Metal/fluorocarbon pyrolants: VI. combustion behaviour and radiation properties of magnesium/poly(carbon monofluoride). *Propellants Explos. Pyrotech.*, **30**, 209.

9. Koch, E.-C. (2010) Metal halocarbon combustion, in *Handbook of Combustion*, New Technologies, Vol. 5 (eds M. Lackner, F. Winter and A.K. Agarwal), Wiley-VCH Verlag GmbH, Weinheim, pp. 355–402.

10. Rozner, A.G. and Helms, H.H. (1972) Smoke generating compositions and methods of use. US Patent 3,634,283, USA.

11. Cudziło, S. and Trzciński, W.A. (2001) Calorimetric studies of metal/polytetrafluoroethylene pyrolants. *Pol. J. Appl. Chem.*, **45**, 25.

12. Cudziło, S. and Trzciński, W.A. (1999) Calorimetry studies of metal/TF pyrolants. 7ᵉ Congres International de Pyrotechnie, Brest France, June 7–11, p. 440.

13. Ault, B. (1980) Infrared matrix isolation study of magnesium metal atom reactions. Spectra of an unsolvated Grignard species. *J. Am. Chem. Soc.*, **102**, 3480.

14. Bare, W.D. and Andrews, L. (1998) Formation of Grignard species from the reaction of methyl halides with laser-ablated magnesium atoms. A matrix infrared study of CH_3MgF, CH_3MgCl, CH_3MgBr and CH_3MgI. *J. Am. Chem. Soc.*, **120**, 7293.

15. Sergeev, G.B., Smirnov, V.V. and Badaev, F.Z. (1982) Low-temperature reaction of magnesium with fluorobenzene. *J. Organomet. Chem.*, **224**, C29.

16. Rieke, R.D. and Hudnall, P.M. (1972) Activated metals. I. Preparation of highly reactive magnesium metal. *J. Am. Chem. Soc.*, **94**, 7178.

17. Beck, C.M., Park, Y.J. and Crabtree, R.H. (1998) Direct conversion of perfluoroalkanes and perfluoroarenes to perfluoro Grignard reagents. *Chem. Commun.*, 693.

18. Howells, R.D. and Gilman, H. (1975) Thermal decomposition of some perfluoroalkyl Grignard reagents. Synthesis of *trans*-1-halo- and *trans*-1-alkylperfluorovinyl compounds. *J. Organomet. Chem.*, **5**, 99–114.

19. Liebman, J.F. and Slayden, S.W. (2008) The thermochemistry of organomagnesium compounds, in *The Chemistry of Organomagnesium Compounds* (eds Z. Rappoport and I. Marek), John Wiley & Sons, Inc., Hoboken, pp. 102–130.

20. Ashby, E.C. and Al-Fekri, D.M. (1990) The reaction of benzotrihalides and benzal halides with magnesium. Synthetic and mechanistic studies. *J. Organomet. Chem.*, **390**, 275.

21. Appelby, I.C. (1971) The hazards of *m*-trifluoromethylphenylmagnesium bromide preparation. *Chem. Ind.*, 120.

22. Knochel, P., Dohle, W., Gommermann, N., Kneisel, F.F., Kopp, F., Korn, T., Sapountzis, I. and Vu, V.A. (2003) Synthese hoch funktionalisierter organomagnesiumreagentien durch halogen-metall austausch. *Angew. Chem. Int. Ed.*, **115**, 4438–4456.

23. Tang, W., Sarvestani, M., Wei, X., Nummy, L.J., Patel, N., Narayanan, B., Byrne, D., Lee, H., Yee, N.K. and Senanayake, C.H. (2009) Formation of 2-trifluoromethylphenyl Grignard reagent via magnesium-halogen exchange: process safety evaluation and concentration effect. *Org. Process Res. Dev.*, **13**, 1426–1430.

24. Davis, S.R. (1990) Theoretical and experimental study of magnesium/polytetrafluoroethylene combustion. Conference, p. 49.

25. Davis, S.R. (1991) Ab initio study of the insertion reaction of Mg into the carbonhalogen bond of fluoro- and chloromethane. *J. Am. Chem. Soc.*, **113**, 4145.

26. Liu, L. and Davis, S.R. (1991) Ab initio study of the Grignard reaction between magnesium atoms and fluoroethylene and chloroethylene. *J. Phys. Chem.*, **95**, 8619.

27. Davis, S.R. and Liu, L. (1994) Ab initio study of the insertion reaction of Mg into a C-F bond of tetrafluoroethylene. *J. Mol. Struct. (THEOCHEM)*, **304**, 227.

28. Hyperchem 8.0. (2007) *Molecular Visualization and Simulation Program Package*, Hypercube, Inc., Gainesville, FL.

29. Koch, E.-C. (2002) Metal-fluorocarbon pyrolants: IV. Thermochemical and combustion behaviour of Magnesium/Teflon/Viton (MTV). *Propellants Explos. Pyrotech.*, **27**, 340–351.

30. Kubota, N. and Serizawa, C. (1986) Combustion of magnesium/polytetrafluoroethylene. 22nd Joint Propulsion Conference, Huntsville AL, June 16–18, AIAA-86-1592.

31. Kubota, N. and Serizawa, C. (1987) Combustion of magnesium/polytetrafluoroethylene. *J. Propul.*, **3**, 303–307.

32. Kubota, N. and Serizawa, C. (1987) Combustion process of Mg/TF pyrotechnics. *Propellants Explos. Pyrotech.*, **12**, 145–148.

33. Hasselman, D.P.H. and Johnson, L.F. (1987) Effective thermal conductivity of composites with interfacial thermal barrier resistance. *J. Compos. Mater.*, **21**, 508–515.

34. Griffiths, V.S., Izod, D.C.A. and O'Sullivan, E. (1970) Some observations of some pyrotechnic compositions. 2nd International Pyrotechnics Seminar, Snowmass at Aspen, July 20–24, p. 450.

35. Kuwahara, T. and Ochiai, T. (1992) Burning rate of Mg/TF pyrolants. 18th International Pyrotechnics Seminar, Breckenridge CO, July 13–17, p. 539.

36. Cudziło, S. and Trzincski, W.A. (1998) Studies of high-energy composites containing polytetrafluoroethylene. International Annual Conference of ICT, Karlsruhe, Germany, p. 151.

7
Ignition of MTV

The ignition of energetic materials unlike the steady-state combustion is a transient process. It is affected by thermal conductivity of the pyrolant, κ; density, ρ; specific heat, c_p; surface area, A; ignition temperature, T_i; $T_a =$ ambient temperature and heat flow into the material, q. The time to ignition as derived by McLain [1] reads

$$t = \frac{\kappa \rho c_p A^2 (T_i - T_a)^2 \text{const}^2}{\pi q^2}$$

The ignition of Magnesium/Teflon/Viton (MTV) by radiative or conductive heat transfer has been investigated in order to understand processes in both bulkhead and laser igniters (Table 7.1). Lombard appears to be the first to have tested radiative ignition of MTV. He used a solar furnace that allowed a maximum irradiance, $H = 100 \text{ W cm}^{-2}$ [2]. Mg/PTFE (polytetrafluoroethylene) samples of unknown stoichiometry were tested at $H_\lambda = 20 \text{ W cm}^{-2}$. The ignition temperature was determined from radiometric measurements and was found to be quite low, 217 °C. This corresponds to 29 s irradiation at the reported irradiance level and translates into a heating rate of 6 K s^{-1}.

Ignition of Mg-based pyrolants with a 300 W CO$_2$ laser has been investigated by Östmark [3, 4]. He found that the radiative ignition of energetic materials is characterized by two asymptotes [3]. At short pulse widths, the ignitability is characterized by the ignition energy density, ε_{crit}; for long pulse widths, the ignition is characterized by the threshold ignition power, P_{crit}. Taking into account the optical absorption affected by different size type Mg powder, one obtains

$$\varepsilon_{crit} = \rho C_p \alpha^{-1}(T_i - T_a) \tag{7.1}$$

$$P_{crit} = 22k\beta\sqrt{\pi (T_i - T_a)} \tag{7.2}$$

where ρ is the density, C_p the specific heat, α the absorption coefficient, β the beam radius, k the thermal conductivity and T_i the ignition temperature. Unspecified consolidated Mg/PTFE samples were irradiated with a 514 nm argon laser, with 65 μm beam diameter and 20 ms pulse duration in nitrogen [5]. For a fixed pellet density, the ignition threshold decreases with increasing nitrogen pressure, and at constant nitrogen pressure, pellets with higher density require a higher ignition energy. Valenta investigated the laser ignition of MTV igniter compositions Types

Metal-Fluorocarbon Based Energetic Materials, First Edition. Ernst-Christian Koch.
© 2012 Wiley-VCH Verlag GmbH & Co. KGaA. Published 2012 by Wiley-VCH Verlag GmbH & Co. KGaA.

Table 7.1 Critical ignition energy data for various MTV [6, 7].

Composition	Critical ignition energy density ($J\ cm^{-2}$)
Type II pellet	13.0–13.2
Type II powder	12.4
Type III pellet	13.0
Type III powder	13.8
Sidewinder pellet	12.7–13.0

II and III both as extruded pellets and granular material [6, 7]. For the experiments, a 400 W CO_2 laser was employed, which allowed a maximum irradiance of $H = 2000\ W\,cm^{-2}$ in the pulse mode or $500\ W\ cm^{-2}$ in the continuous mode. The samples were irradiated at ambient pressure and temperature at four different irradiance levels, $H_\lambda = 70$, 160, 260 and 400 W cm^{-2}, respectively. The most noticeable difference is observed for the pellets of both Types II and III (Figure 7.1). This can be explained by the difference in particle size and surface area for the Mg powder used. Type III Mg has a surface area about 10 times greater than that of Type II Mg. The nearly identical response of Types II and III as granular material may be associated with optical surface effects, compensating the influence of the particle size. As the flux increases, these differences become marginal. It was further noted in the investigation that radiative ignition is preceded by erratic combustion events that become more frequent and subsequently coalesce to give stable combustion.

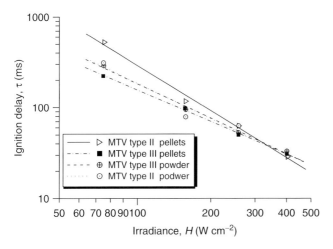

Figure 7.1 Ignition delay for Types II and III MTV composition both in granular and pelletized forms [6, 7].

Type II igniter

53.5 wt% Magnesium, granular chips, Gran 16[a]
30.7 wt% Teflon 7C
15.8 wt% Viton A
Type III Igniter
54.0 wt% Magnesium, atomized Mg spheres[b]
30.0 wt% Halon blend G10/G80
16.0 wt% Viton A

---

[a]100% through 150 μm, <40% retained on 70 μm, <50%
pass 44 μm sieve.
[b]Atomized >90% pass 44 μm sieve.

Sidewinder

54.0 wt% Magnesium[a]
30.0 wt% Halon blend G10/G80
16.0 wt% Viton A

---

[a]Type II Mg, granular Mg chips, granulation 16 (100%
pass 150 μm, 50–55% <44 μm sieve).

Fetherolf et al. investigated the influence of both ambient oxygen pressure on ignition threshold of MTV and boron as a modifier in Mg/PTFE pyrolants (Table 7.4) [8, 9]. On the contrary to the findings under nitrogen, at increased air pressure, a longer ignition delay is observed as shown in Figure 7.2 for MTV with $\xi(Mg) = 50$ wt%.

Figure 7.3 shows the detrimental influence of oxygen partial pressure on ignition delay at 0.1 MPa air and $H = 400$ W cm^{-2}. Ladouceur reasoned that under increased

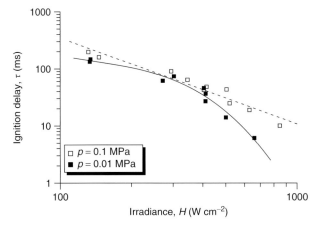

Figure 7.2 Influence of air pressure on ignition threshold of MTV [8].

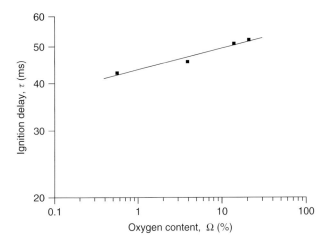

Figure 7.3 Influence of oxygen partial pressure on ignition delay of MTV at 0.1 MPa and $H_\lambda = 400$ W cm^{-2} [8].

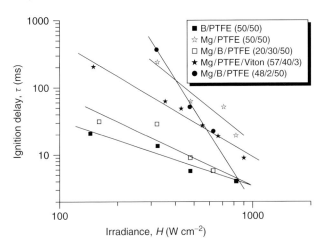

Figure 7.4 Influence of boron content on ignition threshold and ignition delay [9].

oxygen pressure, the formation of CF_4 occurs more readily, which is both quite nonreactive with Mg, thus delaying ignition, and owing to its higher heat capacity compared to that of CF_2 (102 versus 56 J K^{-1} mol^{-1} at 1300 K) also acts as a heat sink and thus lowers flame temperature, leading to a lower heat feedback and thus also reduced rate of combustion (Chapter 8) [10].

An effect on the ignition delay is also affected by the inclusion of amorphous boron [9] (see Figure 7.4), which is known as poor thermal conductor $\lambda(B) = 27$ W m^{-1} K^{-1}, $\lambda(Mg) = 171$ W m^{-1} K^{-1} [11]. Figure 7.5 shows the change of thermal diffusivity as a function of the substitution of Mg by boron in Mg/PTFE.

Chaudhri et al. did radiative ignition experiments on three different MTV samples with successively decreasing Mg particle size (ϕ(Mg): Type 1 > Type 2 >Type 3).

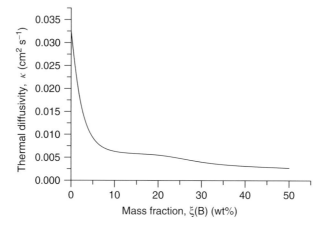

Figure 7.5 Influence of boron content in Mg/PTFE on thermal diffusivity of pyrolant [9].

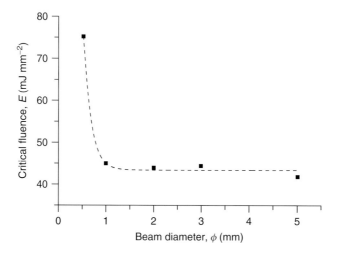

Figure 7.6 Influence of spot size on critical energy fluence to ignition MTV [14].

The actual Mg particle distribution ranged from submicron size to a few microns. From these experiments, it is obvious that laser ignition of MTV is based on absorption of radiation by Mg particles and subsequent heating. The adjacent polymer is decomposed, which is seen on non-ignited samples as darkening of it. This kind of darkening does not occur with pure PTFE samples [12]. However, radiative ignition will not occur below a certain nucleus size believed <1 mm as has been found in experiments [13, 14]. Figure 7.6 shows the steep increase in critical energy fluence for ignition for an unspecified MTV sample exposed to a 1060 nm laser.

The ignition delay for thermal ignition of MTV is depicted in Figure 7.7 for a particular formulation as a function of temperature.

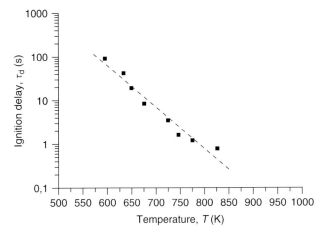

Figure 7.7 Ignition delay for thermal ignition of MTV pellets on a hot plate [13].

References

1. McLain, H.J. (1980) *Pyrotechnics*, The Franklin Institute Press, Philadelphia, PA, p. 184.
2. Lombard, J.M. (1982) Radiometric measurements of igniting thermal flow for various pyrotechnic compositions. 18th International Pyrotechnics Seminar, Steamboat Springs, Colorado, 12–16 July, p. 804.
3. Östmark, H. (1985) Lase as a tool in sensitivity testing of explosives. 8th Symposium on Detonation, USA, July 15–19, pp. 46–53.
4. Östmark, H. (1987) Laser ignition of explosives: ignition energy dependence of particle size. 3e Congres International de Pyrotechnie du Groupe de Travail de Pyrotechnie Spatiale et 12e International Pyrotechnics Seminar, Juan-Les-Pins France, June 8–12, p. 241.
5. Holy, J.A. and Girmann, T.C. (1988) The effects of pressure on the laser initiation of $TiH_x/KClO_4$ and other pyrotechnics. 13th International Pyrotechnics Seminar, Grand Junction, Colorado, July 11–15, p. 449.
6. DeYong, L. and Valenta, F.J. (1989) 14th International Pyrotechnics Seminar, Jersey Channel Islands, September 18–22, p. 123.
7. DeYong, L., Park, B. and Valenta, F. (1990) A Study of the Radiant Ignition

of a Range of Pyrotechnic Materials Using a CO2 Laser. MRL-TR-90-20, Materials Research Laboratory, DSTO, Maribyrnong, Victoria, Australia.
8. Fetherolf, B.L., Chen, D.M., Snyder, T.S., Litzinger, T.A. and Kuo, K.K. (1988) in *Base Bleed: First International Symposium on Special Topics in Chemical Propulsion* (eds K.K. Kuo and J.N. Fleming), Hemisphere Publishing, New York, pp. 43–61.
9. Fetherolf, B.L., Snyder, T.S., Bates, M.D., Peretz, A. and Kuo, K.K. (1989) Combustion characteristics and CO_2 laser ignition behaviour of boron/magnesium/PTFE pyrotechnics. 14th International Pyrotechnics Seminar, Jersey Channel Islands, September 18–22. p. 691.
10. Douda, B.E. (1991) Survey of military pyrotechnics. 16th International Pyrotechnics Seminar, Jönköping Sweden, June 24–28, p. 1.
11. Blachnik, H. (1998) *D'Ans Lax – Taschenbuch für Physiker und Chemiker*, Elemente, Anorganische Verbindungen und Materialien, Minerale, Band **3**, 4th Auflage, Springer-Verlag, Heidelberg.
12. Al-Ramadhan, F.A., Haq, I.U. and Chaudhri, M.M. (1993) Low-energy laser ignition of magnesium-Teflon-Viton

compositions. *J. Phys. D: Appl. Phys.*, **26**, 880–887.

13. Chaudhri, M.M., Al-Ramadhan, F.A. and Haq, I.U. (1993) *Dielectric Breakdown and its Influence on Ignition*, Cavendish Laboratory University of Cambridge.

14. Ramaswamy, A.L., Field, J.E. and Armstrong, R.W. (1995) The laser initiation of a series of energetic materials. 26th International Annual Conference of ICT, Karlsruhe Germany, July 4–7, p. V-18.

8
Combustion

The combustion rate of a pyrolant is determined by ambient pressure, p (kPa), and parameters inherent to a specific pyrolant such as stoichiometry, ξ (wt%), porosity, ϑ (vol%), and specific surface area of its ingredients, Σ (m^2 g^{-1}). The latter factors, in turn, determine density, ρ (g cm^{-3}) and thermal diffusivity, κ (m^2 s^{-1}), of a pyrolant, which are also functionally related to the combustion characteristics.

8.1
Magnesium/Teflon/Viton

Unlike double base and composite propellants, the combustion rate of Magnesium/Teflon/Viton (MTV) does not correlate with the adiabatic combustion temperature, which has its maximum at ξ (Mg) = 0.32. The combustion rate of MTV, however, correlates with the metal content as is observed for many other pyrolants such as Mg/NaNO$_3$ [1], Ti/KNO$_3$ and Zr/KNO$_3$ [2]. Thus, it increases exponentially with ξ (Mg) up to stoichiometries between 55 and 70%. The peak range appears to depend on the surface area of both Mg and polytetrafluorethylene (PTFE) particles applied [3–5].

8.1.1
Pressure Effects on the Burn Rate

As is the case with the many energetic materials, the combustion rate of MTV is affected by ambient pressure. The pressure sensitivity is often described as Vieille's law:

$$u = ap^n$$

where a describes the temperature sensitivity and n is the pressure exponent and pressure p. The pressure sensitivity of MTV has been investigated by Gorbunov, Peretz, Kubota and Kuwahara. In accordance with their results, fuel-rich pyrolants display low-pressure exponents, whereas fuel-lean pyrolants exhibit pronounced sensitivity to pressure changes.

Metal-Fluorocarbon Based Energetic Materials, First Edition. Ernst-Christian Koch.
© 2012 Wiley-VCH Verlag GmbH & Co. KGaA. Published 2012 by Wiley-VCH Verlag GmbH & Co. KGaA.

8.1.2
Particle Size Distribution and Surface Area Effects on the Burn Rate

The influence of both PTFE and Mg particle size distribution on combustion of MTV and its pressure sensitivity has been investigated in great detail by Jackson and Dadley [3], Kubota [6–8] and Kuwahara [9–11]. The data can be summarized as follows:

- high burn rates and low-pressure exponent are realized with

    – fine Mg particle size, small ϕ(Mg) = high surface area
    – large PTFE particle size, large ϕ(PTFE) = low surface area
    – high ξ(Mg)
    – low $\omega = \phi$(Mg)/ϕ(PTFE)(<0.05)
    – low ξ(Viton).

In turn, low burn rates and high-pressure exponent pyrolants are realized with the opposite settings. The ambient pressure combustion behaviour of pyrolants made from 17.6 μm Mg and 5 wt% Viton with different type PTFE particles is depicted in the following pictures. Both PTFE types were further fractionated to obtain defined particle size distributions.

Figure 8.1 shows the influence of PTFE particle sizes on the burn rate. At low ξ(Mg), fine PTFE particles affect high burn rate and coarse PTFE particles yield low burn rates. Although at high Mg content fine particles yield the lowest burn rate and coarse particles cause high burn rates. The inflection point for this effect is roughly at 45 wt% Mg. The same holds true for a different PTFE quality, Fluon G1, as shown in Figure 8.2, with the inflection point at slightly higher mass fraction ξ(Mg) = 50 wt%.

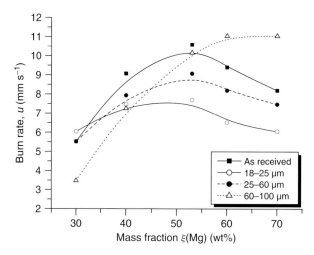

Figure 8.1 Influence of Fluon G2F particle size on the ambient pressure burn rate of MTV [3].

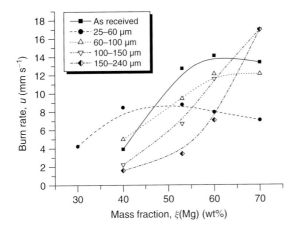

Figure 8.2 Influence of Fluon G1 particle size on the ambient pressure burn rate of MTV [3].

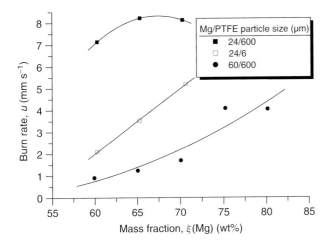

Figure 8.3 Influence of both Mg and PTFE particle size on the burn rate of MTV [9].

Figure 8.3 shows the variation of burn rate at ambient pressure with two grades both Mg and PTFE at 5 wt% level Viton [9].

The influence of the actual Mg surface area is shown in Figure 8.4 for MTV (65/30/5) [9].

The pressure sensitivity for a series of MTV pyrolants is shown in Figures 8.5 and 8.6. Generally, the burn rate increases with increasing pressure. This is indicative of the influence of heat feedback from the gas phase on the burn rate. As the PTFE content decreases, the pressure sensitivity also decreases. Although fuel-rich formulations have a rather linear slope, PTFE-rich formulations display an exponential increase of burn rate with rising pressure, indicating a different combustion mechanism (Tables 8.1 and 8.2).

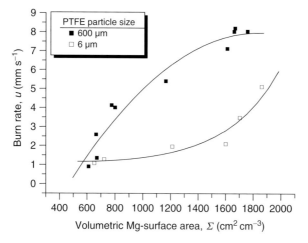

Figure 8.4 Influence of Mg surface area on the ambient pressure burn rate of MTV (65/30/5) [9].

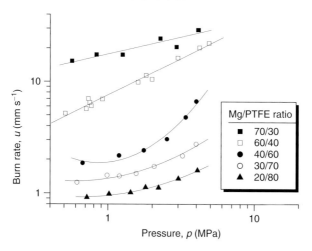

Figure 8.5 Influence of stoichiometry on pressure sensitivity of the burn rate of MTV with 3 wt% Viton with Mg (22 μm) and PTFE (25 μm).

The influence of different Mg particle sizes on fuel-rich MTV (70/18/12) is shown in Figure 8.7 [11]. Very fine Mg yields the highest burn rate owing to scatter the linear fit asserts a negative pressure exponent (Table 8.3). At lower pressures, 49 μm Mg particles show significantly low burn rate that will increase with increasing pressure and coincide with 19 μm material at ∼4 MPa. Similar effects are observed for the pyrolant using the 79 μm Mg. However, it does not reach the 19 μm material level.

The combustion behaviour of fuel-lean Mg/PTFE pyrolants (23/77) at reduced pressure (2–26 kPa) indicates an incomplete reaction [12]. From another study,

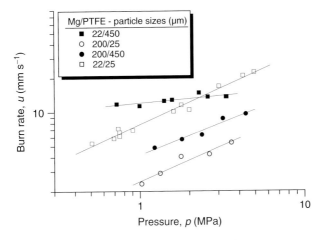

Figure 8.6 Influence of both Mg and PTFE particle size distribution on pressure sensitivity of a 60/40/3 MTV formulation particle sizes indicated [7].

Table 8.1 The Vieille parameter for Figure 8.5.

Parameter	70/30	60/40
a	17.541	7.474
n	0.3278	0.6867

Table 8.2 The Vieille parameter for Figure 8.6.

Parameter	22/450	22/25	200/25	200/450
a	12.056	7.627	2.509	4.126
n	0.1245	0.6803	0.5897	0.5808

low-pressure combustion behaviour in both air and nitrogen has been investigated (Figure 8.8 and Table 8.4) [13, 14].

The authors argue that the lower burn rate in air is due to competing oxidation that is less exothermic than fluoridation and, hence, influences the heat balance at the surface of the pyrolant. At higher pressures, these effects are compensated by better heat transfer as can be seen by the convergent lines in Figure 8.8 [13, 14]. Ladouceur explained the lower burn rate with preferred CF_4 formation in oxygeneous atmosphere. CF_4 reacts slower with Mg than that with CF_2 and, thus, is held responsible for the observed lower burn rate [15]. The pressure dependence of a stoichiometric binary Mg/PTFE (32/68) formulation made with fine particles

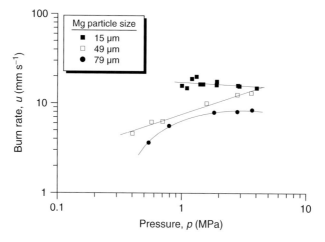

Figure 8.7 Influence of Mg particle size on pressure sensitivity of MTV (70/18/12) [11].

Table 8.3 The Vieille parameter for Figure 8.7.

Parameter	15 μm	49 μm
a	17.406	7.662
n	−0.0933	0.4480

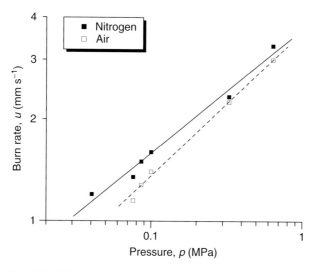

Figure 8.8 Influence of pressure and the type of atmosphere on the combustion rate of (50/50/x) MTV using 20 μm Mg [13, 14].

Table 8.4 The Vieille parameter for Figure 8.8.

Parameter	Air	Nitrogen
a	0.188	0.265
n	0.429	0.386

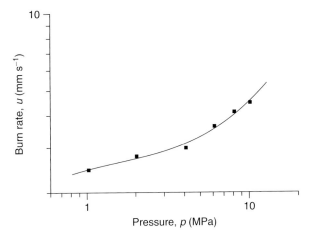

Figure 8.9 Pressure dependence of combustion rate Mg/PTFE (32/68) [4].

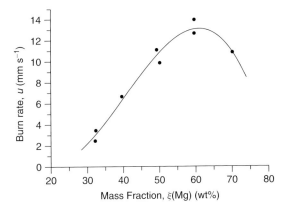

Figure 8.10 Combustion rate of Mg/PTFE as a function of stoichiometry at 6.1 MPa [4].

($\sim$4 μm Mg) is shown in Figure 8.9. It shows the exponential slope typical for PTFE-rich formulations [4].

The variation of combustion rate with stoichiometry at 6 MPa is shown in Figure 8.10.

The influence of grain temperature on the pressure-dependent burn rate of MTV has been investigated in [16–18] (Figure 8.11). It clearly indicates the importance

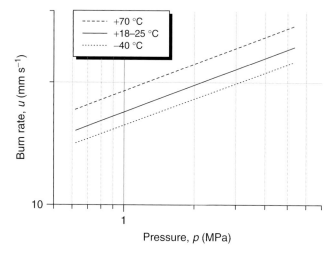

Figure 8.11 Influence of pyrolant temperature on pressure dependence of MTV (58/38/4) [16–18].

Table 8.5 The Vieille parameter for Figure 8.11.

Parameter	−40 °C	+25 °C	+70 °C
a	15.7	16.9	19.2
n	0.22	0.22	0.22

of the condensed phase heat balance on the burn rate. The pressure exponent basically does not change, but the temperature coefficient is altered (Table 8.5).

The pressure dependence for pyrolants using two different types of Mg (49 and 78 µm) is shown in Figure 8.12. Although fine Mg yields a burn rate increase at 70 wt% Mg, it does effect a decrease in burn rate at 80 wt%, even though the pressure exponent is very low ($n = 0.06$) (Table 8.6) [10].

The pressure dependence for an igniter formulation

Type I MTV

---

54 wt% Magnesium, type III, granulation 16, with 50–55 wt% is 325 mesh
30 wt% Teflon, L-P-403 with 15 wt% type II and 15% type IV
16 wt% Viton, WS 7682

---

up to 8 MPa is shown in Figure 8.13 [19]. Its sigmoidal shape of the pressure curve is indicative of two mechanistic transitions as proposed in [20].

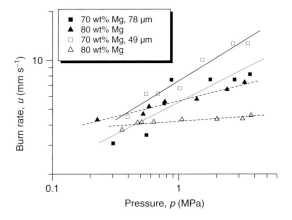

Figure 8.12 Pressure dependence of MTV with 49 μm Mg [10].

Table 8.6 The Vieille parameter for Figure 8.12.

Parameter	50 wt% Mg (49 μm)[a]	70 wt% Mg (78 μm)	70 wt% Mg (49 μm)	80 wt% Mg (78 μm)	80 wt% Mg (49 μm)
a	2.9	5.8	7.4	5.6	4.2
n	0.7	0.32	0.47	0.20	0.06

[a] Not shown in Figure 8.12.

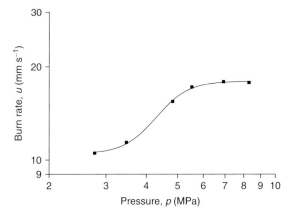

Figure 8.13 Pressure dependence of the burn rate of type I MTV according to Ref. [19].

8.2
Porosity

Other inherent factors influencing the burn rate include both the density and porosity of the pyrolant grain. It is a known fact that the burn rate, u, of MTV and its derivatives will increase with decreasing density, ρ_{exp} [21]:

$$u \approx \frac{\rho_{max}}{\rho_{exp}} \qquad (8.1)$$

If the porosity increases, the permeability of the grain for gaseous reaction products increases as well. Cudziło investigated the effect of density on the burn rate of both binary Mg/PTFE pyrolants Mg_4Al_3/PTFE and could demonstrate a significant effect of porosity on the combustion rate (see below) [22, 23].

8.3
Burn Rate Description

A number of burn rate models have been developed for MTV by Cudziło, Kubota, Kuwahara and Koch.

Kubota's model is based on the assumption that the heat generation with MTV occurs exclusively in the gas phase [8]. Thus, heat transfer from the gas phase to the burning surface, $\dot{q}_{g,s}$ (W mm^{-2}), is the determining factor for the burn rate [6–8]:

$$u = \frac{\dot{q}_{g,s}}{\rho_p(\Delta T C_p - Q_s)} (\text{mm s}^{-1}) \qquad (8.2)$$

where ρ_p is the density of the pyrolant; C_p the specific heat of the pyrolant; $Q_s = \Delta_v H(\text{PTFE}) + \Delta_m H(\text{Mg})$; $\Delta_v H(\text{PTFE})$ the dissociation enthalpy of PTFE; $\Delta_m H(\text{Mg})$ the fusion enthalpy of Mg and ΔT the temperature difference between the gas phase and burning surface.

A similar yet different burn rate model has been applied by Kuwahara et al. [9]. They assumed that

- The reaction rate of the individual Mg particles, u_{Mg}, determines the burn rate, u.
- A Mg particle has a uniform temperature and its surface area reacts with the PTFE to yield a reaction layer δ, which is mainly solid MgF_2.
- The ejection of an Mg particle from the condensed phase takes place after the reaction layer reaches a value of δ (for definition, see Figure 6.4 on page 72)
- Reaction rate of the Mg particle, u_{Mg}, is determined by the particle diameter, ϕ_{Mg}.

The schematic combustion shown in Figure 6.4 on page 72 accounts for Kuwahara's model:

$$u = \frac{1}{t_{Mg} \cdot \mu} \qquad (8.3)$$

$$t_{Mg} = \frac{\delta_{MgF_2}}{u_{Mg}} \qquad (8.4)$$

$$\mu = \frac{1}{\phi_{Mg}} \cdot \sqrt[3]{\frac{\rho_p \xi(\text{Mg})}{\left(\frac{\pi}{6\rho(\text{Mg})}\right)}} \qquad (8.5)$$

Substituting both Eqs. (8.5) and (8.4) in Eq. (8.2) yields

$$u = \frac{1}{\delta_{MgF_2}} \cdot \sqrt[3]{\frac{1}{\rho_p \xi(Mg)}} \cdot \sqrt[3]{\frac{\pi}{6\rho(Mg)}} u_{Mg}\phi_{Mg}$$

where $\xi(Mg)$ is the mass fraction of Mg, ρ_p the density of the pyrolant grain, $\rho(Mg)$ the density of Mg and δ the thickness of MgF$_2$ layer.

Unlike Kubota [8], Kuwahara [9] took into account energy-releasing reactions in the condensed phase as well. However, his analytical model requires only burn times and does not need to solve the heat and mass balance equations for this case.

Cudziło makes an even broader approach [22, 23] and takes into account both conductive and radiative heat flux from the gas-phase reaction and heat generated in the condensed phase (Figure 6.12 page 76). His model is based on a burn rate model that has been developed earlier for Mg/NaNO$_3$ [24, 25]:

$$u\rho_p \left[\xi_{PTFE} Q_{PTFE}^{m,dec.} + (1 - \xi_{PTFE})Q_{Mg}^m + C_p(T_f - T_0) \right] = Q_k + Q_r + Q_s$$

where $Q_{PTFE}^{m,dec.}$ is the sum of both enthalpy of fusion and decomposition of PTFE; Q_{Mg}^m the enthalpy of fusion of Mg; T_f the burning surface temperature; T_0 the ambient temperature; Q_k the conductive heat flux; Q_r the radiative heat flux and Q_s the heat flux from condensed phase reaction.

Koch in his model has made the similar assumptions as both Cudziło and Kuwahara [26]. He also showed that models based on exclusive radiative transfer (Q_r) from the gas phase are insufficient to account for the observed burn rate behaviour.

8.4
Combustion of Metal–Fluorocarbon Pyrolants with Fuels Other than Magnesium

8.4.1
Magnesium Hydride

Magnesium hydride has been proposed as a slow burning fuel in tracer and illuminant compositions [27]. The slow burn rate of MgH$_2$-based pyrolants is mainly due to the low thermal conductivity that is in the order of $\kappa = 4 \times 10^{-2}$ W m^{-1} K^{-1} (compared to 18 W m^{-1} K^{-1} for Mg), which, on the other hand, also allows for a much easier ignition as heat accumulates better than that in magnesium-based pyrolants. In addition, the release of hydrogen facilitates combustion in oxygeneous environments. The combustion of binary MgH$_2$/PTFE pyrolants has been investigated by Koch *et al.* [28]. Steady-state combustion occurs between 28 and 68 wt% MgH$_2$ (Figure 8.14).

At stoichiometries above 60 wt% MgH$_2$, the combustion wave proceeds in a different manner through the column as at lower MgH$_2$ values. This is recognized by intense sparkling and explosion of particles and a less pronounced combustion flame (see Figure 9.15 on page 129).

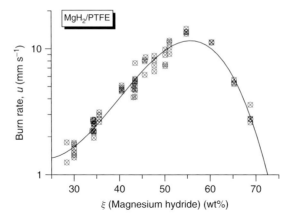

Figure 8.14 The burn rate of MgH$_2$/PTFE pyrolant at ambient pressure [28].

8.4.2
Alkali and Alkaline Earth Metal

8.4.2.1 Lithium

Mixtures of granular lithium with perhalogenated carbon compounds such as poly-chlorotrifluoroethylene, (ClFCCF$_2$)$_n$, (PCTFE), are reported to yield low combustion temperatures when reacted in confinement (constant volume). The measured combustion temperatures are given below in Table 8.7. Thermochemical codes fail to converge for these fuel-rich stoichiometries as the reaction products are in the condensed phase exclusively.

The combustion of closed-vessel intimate Li/PTFE sandwich arrangements has been reported by Smith [30]. In this investigation, lithium rods (length 216 mm, outer diameter: 26 mm) were machined to held various diameter rods from either machined PTFE (high density = 98% Theoretical maximum density (TMD)) or pressed PTFE powder (low density <50% TMD). The Li tube/PTFE rod composite was placed in a matching stainless steel tube and evacuated to 0.1 kPa. The composite was ignited on top of the PTFE rod with a pyrolant. The propagation

Table 8.7 The reaction temperature of binary Li/PCTFE pyrolants.

ξ(Li)/wt%	Measured temperature (°C)
75	650
80	590

From Ref. [29].

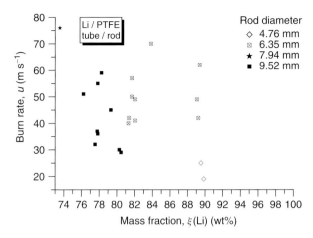

Figure 8.15 Speed of combustion for various Li/PTFE stoichiometries and rod diameters [30].

of the combustion front was determined by tracing the temperature rise along the stainless steel tube with probes attached on the outside. The maximum pressure for reactions with stoichiometries in the range of $\xi(Li) = 74–90$ wt% was between 1.5 and 3.0 MPa during combustion. After cooling of the vessel, the remainder pressure was back at the initial value (0.1 kPa) indicative that only condensed products at STP are formed. The propagation velocities range between $u = 20$ and 75 m s^{-1} for all stoichiometries investigated indicative of rather low sensitivity of the deflagration process to stoichiometry in this compositional range (Figure 8.15).

A combustion model has been developed for the Li/PTFE sandwich configuration burn rate [30]. It is concluded that condensed phase heat transfer is the sole mechanism by which the reaction propagation is controlled. This explains the relatively high values obtained for fuel-rich stoichiometries and the large scatter observed that is due to different sample density/porosity and thus varying thermal diffusivity, α.

8.4.2.2 Magnesium–Aluminium Alloy

The combustion of binary Mg_4Al_3/PTFE has been studied by Cudziło [22, 23] and Minghua Chen [31]. The burn rate increases with decreasing density and increasing metal content (Figure 8.16). The Mg_4Al_3 particle size used by Cudziło was 75–105 μm.

8.4.3
Titan

As other metallic fuels Ti/PTFE compositions show a similar functional relationship between metal content and burn rate [4]. Figure 8.17 shows the burn rate for

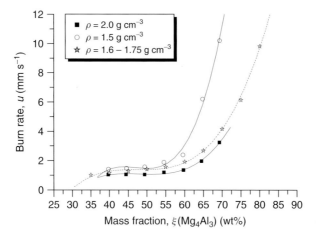

Figure 8.16 The burn rate of binary Mg_4Al_3/PTFE mixtures at varying densities.

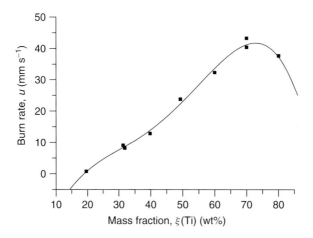

Figure 8.17 The burn rate of binary Ti/PTFE with 10 μm particle size at 6.1 MPa [4].

stoichiometries with 12 wt% Viton each [11]. Figure 8.18 shows the effect of stoichiometry and particle size on the pressure-dependent burn rate (Table 8.8).

The solid-state reaction between titanium and carbon black is commencing not below 1200 K; hence, it has been proposed to modify Ti/C pyrolant with PTFE to lower the activation energy, and facilitate ignition [32, 33]. The burn rate and pressure dependence of Ti/C pyrolant having a constant fraction of 9.1 wt% PTFE with very small amount Viton binder ∼0.1 wt% have been investigated, which are shown in Figures 8.19 and 8.20. The rate of combustion increases with increasing metal content and is faster with small Ti particles. The pressure exponents for both 60 and 80 wt% Ti are very close, indicating a similar dependence on the gas-phase reaction (Table 8.9).

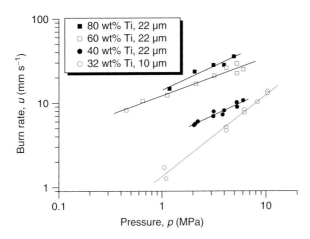

Figure 8.18 The burn rate of Ti/PTFE/Viton with 10 and 22 μm Ti particle size [11].

Table 8.8 The Vieille parameter for Figure 8.18.

Parameter	32 wt% Ti	40 wt% Ti	60 wt% Ti	80 wt% Ti
a	1.16	4.0	12.4	14.6
n	1.04	0.48	0.43	0.53

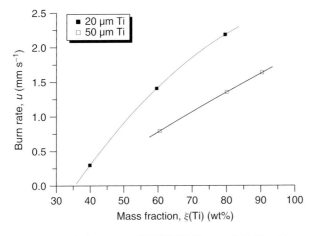

Figure 8.19 The burn rate of Ti/C/PTFE/Viton at 0.1 MPa with two-particle qualities.

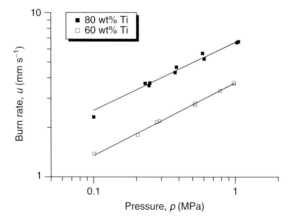

Figure 8.20 The burn rate of two Ti/C/PTFE/Viton formulations as a function of pressure.

Table 8.9 The Vieille parameter of Ti/C/PTFE/
Viton pyrolants for Figure 8.20.

Parameter	60 wt% Ti	80 wt% Ti
a	3.7	6.6
n	0.442	0.414

8.4.4
Zirconium

The combustion rate of Zr increases with increasing metal content but is significantly lower at given particle size than with homologous titanium. Figure 8.21

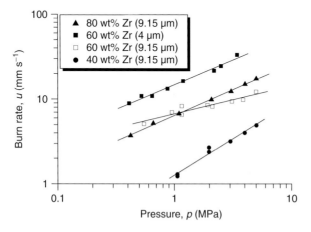

Figure 8.21 The burn rate of Zr/PTFE/Viton pyrolant with 12 wt% Viton and 9 and 4 μm zirconium particles [11].

Table 8.10 The Vieille parameter for Figure 8.21.

Parameter	40 wt% Zr 9 μm	60 wt% Zr 9 μm	60 wt% Zr 4 μm	80 wt% Zr 9 μm
a	1.4	6.7	14.5	6.3
n	0.80	0.33	0.62	0.76

shows the burn rate a three different stoichiometries and in one case with two different Zr particle qualities as a function of pressure (Table 8.10) [11].

8.4.5
Zinc

The burn rates for Zn/PTFE are low when compared to other pyrolants and range from 0.2 to $1 \, \text{mm s}^{-1}$ for fuel-rich stoichiometries (S. Cudziło, private communication) [34] (Figure 8.22). The combustion proceeds in a spinning mode and seems restricted to the condensed phase with very little dark red flame.

Cudziło *et al.* have determined thermal diffusivity and specific heat for the range of temperature between 30 and 290 °C (Table 8.11) [35, 36].

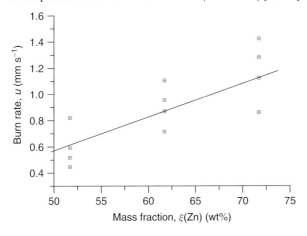

Figure 8.22 The burn rate for Zn/PTFE at ambient pressure at various pressing densities.

Table 8.11 Thermophysical/thermochemical properties of binary Zn/PTFE pyrolants for the range $T = 30-290 \, °C$ [35, 36].

$\xi(Zn)$ (wt%)	Density (g cm^{-3})	C_p (J g^{-1} K^{-1}) $= \alpha + \beta \times T + \gamma \times T^2$			K(m^2 s^{-1}) $= \mu + \nu \cdot \ln T$		α (K^{-1})
		α	β	γ	μ	ν	
56.7	3.616	0.6270	1.8786×10^{-4}	1.8143×10^{-6}	4.2734	−0.2691	81.8
76.7	4.706	0.5124	2.0986×10^{-4}	8.1429×10^{-7}	9.2771	−0.8825	75.9

8.4.6
Boron

Shidlovskii *et al.* investigated the combustion behaviour of binary boron/PTFE pyrolants. The burn rate at 0.6 MPa as a function of stoichiometry is shown in Figure 8.23 [4]. It peaks at ~35 wt% boron, showing that it does not correlate with the peak adiabatic combustion temperature that is obtained at 12.5 wt%.

The pressure dependence of three different binary pyrolants is shown in Figure 8.24.

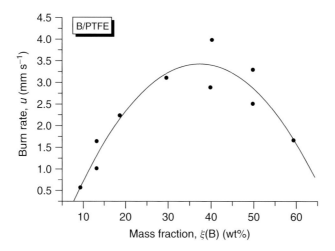

Figure 8.23   The burn rate of binary B/PTFE (boron particle size: ~3 μm) at 6 MPa [4].

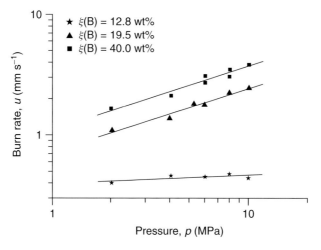

Figure 8.24   Pressure dependence of the burn rate of binary B/PTFE (boron particle size: ~3 μm) [4].

8.4.7
Magnesium Boride, MgB$_2$

A large range of stable combustion stoichiometries is obtained with MgB$_2$/PTFE (40–95 wt%). The burn rate peaks at ~75 wt% and is a little slower than that with Mg$_2$Si (Figure 8.25). Again it does not correlate with the calculated flame temperature of the primary combustion zone [28].

8.4.8
Aluminium

The burn rate of thermally ignited loose Al/PTFE ($\ll$50% TMD) both unconfined and confined in tubes has been investigated by numerous researchers [37–40]. The results for both micrometric and nano-metric aluminium blended with 200 nm PTFE particles are shown in Figure 8.26. The burn rate was obtained from loose powder filled in a semicircular groove (3.175 mm in diameter) in an acrylic block ignited by Nichrome$^{®}$ wire. The burn rate of the nano-pyrolant is approximately an order of magnitude greater than that of the micro-pyrolant.

At stoichiometries between 10 and 50 wt% aluminium, the Al/PTFE reaction yields significant amount of gaseous products such as AlF, AlF$_2$ and AlF$_3$ (see

Table 8.12 The Vieille parameter for B in Figure 8.24.

Parameter	12.8 wt% B	19.5 wt% B	40 wt% B
a	0.398	0.693	1.046
n	0.068	0.548	0.557

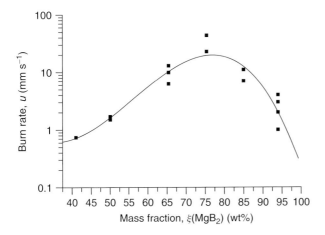

Figure 8.25 The combustion rate of MgB$_2$/PTFE [28].

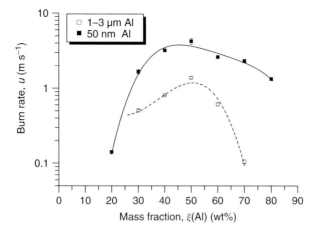

Figure 8.26 The burn rate for loose powder Al/PTFE with both 1–3 μm and 50 nm Al and 200 nm PTFE.

Figure 5.29 page 63). Unlike gasless pyrolant systems that essentially prove pressure independent such as Al/MoO₃, confinement has an accelerating effect on the combustion of Al/PTFE as shown in Figure 8.27. Hence, high burn rates under confined conditions appear to coincide with the stoichiometry for maximum adiabatic combustion temperature (see Figure 5.28 on page 63) and gas volume.

The combustion rate of pressed nano-AlTV as a function of stoichiometry at 1.9 MPa is shown in Figure 8.28. The Al-content/burn rate curve of the consolidated AlTV displays a similar shape with similar maximum as that of the loose powder burns.

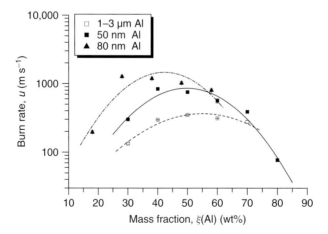

Figure 8.27 The confined burn rate for Al/PTFE/Viton (80/200 nm) and two binary Al/PTFE mixtures (50/200 nm) and (1–3 μm/200 nm) with error bars.

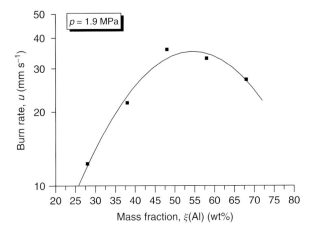

Figure 8.28 Influence of stoichiometry on pressed Al/PTFE/Viton (80/200 nm) burn rate (6.35 mm pellet diameter).

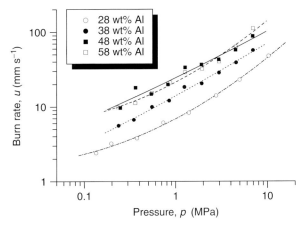

Figure 8.29 The combustion rate for pelletized Al/PTFE/Viton (80/200 nm) as a function of pressure (6.35 mm diameter) with second-order polynomial fit.

The pressure dependence as a function of stoichiometry of consolidated AlTV is shown in Figure 8.29. The data do not fit the Vieille power law well. This is highlighted by the second-order polynomial fit superimposed on each data set. Except for the 48 wt% Al data set, each set displays a sigmoidal shape that has been described before for fuel-lean MTV (see e.g. Figure 8.9). Generally, the pressure dependence varies with pressure for these stoichiometries. This has been explained with coincidence of the measurement conditions with the low-pressure transition range in the Ward–Son–Brewster combustion model. That is transition from the condensed phase-controlled reaction to gas-phase controlled regime (Table 8.13) [20].

Table 8.13 The Vieille parameter for nano-AlTV for Figure 8.29.

Parameter	28 wt% Al	38 wt% Al	48 wt% Al	58 wt% Al
a	8.036	14.080	23.700	21.134
n	0.662	0.678	0.594	0.676

Farnell *et al.* investigated the low-pressure effect of high altitude (12 200 m) effect on the combustion of micrometric Al/PTFE compositions [41, 42].

8.4.9
Silicon

Pyrolants made from micrometric material ($\phi(\text{Si}) < 100\,\mu\text{m}, \phi(\text{PTFE}) \sim 40\,\mu\text{m}$) offer a narrow stoichiometry range ($\xi(\text{Si}) = 51$–$66\,\text{wt\%}$) with stable but slow combustion ($u < 2\,\text{mm s}^{-1}$). The burn rate of Si/PTFE decreases with increasing sample density being indicative of a strong effect of the gas phase on the burn rate. Figure 8.30 shows the variation of burn rate with density. The specific heat for a composition containing 56 wt% Si has been determined to $c_p = 0.854\,\text{J g}^{-1}\,\text{K}^{-1}$ [35]. The Arrhenius parameters for the reaction of Si/PTFE (50/50) ($\rho = 2.02\,\text{g cm}^{-3}$) are $\ln A = 35.77$ and $E_A = 287.5\,\text{kJ mol}^{-1}$, and the rate constant for the decomposition reaction at 850 K, $k_{850} = 0.0074\,\text{s}^{-1}$ [43]. Yarrington *et al.* [39, 40] prepared ternary pyrolants from amorphous nano-scale silicon ($\phi(\text{Si}) < 50\,\text{nm}, \Sigma \geq 80\,\text{m}^2\,\text{g}^{-1}, \phi(\text{PTFE}) \sim 200\,\text{nm}, \Sigma \geq 5$–$10\,\text{m}^2\,\text{g}^{-1}$) and additional fluoropolymer binder (Fluorel FC 2175). They obtained similar results for

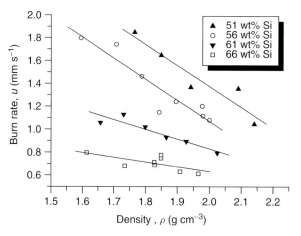

Figure 8.30 Influence of density on the burn rate of different stoichiometry Si/PTFE pyrolant. After Ref. [44].

the burn rate as with the micrometric materials reported above but were able to extend the stable combustion range down to $\xi(Si) = 24$ wt% (Figure 8.31). They also explored the pressure sensitivity of these pyrolants in the range between 0.1 and 7 MPa under argon and found a quite constant pressure exponent of $n = 0.77 \pm 0.05$ (Figure 8.32). Yarrington *et al.* also investigated the combustion of loose pyrolant ($\rho = 13–15\%$TMD) confined in Plexiglas tubes. Under these conditions, burn rates between $u = 250$ and 400 m s^{-1} have been realized (Figure 8.33) [39, 40].

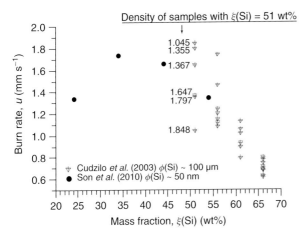

Figure 8.31 The burn rate of nano-Si/PTFE/Viton and micro-Si/PTFE. After Refs. [39, 40, 44].

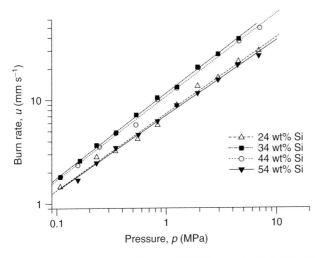

Figure 8.32 The burn rate of nano-Si/PTFE/Viton. After Refs. [39, 40].

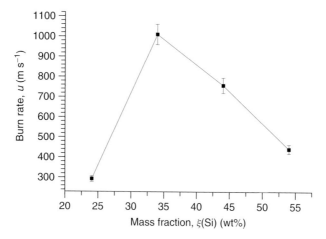

Figure 8.33 The burn rate of low-density nano-SiTV ($\rho =$ 13–15%TMD) at ambient pressure.

8.4.10
Silicides

8.4.10.1   Dimagnesium Silicide, Mg$_2$Si

Koch *et al.* prepared binary pyrolants based on irregular crystalline Mg$_2$Si (Figure 8.34) and PTFE [28].

The combustion rate of Mg$_2$Si/PTFE [28] was found stable to proceed for stoichiometries ranging from 57 to 92 wt%. At ratios below 57 wt% Mg$_2$Si, pyrolants display high frequent oscillatory combustion and eventually extinguish. The burn rate increases with decreasing density pointing towards the influence of gas-phase reactions on the combustion mechanism. Between 70 and 80 wt% magnesium silicide, very high burn rate ranges (>50 mm s^{-1}) are obtained. Figure 8.35 shows the burn rate as a function of stoichiometry.

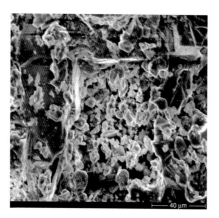

Figure 8.34 The scanning electron microscopic image of commercial Mg$_2$Si.

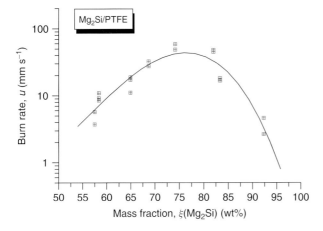

Figure 8.35 The burn rate of $Mg_2Si/PTFE$ at ambient pressure for sample densities between $\rho = 1.48$ and $2.12\,g\,cm^{-3}$.

At stoichiometries $\sim$75 wt% Mg_2Si, the combustion phenomenology changes. Below that, linear combustion rate occurs with significant gas-phase combustion. Above the point, fast glowing of the pyrolant strand with eventual terminal sparkling or explosion is encountered. Thus, the determination of the burn rate for values above 75 wt% Si is slightly ambiguous.

8.4.10.2 Calcium Disilicide

Pyrolants based on calcium disilicide have been investigated by Cudziło [34, 35, 44, 45] and the author. The combustion of a pressed stoichiometric mixture is shown in Figure 8.36, which shows the development of steady-state combustion after ignition of the pellet by a butane torch. Some experimentally determined thermophysical and thermochemical properties of various $CaSi_2/PTFE$ based pyrolants are listed in Table 8.14.

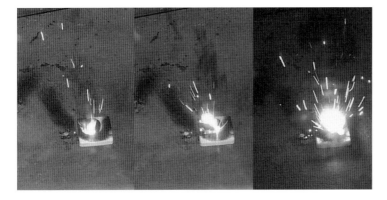

Figure 8.36 Combustion sequence of $CaSi_2/PTFE$ (28/72).

Table 8.14 Thermophysical/thermochemical properties of binary CaSi$_2$/PTFE pyrolants.

ξ(Si) (wt%)	Density (g cm^{-3})a	c_p (Jg^{-1}K^{-1})	λ (W m^{-1}K^{-1})	κ (mm^{2}s^{-1})	E_A (kJ mol^{-1})	ln A	$k_{850 K}$ (s^{-1})	α (K^{-1})	ΔH_c (J g^{-1})
27.8	2.327	0.683	1.049	0.662	–	–	–	–	–
50.0	–	–	–	–	356.6	35.77	0.0074	–	–
52.8	2.370	0.570	1.800	1.335	–	–	–	49.5	–
57.8	–	–	–	–	–	–	–	–	4800

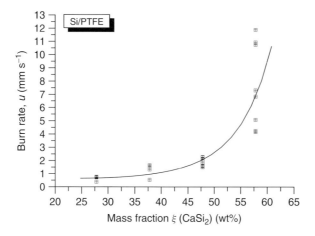

Figure 8.37 Influence of stoichiometry on the burn rate (density variations not indicated).

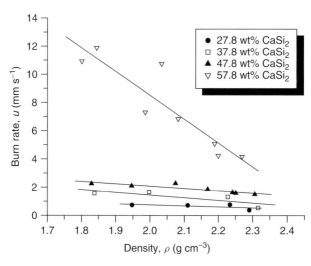

Figure 8.38 Influence of pyrolant density and stoichiometry on the burn rate of CaSi$_2$/PTFE.

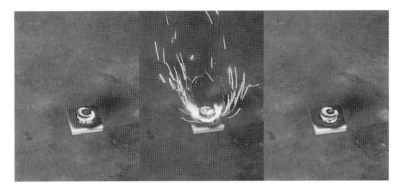

Figure 8.39 Oscillatory combustion of ZrSi$_2$/PTFE (28/72).

CaSi$_2$/PTFE undergoes stable combustion in the range between 28 and 58 wt% CaSi$_2$ (Figure 8.37) [44]. Especially fuel-rich pyrolant (58 wt% Si) is quite sensitive to density and shows variation of burn rate by an order of 300% between low and high density ($\rho = 2.25$ vs 1.80 g cm^{-3}) (Figure 8.38).

8.4.10.3 Zirconium Disilicide
Koch investigated ZrSi$_2$ (325 mesh) as a possible fuel in pyrolants. However, only a binary near-stoichiometric composition (32.9/67.1) ($\rho = 2.55$ g cm^{-3}) was ignitable but did not sustain combustion as shown in Figure 8.39.

DSC analysis revealed two consecutive exothermic reactions with onset at both 472 and 512 °C. The combined heat release is equal to 130 J g^{-1}, which is about half the energy released in the stoichiometric CaSi$_2$/PTFE pyrolant.

8.4.11
Tungsten–Zirconium Alloy

Hahma developed high-density pyrolants for autophagous nose cones for advanced infrared decoy flares with improved kinematic separation behaviour [46]. Typical payloads are depicted:

Hahma 1	Hahma 2
20 wt% Zirconium–tungsten alloy (50/50)	18 wt% Zirconium–tungsten alloy
10 wt% Hostaflon TF 9205	10 wt% Hostaflon TF 9205
70 wt% Lead dioxide	82 wt% Bismuth trioxide
Burn rate[a]	
2.4 mm s^{-1}	0.9 mm s^{-1}.

[a](personal communication A. Hahma).

Table 8.15 Actual composition and TMD of binary Al/PTFE mixtures taking into account 20 wt% of Al powder being Al_2O_3 [47].

Stoichiometric ratio σ (−)	Aluminium (wt%)	PTFE (wt%)	Theoretical maximum density (g cm⁻³)	Actual density range (% TMD)
1	34.88	65.12	2.361	14.3–42.2
1.5	44.42	55.58	2.424	11.5–15.2
2	51.57	48.43	2.474	14.1–15.8

8.5
Underwater Combustion

Binary pyrolants based on nano-metric Al (84 nm) and PTFE at low density (<20% TMD) will ignite and deflagrate underwater [47]. Stable deflagrations have been observed at compositions ranging from 30, 39 to 46 wt% Al, which translates into stoichiometric ratios, σ, of 1, 1.5 and 2 (Table 8.15):

$$4n\, Al + 3(C_2F_4)_n \longrightarrow 4AlF_3 + 6C \tag{8.6}$$

The stoichiometric ratio, σ, is defined as

$$\sigma = \frac{\left(\frac{\xi(Al)}{\xi(PTFE)}\right)}{0.3598} \tag{8.7}$$

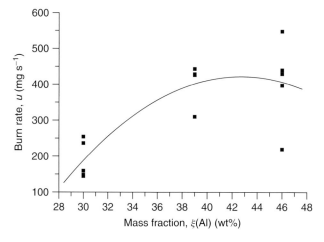

Figure 8.40 The burn rate of submerged nano-Al/PTFE [47].

Figure 8.41 Stoichiometric (a) and fuel-rich
(b) combustion plume. Reproduced with kind permission by
Prof. M. Pantoya.

(a)   (b)

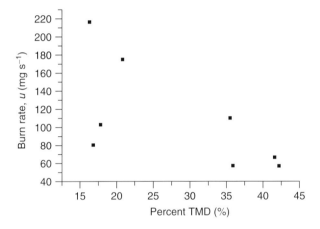

Figure 8.42 The burn rate of stoichiometric Al/PTFE as a function of density.

With increasing ξ(Al), the burn rate increases as shown in Figure 8.40. This is in line with the deflagration behaviour of powdered nano-Al/PTFE in air described above. At a stoichiometric ratio $\sigma = 1$, significant amounts of carbon are formed, which extinct the radiation from the combustion plume (Figure 8.41).

For the balanced composition with 30 wt% Al, the effect of density on combustion behaviour has been investigated. At low densities, a fast consumption with a single large plume is observed, whereas at increasing density the burn rate slows down (Figure 8.42) and a rather successive combustion of the pyrolant with concomitant release of carbon shells occurs [47].

References

1. Singh, H., Somayajulu, M.R. and Rao, R.B. (1989) A study on combustion behavior of magnesium-sodium nitrate binary mixtures. *Combust. Flame*, **76**, 57–61.

2. Miyata, K. and Kubota, N. (1996) Combustion of Ti and Zr particles with KNO_3. *Propellants Explos. Pyrotech.*, **21**, 29–35.

3. Jackson, D. and Dadley, D.A. (1969) Investigation into the Manufacture and Properties of Magnesium/Fluorocarbon Compositions for Pyrotechnic Applications, RARDE Memorandum 31/69, Fort Halstead, Kent.

4. Shidlovskii, A.A. and Gorbunov, V.V. (1978) Combustion of PTFE mixed with boron, titanium, or magnesium. *Fiz. Goren.*, **14**, 157–159.

5. Hall, A.R. (1980) Some Combustion Characteristics of a Magnesium/Fluorocarbon Pyrotechnic Composition, SR 886b, Propellants, Explosives and Rocket Motor Establishment, Westcott, Aylesbury, Bucks.

6. Kubota, N. and Serizawa, C. (1986) Combustion of magnesium/polytetrafluoroethylene. 22nd Joint Propulsion Conference, Huntsville, USA, June 16–18, AIAA-86-1592.

7. Kubota, N. and Serizawa, C. (1987) Combustion of magnesium/polytetrafluoroethylene. *J. Propulsion*, **3**, 303–307.

8. Kubota, N. and Serizawa, C. (1987) Combustion process of Mg/TF Pyrotechnics. *Propellants Explos. Pyrotech.*, **12**, 145–148.

9. Kuwahara, T. and Ochiai, T. (1992) Burning rate of Mg/TF pyrolants. 18th International Pyrotechnics Seminar, Breckenridge, Colorado, July 13–17, p. 539.

10. Kuwahara, T., Matuso, S. and Shinozaki, N. (1997) Combustion and sensitivity characteristics of Mg/TF pyrolants, propellants explos. *Pyrotech*, **22**, 198–202.

11. Kuwahara, T. and Takizuka, M. (1998) Combustion and sensitivity characteristics of metal/Teflon (Mg/Ti/Zr/Teflon) pyrolant. *Kayaku Gakkaishi*, **59** (1), 18–23.

12. Speckart, E.O. (1962) The Pressure Dependence of the Solid State Reaction Between Magnesium and Teflon, Naval Postgraduate School, Monterey, CA.

13. Chen, D.M., Hsieh, W.H., Snyder, T.S., Yang, V. and Kuo, K.K. (1988) Study of the thermophysical properties and combustion behavior of metal-based solid fuels. AIAA, ASME, SAE, and ASEE, 24th Joint Propulsion Conference, Boston, 11–13 July, pp. 11.

14. Chen, D.M., Hsieh, W.H., Snyder, T.S., Yang, V., Litzinger, T.A. and Kuo, K.K. (1991) Combustion behavior and thermophysical properties of metal-based solid fuels. *J. Propul.*, **7**, 250–257.

15. Ladouceur. H.D (2005) An overview of the known chemical kinetics and transport effects relevant to Mg/PTFE combustion. 2nd Workshop on Pyrotechnic Combustion Mechanisms, Karlsruhe, Germany, June 27, 2005.

16. Peretz, A. and Cohen, J. (1980) Development of a Magnesium-Teflon-Viton composition for propulsion system igniter. *Isr. J. Technol.*, **18**, 112–114.

17. Peretz, A. (1982) Investigation of pyrotechnic MTV compositions for rocket motor igniters. 18th Joint Propulsion Conference, Cleveland, June 21–23, AIAA-82-1189.

18. Peretz, A. (1984) Investigation of pyrotechnic MTV compositions for rocket motor igniters. *J. Spacecr.*, **21**, 222–224.

19. Valenta, F.J. (1988) MTV as a pyrotechnic composition for solid propellant ignition. 13th International Pyrotechnics Seminar, p. 811.

20. Ward, M.J., Son, S.F. and Brewster, M.Q. (1998) Role of gas- and condensed-phase kinetics in burning rate control of energetic solids. *Combus. Theory Model.*, **2**, 293–312.

21. Frolov, Y.V. and Korostelev, V.G. (1989) Combustion of gas-permeable porous systems. *Propellants Explos. Pyrotech.*, **14**, 140–149.

22. Cudziło, S. and Trzcinski, W.A. (1996) Badanie procesu spalania mieszanin metal-politetrafluoroetylene. *Biul. WAT*, **XLV**, 131–148.

23. Cudziło. S. and Trzcinski, W.A. (1998) Studies of high energy composites containing polytetrafluoroethylene. 29th International Annual Conference of ICT, Karlsruhe, Germany, June 30 – July 3, p. P151.

24. Kashparov, L.Y., Klyachko, L.A., Silin, N.A. and Shakhidzhanov, E.S. (1994) Burning of mixtures of magnesium with sodium nitrate. I. Burning velocity of two-component mixtures of magnesium with sodium nitrate. *Combust., Explos. Shock*, **30**, 608–615.

25. Klyachko, L.A. (1971) Burning of a particle of low-boiling metal moving relative to a gaseous oxidizer. *Combust. Explos. Shock*, **7**, 199–202.

26. Koch, E.-C. (2002) Metal-fluorocarbon pyrolants: IV. Thermochemical and combustion behaviour of Magnesium/Teflon/Viton (MTV). *Propellants Explos. Pyrotech.*, **27**, 340–351.

27. Ward, J.R. (1981) MgH_2 and $Sr(NO_3)_2$ pyrotechnic composition. US Patent 4,302,259, USA.

28. Koch, E.-C., Weiser, V. and Roth, E. (2011) Combustion behaviour of binary pyrolants based on MgH_2, MgB_2, Mg_3N_2, Mg_2Si, and polytetrafluoroethylene. EUROPYRO 2011, Reims, France, May 16–19.

29. Biermann, U.K.P. and Schroder, J. (1976) Method of Manufacturing reaction mixtures of finely divided metals or alloys and solid perhalogenated Carbon Compounds. US Patent 3,963,541, Germany.

30. Smith, R.B. (1975) *Propagation of a Chemical Reaction Through Heterogeneous Lithium-Polytetrafluoroethylene Mixtures*, Pennsylvania State University, Philadelphia, PA.

31. Chen, M., Jiao, Q., Wen, Y. and Lu, B. (2005) The research on emission performance of Mg_4Al_3/PTFE infrared composition. *Laser Infrared*, **35**, 500–503.

32. Takizuka, M., Onda, T., Kuwahara, T. and Kubota, N. (1998) Thermal Decomposition Characteristics of Ti/C/TF Pyrolants, AIAA Paper 98-3826.

33. Takizuka, M., Onda, T., Kuwahara, T. and Kubota, N. (1999) Combustion of TiC pyrolants. *J. Pyrotech.*, **10**, 45–48.

34. Koch, E.-C. (2003) Pre-ignition reactions in metal/organohalogen pyrolants. 35th International Annual ICT Conference, Karlsruhe, Germany, June 29–July 2, p. 126.

35. Panas, A.J., Cudziło, S. and Terpilowski, J. (2002) Investigation of the thermophysical properties of metal-polytetrafluoroethylene pyrotechnic compositions. *High Temp. High Press.*, **34**, 691–698.

36. Wilker, S., Pantel, G., Cudziło, S., Panas, A. and Terpilowski, J. (2003) Heat conductivity measurements of metal-PTFE compositions – A comparison of different methods. International Annual Conference of ICT, Karlsruhe, Germany.

37. Watson, K.W. (2007) Fast reaction of nano-aluminium: a study on oxidation versus fluorination. Thesis. Texas Tech University.

38. Watson, K.W., Pantoya, M.L. and Levitas, V.I. (2008) Fast reactions with nano- and micrometer aluminum: a study on oxidation versus fluorination. *Combust., Flame*, **155**, 619–634.

39. Yarrington, C.D., Lothamer, B., Son, S.F. and Foley, T.J. (2009) Combustion properties of Silicon/Teflon/Viton and Aluminum/Teflon/Viton composites. 47th AIAA Aerospace Sciences Meeting, Orlando, January 5–8, AIAA-2009-760.

40. Yarrington, C.D., Son, S.F. and Foley, T.J. (2010) Combustion of Silicon/Teflon/Viton and Aluminum/Teflon/Viton energetic composites. *J. Propul. Power*, **26**, 734–743.

41. Farnell, P.L. and Taylor, F.R. (1984) Combustion propagation rates of metal Fuel/Oxidant/Binder systems at simulated high altitudes. 9th International Pyrotechnic Seminar, Colorado Springs, CO, August 6–10, p. 151.

42. Farnell, P.L., Campbell, C. and Taylor F.R. (1989) Effect of reduced atmospheric pressure on the performance characteristics of pyrotechnic compositions containing aluminum. ARAED-TR-89024, Picatinny Arsenal, Dover, NJ.

43. Ksiazczak, A. and Boniuk, H.C. (2003) Thermal decomposition of PTFE in the presence of silicon, calcium silicide, ferrosilicon and iron. *J. Therm. Anal.*, **74**, 569–574.

44. Cudziło, S., Huczko, A., Lange, H., Panas, A.J. and Trzcinski, W.A. (2003) Self-propagating synthesis of ceramics in B/PTFE and Si/PTFE compositions. 8e Congres International de Pyrotechnie, Saint-Malo, France, June 23–27, p. 547.

45. Trzcinski, W.A., Cudziło, S., Szala, M. and Gut, Z. (2007) Investigation

of the combustion of calcium sili-
cide/polytetrafluoroethylene mixtures.
Arch. Combust., **27** (3–4), 69–79.

46. Hahma, A. (2011) Missile with a
pyrotechnic charge EP2295927A2,
Deutschland.

47. Shawn, S.C., Pantoya, M., Prentice,
D.J., Steffler, E.D. and Daniels, M.A.
(2009) Nanocomposites for underwater deflagration. *Adv. Mater. Processes*,
33–35.

9
Spectroscopy

9.1
Introduction

Combustion reactions are general accompanied by intense emissions in the infrared (IR, $\lambda_{IR} = 0.8 - 14\,\mu m$) and visible ($\lambda_{VIS} = 300 - 700\,nm$) ranges. Sometimes significant emission occurs in the ultraviolet range as well. IR emissions are due to vibrational and rotational transitions of molecules, ultraviolet–visible (UV–VIS) and near-infrared (NIR) emissions are due to electronic transitions in molecules or atoms. The emission spectra of combustion flames are determined by their temperature and the chemical species and their concentration present in the flame [1, 2]. The three major sources of radiation are dispersed condensed particles that emit continuum radiation [3] in both UV–VIS and IR, hot gaseous molecules that emit band radiation also in both UV–VIS and IR [3, 4] and finally vaporized metals that yield line radiation in the UV–VIS and NIR [5]. On top of this, above certain temperatures, ionization of species can take place leading to broadband recombination radiation mainly in the UV–VIS [6]. Thus, UV–VIS and IR emission spectra of flames reveal species involved in the gas-phase combustion reactions, which eventually allow resolving details of the combustion mechanism [7].

Unlike hydrocarbon combustion flames that are dominated by emissions from $C_{(s)}$, CO, CO_2, OH and H_2O [2, 3], the combustion flames from metal/fluorocarbon pyrolants show distinct differences because of the absence of oxygen from the primary reaction zone involving solely fluorine, metal and carbon.

Figure 9.1 depicts a combustion flame and typical morphology of a Mg/PTFE (polytetrafluoroethylene) flame. The luminous cone designated 'a' is dominated by continuum radiation and both fluorocarbon species. The outer aerobic combustion zone 'b' is less optically dense and shows mainly molecular radiation of MgF, MgO, C_2, CO and CO_2.

The most characteristic feature of any metal fluorocarbon flame is a strong continuum-like emission commencing at about 250 nm and ranging to $\sim$10 μm, which is due to radiation from thermally excited carbon particles formed by reduction of fluorocarbon species in the inner flame zone. In addition, gaseous metal, carbon and fluorine species yield particular emissions in both UV and IR. In the outer diffusion zones, atmospheric constituents such as oxygen and

Metal-Fluorocarbon Based Energetic Materials, First Edition. Ernst-Christian Koch.
© 2012 Wiley-VCH Verlag GmbH & Co. KGaA. Published 2012 by Wiley-VCH Verlag GmbH & Co. KGaA.

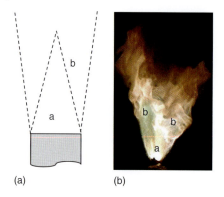

(a)   (b)

Figure 9.1 Morphology of Mg/PTFE combustion flame (A) and colour photograph (B). (Reproduced with kind permission from Volker Weiser.)

water do take part in the combustion reaction and eventually cause characteristic emissions. To account for this, combustion flames are also photographed at a low aperture and with short exposure to avoid saturation and allow for resolution of differently tinted zones.

9.2
UV–VIS Spectra

The most abundant species in metal fluorocarbon pyrolant flames in both UV–VIS range and their maximum wavelength are given in Table 9.1 [4, 8].

UV–VIS spectra can be acquired with any common diode-array-type spectrometer.

Figure 9.2 shows the general measurement setup for UV–VIS emission used for determination of the majority of spectra reported in the following. The focus of the

Table 9.1 UV–VIS spectroscopic properties of most abundant species in magnesium fluorocarbon flames and their most relevant transitions [4, 8].

Species	T_0, λ (nm)	Species	T_0, λ (nm)	Species	T_0, λ (nm)	Species	T_0, λ (nm)
C_2	214	CF	187	Mg	285	MgO	606
	231		190		383		500
	251		203		383		386
	343		233		384		381
	516	CF_2	269		517		377
	1210		504		518		372
CO	283	MgF	235	OH	278	MgOH	370
	297		269		309		380
	313		359	COF_2	216		383
	484	MgCl	377		218	MgH	521

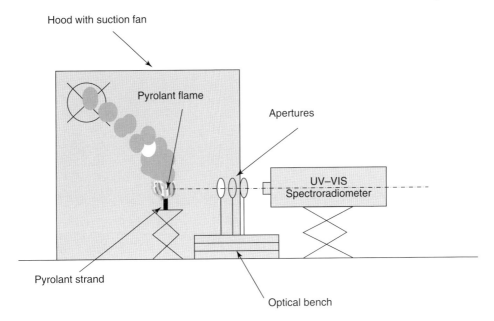

Hood with suction fan

Pyrolant flame

Apertures

UV–VIS
Spectroradiometer

Pyrolant strand

Optical bench

Figure 9.2 Measurement setup for UV–VIS spectroscopic emission.

spectrometer is set above the pyrolant strand to aim at the gas phase. Experiments are conducted in a laboratory hood or combustion bomb [9] equipped with a strong suction fan. Many of the spectra reported in the following have been obtained with a grating spectrometer with 250 mm focus and UV-sensitized silicon CCD (ANDOR) with a resolution of 1024 points and minimum exposure time of 10 ms per spectrum. The spectrometer is calibrated with Hg lamps.

9.2.1
Polytetrafluoroethylene Combustion

In the preignition zone of PTFE-based pyrolants, PTFE undergoes fusion and finally depolymerizes to yield gaseous tetrafluoroethylene, C_2F_4 (TFE) (see Figure 6.4 in Chapter 6). Spectroscopy of dissociated PTFE in a vacuum plasma plume has been reported in Ref. [10].

TFE further decomposes and eventually undergoes reaction with both metal species and air. Figure 9.3 depicts the spectrum of a combustion flame of TFE premixed with oxygen (45/55) [11]. It nicely displays the A^1B_1-X^1A_1-band system of CF_2 superimposed from a strong carbon continuum. Although COF_2 formation is anticipated in the same reaction (see Equation 9.1) [11], it is not seen in the probed wavelength range [8].

$$C_2F_4 + O_2 \longrightarrow CF_2 + COF_2 \tag{9.1}$$

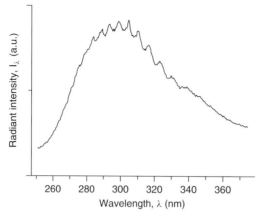

Figure 9.3 UV spectrum of a premixed C_2F_4/O_2 (45/55) flame showing the $A^1B_1-X^1A_1$ band system of CF_2; not corrected for intensity, and therefore, the overall maximum at $\lambda = \sim300$ nm has no physical meaning. (Data from Ref. [11].)

9.2.2
Magnesium/Fluorocarbon Pyrolants

Griffiths *et al.* have been the first to investigate the UV–VIS spectra of Mg/PTFE flames at various stoichiometries and various pressures [12]. They identified $Mg_{(g)}$, $MgF_{(g)}$, $MgO_{(g)}$ and C_2 species in these flames depending on the pressure regime. Table 9.2 shows the signals found for a stoichiometric Mg/PTFE composition (32/68 wt%) at different pressures.

Table 9.2 Spectroscopic properties of Mg/PTFE (32/68 wt%) at different pressures.

Species	Wavelength (nm)	Pressure (kPa)		
		101 − 33	19 − 6.6	6.6 − 2.6
$MgF_{(g)}$	359.0	x	x	x
	369.0	x[a]	o	o
$Mg_{(g)}$	285.2	x[a]	o	o
	309.7	o	x	x
	383.7	x	x	x
	457.1	x	x	o
	518.4	x	x	x
$MgO_{(g)}$	370–499	x	o	o
C_2	516	o	x	x

x, detected; o, not observed.
[a] Signal in absorption.
After Ref. [12].

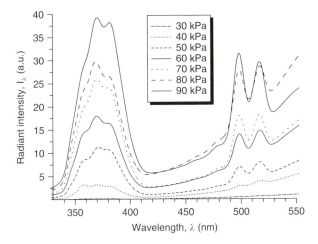

Figure 9.4 UV–VIS spectra of MTH at various levels of air
pressure. (Reproduced with kind permission from Volker
Weiser and Evelin Roth [13].)

MgO signals disappear at low pressures because of the reduced oxygen partial
pressure and hence the reduced probability of reaction of magnesium with oxygen.
Weiser *et al.* in their study of MTH strands in a combustion bomb observed
decrease in MgO signal intensity in the triplet region in favour of the MgF signal
at 359 nm [13] (Figure 9.4).

Figure 9.5 depicts the combustion flame of MTH strands at various pressures. The
pictures at higher pressure show significant greater area of blue chemiluminescence
evidencing MgO emission.

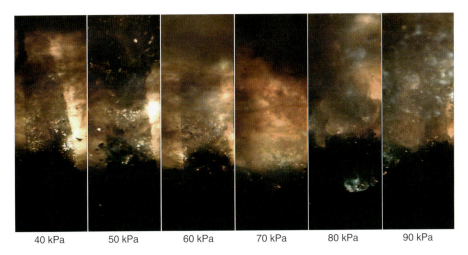

Figure 9.5 Combustion of MTH at reduced air pressure.
(Reproduced with kind permission from Weiser *et al.* [13].)

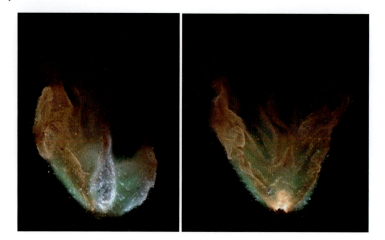

Figure 9.6 Short-exposure photographs (1/8000 s, aperture: 16) of MTH. (Reproduced with kind permission from Volker Weiser.)

Small aperture (16) short exposures (1/8000 s) of fuel-rich Mg/PTFE/Hycar® combustion flames at atmospheric pressure are depicted in Figure 9.6 [14]. They show the typical morphological features of pyrolant flames described above. With the reported photographic settings applied, the combustion flame nicely exhibits differently tinted zones (orange, bluish green and bluish white) indicating the presence of the respective emitting chemical species. Thus, thermally excited carbon particles yield orange luminescence. Bluish green luminescence is very likely due to MgO emission band at 499 nm. Bluish white emission is concluded to be due to Mg emission at 380–400 nm.

Figure 9.7 shows the UV–VIS emission spectrum of magnesium/poly(carbon monofluoride)/polychloroprene in air between 250 and 900 nm [14, 15]. It is characterized by a continuum superimposed from $Mg_{(g)}$, $MgF_{(g)}$, $MgO_{(g)}$, C_2 and trace impurities such as $Na_{(g)}$ and $K_{(g)}$. The peak at $\lambda = \sim800$ nm has no physical meaning, as the spectrometer was not intensity calibrated before measurement. A more narrow spectral range is depicted in Figure 9.8 and displays emission traces of both combustion of magnesium ribbon in air and the above pyrolant. The most prominent differences between the combustion of pure Mg and pyrolant combustion are the continuum due to carbon and signals related to MgF in the Mg-triplet range.

Below 300 nm, magnesium combustion in air yields a partially self-absorbed singlet around 285 nm [16, 17]. In case of the pyrolant, additional signals for CO and transient MgF and CF_2 are discernible as well (Figure 9.9). In regions well above the primary combustion zone, the composition of the flame changes and correspondingly the spectra change. This is seen in Figure 9.10 that depicts an increase in CO and a reduction in Mg and MgF, as well as the disappearance of CF signals as a function of distance over strand surface, indicating afterburn reactions

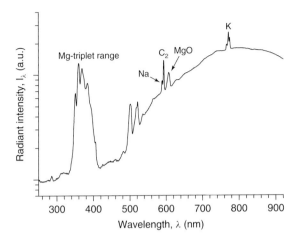

Figure 9.7 The 250–900 nm UV–VIS spectra of magnesium pyrolant combustion in air (not baseline corrected; therefore, the overall maximum at $\lambda = \sim$800 nm has no physical meaning). (Reproduced with kind permission from Volker Weiser and Evelin Roth [14, 15].)

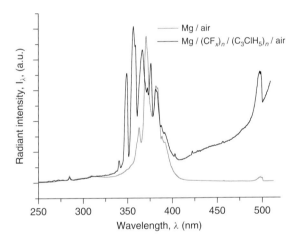

Figure 9.8 The 250–510 nm UV–VIS spectra of magnesium combustion in air and combustion of a Mg-based pyrolant. (Reproduced with kind permission from Volker Weiser and Evelin Roth.)

of carbon and precipitation of $MgF_{2(l,s)}$ and $MgO_{(l,s)}$, which are not active in this spectral range.

In the 350–400 nm range (Figure 9.11), Mg displays strong atomic lines as a triplet at 383–384 nm [14, 15]. Both MgO and MgOH also yield strong molecular bands at 360–365, 367–375, 380–387 nm as well as 368–373 and 377–392 nm.

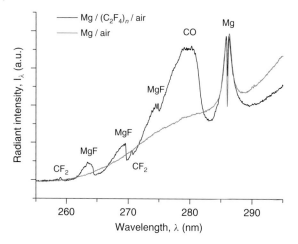

Figure 9.9 UV–VIS (255–295 nm) spectra of magnesium combustion in air and combustion of Mg-based pyrolant in air. (Reproduced with kind permission from Volker Weiser and Evelin Roth.)

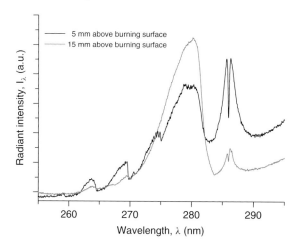

Figure 9.10 UV–VIS spectra (255–295 nm) of combustion of Mg-based pyrolant in air showing the effect of carbon afterburn as a function of height above the burning surface. (Reproduced with kind permission from Volker Weiser and Evelin Roth.)

The pyrolants further yield MgF with signals at 349, 359 and 368 nm as well as a C_2 signal at 343 nm.

The short exposures of Magnesium/Teflon/Viton (MTV) combustion show distinct areas with green and blue luminescence (Figure 9.6). The bluish green luminescence is seen all over the flame volume. This emission can be attributed to the Mg triplet centred at 518 nm. The light blue luminescence in contrast is

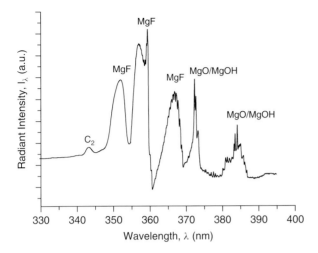

Figure 9.11 UV–VIS spectra (350–395 nm) of Mg-based pyrolant in air. (Reproduced with kind permission from Volker
Weiser and Evelin Roth.)

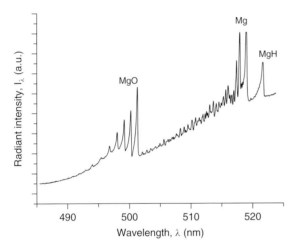

Figure 9.12 UV–VIS spectra ($\lambda = 485$–525 nm) of combustion of Mg-based pyrolant in air. (Reproduced with kind
permission from Volker Weiser and Evelin Roth.)

discernible only in the outer layers of the plume and is assigned to MgO that yields
a rotationally resolved signal at $\lambda = 499$ nm (Figure 9.12). A signal at $\lambda = 516$ nm
is assigned to C_2. The array of signals between 500 and 516 nm is assigned to the
$A^2\Pi - X^2\Sigma^+$ transition of MgH with peak at 521 nm [18]. Another band system
centred at $\lambda = \sim535$ nm is tentatively assigned to CaF, which originates from Ca
stearate contamination of the twin screw-extruded MTH pyrolant (Figure 9.13)
[14, 15].

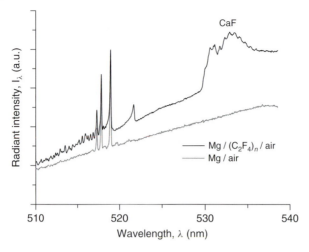

Figure 9.13 UV–VIS spectra (510–540 nm) of combustion of Mg in air and Mg-based pyrolant in air. (Reproduced with kind permission from Volker Weiser and Evelin Roth.)

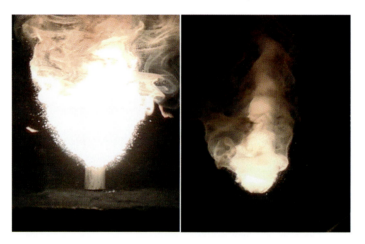

Figure 9.14 Short-exposure photographs (1/8000 s, open aperture, 0 dB) of MgH$_2$/PTFE at lean fuel content [19].

9.2.3
MgH$_2$, MgB$_2$, Mg$_3$N$_2$, Mg$_2$Si/Mg$_3$Al$_2$/Fluorocarbon Based pyrolants

Fluorocarbon-based pyrolants containing Mg-based fuels with no metallic magnesium yield combustion flames with distinct spectral differences to Mg/PTFE. Koch *et al.* have investigated the combustion properties of a series of experimental binary pyrolants based on, MgH$_2$, MgB$_2$, Mg$_3$N$_2$, Mg$_2$Si and PTFE at ambient pressure in air (Table 9.3) [19]. Stills from MgH$_2$/PTFE combustion at both fuel-lean and fuel-rich conditions are depicted in Figures 9.14 and 9.15. Figure 9.14 shows a

Table 9.3 Spectroscopic properties of various stoichiometric pyrolants at ambient pressure under air [19].

Species	Wavelength (nm)	Mg/PTFE (32/68)	MgH$_2$/PTFE (34/66)	MgB$_2$/PTFE (48/52)	Mg$_3$N$_2$/PTFE (57/43)	Mg$_2$Si/PTFE (44/56)
MgF	264	x	o	o	o	o
	268	x	o	o	o	o
	275	x	o	o	o	o
	349	x	x	x	x	x
	358	x	x	x	x	x
	367	x	x	x	x	x
Mg	285	x	o	o	o	x
	384	x	o	o	o	x
	499	x	o	o	o	x
	518	x	o	o	o	x
MgO	370–499	x	o	o	o	x
CF	271	x	o	o	o	–
CO	283	x	o	o	o	–
Other	–	–	–	BO$_2$ multiplet	–	–

x, detected; o, not observed.

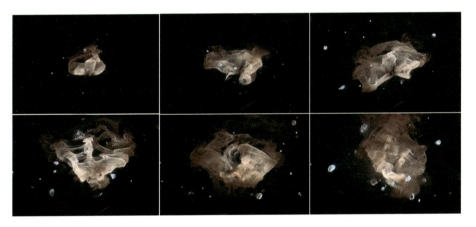

Figure 9.15 Short-exposure photographs (1/8000 s, aperture: 16) of MgH$_2$/PTFE at high fuel content [19].

luminous zone that unlike Mg/PTFE/Viton® shows no distinct colour. This is confirmed by the UV–VIS spectrum in Figure 9.16 that virtually shows only a weak continuum. Opposed to that the combustion of fuel-rich MgH$_2$/PTFE shows ejection of MgH$_2$ particles with distinct blue vapour-phase combustion aureoles (MgO) superimposed on a background of thermally excited carbon particles glowing orange red (Figure 9.15).

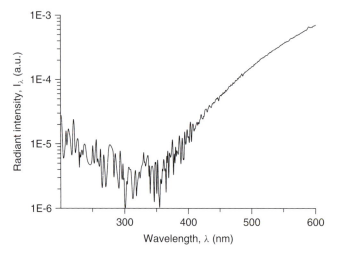

Figure 9.16 UV–VIS spectrum (200–600 nm) of combustion of MgH$_2$/PTFE in air. (Reproduced with kind permission from Volker Weiser and Evelin Roth [19].)

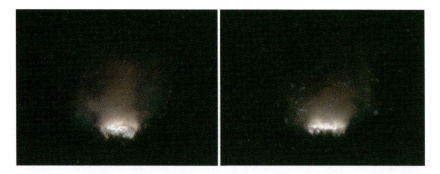

Figure 9.17 Short-exposure photographs (1/8000 s, aperture: 16) of MgB$_2$/PTFE at lean fuel content [19].

Combustion of fuel-lean MgB$_2$-based pyrolant in air yields a green seamed luminous flame (Figure 9.17). The green luminescence origins from boron oxidation as is obvious from the corresponding UV–VIS spectrum (Figure 9.18) that depicts BO$_2$ bands and MgF signals. Combustion of fuel-rich pyrolant depicts ejection of luminous particles displaying both boron and magnesium combustion (Figure 9.19).

The fuel-lean combustion flames of Mg$_3$N$_2$/PTFE (Figure 9.20) show a luminous flame and a faint blue seam. The *de novo* formation of Mg$_3$N$_2$ adjacent to the pyrolant strand is observed as an inclined translucent filamentous structure. In the fuel-rich flame, an expanded flame with intense afterburn of Mg is observed (Figure 9.21). This impressively depicts the effect of nitrogen release on the size of

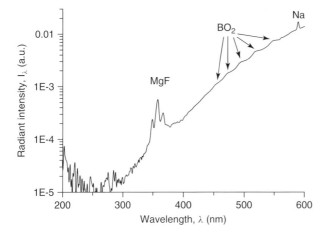

Figure 9.18 UV–VIS spectrum (200–600 nm) of combustion of fuel-lean MgB_2/PTFE in air. (Reproduced with kind permission from Volker Weiser and Evelin Roth [19].)

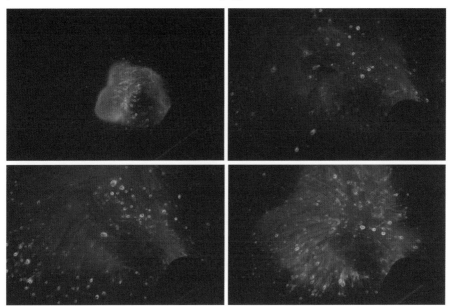

Figure 9.19 Short-exposure photographs (1/8000 s, aperture: 16) of MgB_2/PTFE at rich fuel content [19].

the plume. The MgF triplet at 359 nm. is visible as the only remarkable feature in the UV–VIS spectrum (Figure 9.22).

Fuel-lean pyrolant flames based on Mg_2Si/PTFE (Figure 9.23) even at short exposure and smallest aperture show a dazzling light indicating a high plume temperature. The UV–VIS spectrum (Figure 9.24) shows strong signals for Mg,

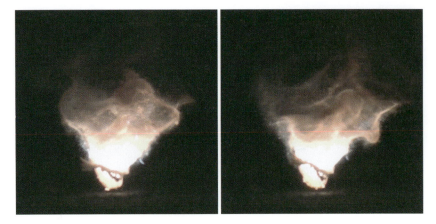

Figure 9.20 Short-exposure photographs (1/8000 s, aperture: 16) of Mg_3N_2/PTFE at lean fuel content [19].

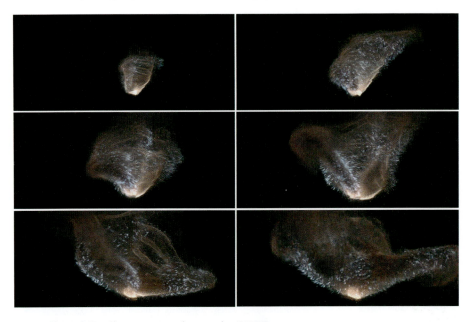

Figure 9.21 Short-exposure photographs (1/8000 s, aperture: 16) of Mg_3N_2/PTFE at rich fuel content [19].

MgF and MgO in the 340–395 nm range, and also a weak Mg signal at 282 nm in emission is obtained. This could possibly call for a reaction between Si and fluorine and thus explain the Mg surplus. However, no signals are detected for SiF that would give intense bands near 437 nm [20, 21]. The fuel-rich combustion pictures show dominating MgO luminescence and very little carbon particle emission (Figure 9.25).

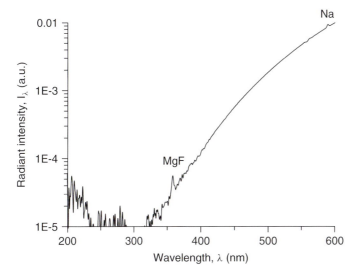

Figure 9.22 UV–VIS spectrum (200–600 nm) of combustion of fuel-lean Mg_3N_2/PTFE in air. (Reproduced with kind permission from Volker Weiser and Evelin Roth [19].)

Figure 9.23 Short-exposure photographs (1/8000 s, aperture: 16) of Mg_2Si/PTFE at lean fuel content [19].

9.2.4
Silicon/PTFE Based Pyrolants

Cudziło has measured the VIS–NIR spectrum of Si/PTFE (36/64). The spectrum is a greybody curve and superimposed by two local maxima at $\lambda = 850$ and 960 nm, which are of unknown origin [22] (Figure 9.26).

The combustion flame of pulverized mesoporous silicon mixed with perfluoroeicosane ($C_{20}F_{48}$) (PFE) is depicted in Figure 9.27 [23]. The ignition of a mesoporous silicon waver infiltrated with polyfluoropolyether (PFPE) is seen in Figure 9.28 [24].

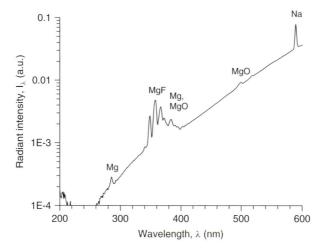

Figure 9.24 UV–VIS spectrum (200–600 nm) of combustion of fuel-lean Mg₂Si/PTFE in air. (Reproduced with kind permission from Volker Weiser and Evelin Roth [19].)

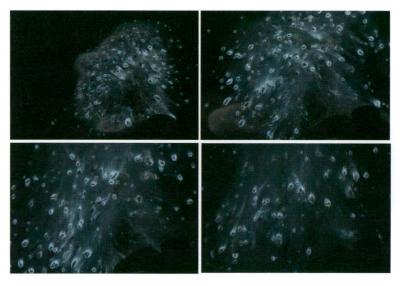

Figure 9.25 Short-exposure photographs (1/8000 s, aperture: 16) of Mg₂Si/PTFE at rich fuel content [19].

9.2.5
Boron/PTFE/Viton Based Pyrolants

Boron/PTFE/perchlorate based payloads display a beautiful deep green combustion flame such as is depicted in Figure 9.29 [25]. The UV–VIS spectrum (Figure 9.30) shows the typical BO₂ multiplet responsible for the green emission. However, two

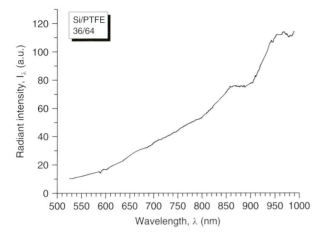

Figure 9.26 VIS–NIR spectrum (500–1000 nm) of combustion of Si/PTFE in air. (After Ref. [22].)

Figure 9.27 Combustion sequence of powdered meso-porous silicon mixed with perfluoroeicosane. (Reproduced with kind permission from Volker Weiser.)

non-resolved heads at both 320 and 360 nm could be due to BF species [4, 21], as these heads do not appear in either B/KNO$_3$ or B/AP combustion flames.

9.3
MWIR Spectra

The most dominating species present in Mg/fluorocarbon flames in the IR are carbon particles. A few other species also appear. These are given together with the band strengths, Π, in Table 9.4.

Figure 9.28 Combustion still of mesoporous silicon infiltrated with PFPE. (Reproduced with kind permission from Prof. Steve Son and Aaron B. Mason [24].)

Figure 9.29 Combustion flame of Boron/PTFE/Viton/KClO$_4$. (Reproduced with kind permission from Weiser *et al.* [25].)

9.3.1
Polytetrafluoroethylene Combustion

The FTIR spectrum (Figure 9.31) of a premixed TFE/O$_2$ ambient pressure (100 kPa) combustion flame shows COF$_2$ and CO$_2$ distinct emissions but no signs of TFE, indicating a fast combustion kinetics.

9.3.2
Magnesium/Fluorocarbon Combustion

The IR emission spectrum of combustion of Mg/PTFE (5/95) under pressurized oxygen atmosphere (200–300 kPa) is depicted in Figure 9.32 (H. D. Ladouceur,

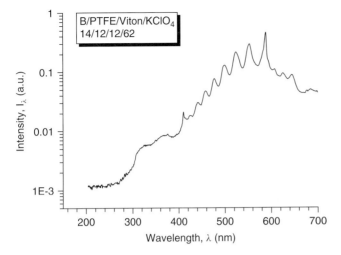

Figure 9.30 UV–VIS spectrum of Boron/PTFE/Viton/KClO$_4$.
(Reproduced with kind permission from Volker Weiser and
Evelin Roth [25].)

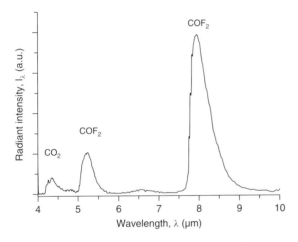

Figure 9.31 FTIR spectrum of a premixed TFE/O$_2$ flame
(50/50). (After data obtained from Ref. [11].)

personal communication) [29]. Unlike ambient pressure combustion of MTV
(Figure 9.33), carbon emission is reduced to a minimum, thus only very weak
continuum level is obtained. The spectrum in Figure 9.33 shows both HF emission
and CO$_2$ and COF$_2$. The ratio of the latter is different than that in Figure 9.32 due
to the formation of both MgF and MgF$_2$, and thus the lower COF$_2$ content. The
signals of COF$_2$ are superimposed from absorptions at 5.22 and 7.8 µm, possibly
indicating the trifluorovinyl radical (C$_2$F$_3$) [27, 30], which may have been formed
via either fluorine atom dissociation or collision of TFE with Mg.

Table 9.4 Vibrational wavelengths and band strengths of important species abundant in metal-fluorocarbon flames.

	λ (μm)	Π (km mol^{-1})	References
CO	4.67	250.36	[3]
	2.35	1.97	
	1.57	0.011	
CO$_2$	4.9	0.6	[3]
	4.3	2700	
	2.7	67	
	2.0	1.6	
CF$_2$	14.97	3	[28]
	9.07	160 ($\pm$80)	
	8.20	90 ($\pm$25)	
CF$_3$	7.99	vs[a]	–
	9.20	s[b]	
COF$_2$	5.19	$\sim$700	–
	8.01	$\sim$1000	
CF$_4$	7.80	1159 ($\pm$7)	[26]
C$_2$F$_3$	8.16	119 ($\pm$39)	[27]
	7.75	106 ($\pm$41)	
C$_2$F$_4$	7.50	–	–
	8.42		
HF	2.53	389	[3]
	1.29	12.30	
H$_2$O	6.3	300	[3]
	2.7	220	
	1.87	24	
	1.38	18	
	1.14	1	

[a]vs = very strong
[b]s = strong

The midwave ambient atmosphere and pressure IR spectrum of a Mg/PTFE/Viton combustion flames is dominated by a strong carbon continuum superimposed from a series of molecular emitters such as H$_2$O, HF and CO$_2$ both in emission and absorption (Figure 9.33). As carbon dioxide emission is always partially absorbed by the cooler combustion gases, the typical so-called blue spike and red wing appear. At greater distances between flare and spectrometer, the absorption of radiation because of the atmospheric constituents becomes relevant as can be seen from the overall spectrum depicted in Figure 9.34, which has been recorded in 10 m distance to the flare.

It nicely shows the blackbody-like trace with a peak value of $\lambda = 1.9\,\mu$m and distinct atmospheric absorptions due to water and CO$_2$.

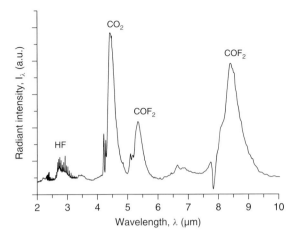

Figure 9.32 FTIR spectrum of a Mg/PTFE = 5/95 wt% flame in pure O_2 at 200–300 kPa. (After data obtained from H. D. Ladouceur, personal communication, Ref. [29].)

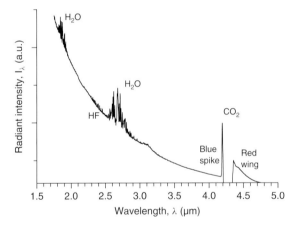

Figure 9.33 Close-range (1 m) FTIR spectrum (1.8–6 μm) of combustion of magnesium/PTFE/Hytemp in ambient pressure air. (Reproduced with kind permission from Volker Weiser and Evelin Roth.)

9.3.3
MgH_2, MgB_2, Mg_3N_2, Mg_2Si/Fluorocarbon Based Pyrolants

The circular variable filter (CVF)-IR-emission spectra of the aforementioned pyrolants based on MgH_2, MgB_2, Mg_3N_2 and Mg_2Si do not exhibit any features related to the included elements, in particular, no FBO or SiO are observed (Figure 9.35) [19]. This may be due to the general lower resolution of CVF spectra compared to FTIR spectra. However, the spectra display differences with respect to overall intensity as is discussed in section radiometric properties in Table 9.5.

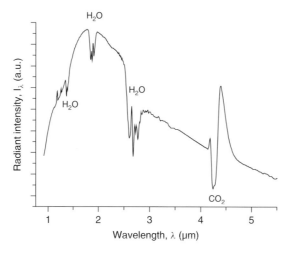

Figure 9.34 Mid-range (10 m) FTIR spectrum (0.9–6 μm) of combustion of magnesium/polytetrafluoroethylene/Viton in air.

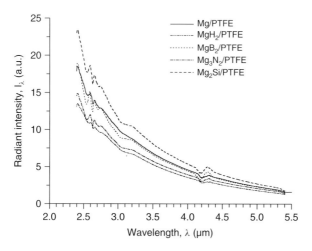

Figure 9.35 Close-range (1 m) CVF-IR spectrum (2.4–6 μm) of combustion of Mg/PTFE, MgH$_2$/PTFE, MgB$_2$/PTFE, Mg$_3$N$_2$/PTFE and Mg$_2$Si/PTFE in air. (Reproduced with kind permission from Volker Weiser and Evelin Roth [19].)

9.3.4
Si/Fluorocarbon Based Pyrolants

Cudziło investigated the spectral emittance of various pyrolants including Si/PTFE (S. Cudziło, private communication). The spectrum is mainly a greybody trace superimposed from a strong CO$_2$ peak at 4.3 μm due to the afterburn of carbon (Figure 9.36).

Table 9.5 Normalized radiant intensity $I_{1.8-4.8\,\mu m}$ (W sr^{-1}) for investigated pyrolants [19].

Pyrolant	Mg/PTFE (32/68)	MgH$_2$/PTFE (34/66)	MgB$_2$/PTFE (48/52)	Mg$_3$N$_2$/PTFE (57/43)	Mg$_2$Si/PTFE (44/56)
Normalized intensity	1.000	0.756	0.965	0.835	1.179

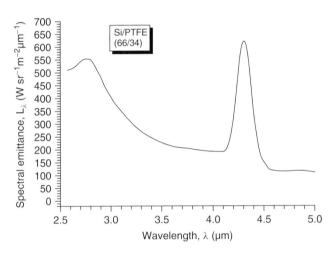

Figure 9.36 Close-range (1 m) CVF-IR spectrum (2.4–6 μm) of combustion of Si/PTFE in air. (Reproduced with kind permission from Prof. Dr. Stanislaw Cudziło.)

9.3.5
Boron/PTFE/Viton Based Pyrolants

The IR spectrum of B/PTFE/Viton/KClO$_4$ flames exhibits specific signals for both HF and BOF [25]. Figure 9.37 covers a spectrum between 4500 and 2500 cm^{-1}. The fundamental HF emission is seen around 3962 cm^{-1}, HOBO and water coincide at 3700 cm^{-1}. Figure 9.38 shows the wave number range between 2500 and 1000 cm^{-1}. The selective band heads are due to three boron compounds, FBO [31], HOBO [31] and B$_2$O$_2$ [32].

9.4
Temperature Determination

As mentioned in the beginning of this chapter, spectra are determined by the nature and concentration of chemical species present in the flame and the temperature of both gas phase and particles. Thus, it can be concluded that determination of the temperature of both gas phase and particles is possible [2].

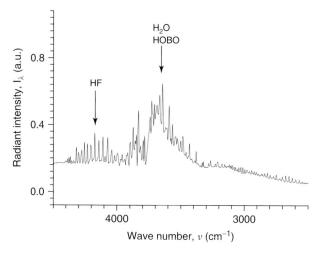

Figure 9.37 FTIR spectrum of B/PTFE/Viton/KClO₄. (Reproduced with kind permission from Volker Weiser and Evelin Roth [25].)

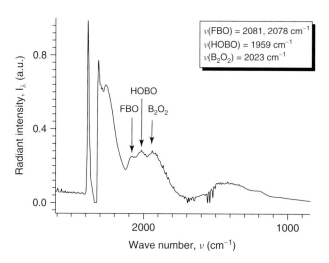

Figure 9.38 FTIR spectrum of B/PTFE/Viton/KClO₄. (Reproduced with kind permission from Volker Weiser and Evelin Roth [25].)

9.4.1
Condensed-Phase Temperature

The NIR spectral range of metal fluorocarbon pyrolant combustion flames is mainly a continuum due to hot carbon particle emission. Thus, the shape and slope of the curve can be exploited to determine the combustion temperature by curve fitting

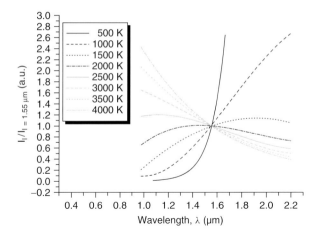

Figure 9.39 Slope of Planck curve for various temperatures after normalized to radiant intensity at 1.55 μm. (After Ref. [33].)

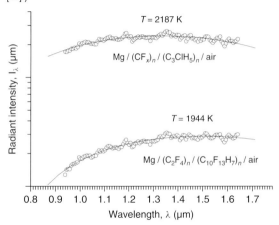

Figure 9.40 NIR spectrum (350–395 nm) of combustion of both MTV and MPP in air. (Reproduced with kind permission from Volker Weiser and Evelin Roth.)

with a Planck function varying temperature at assigned emissivity.

$$M_\lambda = \frac{2\pi c_1}{\lambda^5 \left(e^{\frac{c_2}{\lambda T}} - 1 \right)} (\mathrm{W m^2}) \tag{9.2}$$

Figure 9.39 depicts traces for various blackbodies at temperature ranging from 500 to 4000 K normalized to the maximum intensity at 1.55 μm. This shows the versatility of this wavelength range to determine the temperature as slight changes in temperature causes a significant alteration in the slope and the algebraic sign of this part of the blackbody curve [33].

Table 9.6 Temperatures of various stoichiometric PTFE-based combustion flames in 0.1 MPa air compared to the calculated adiabatic temperatures [19] in air.

	Mg$_2$Si	MgB$_2$	Mg	MgH$_2$	Mg$_3$N$_2$
Theory (K)	2734	2754	2689	2591	2349
Experiment (K)	2052	1926	2023	1730	1718
Δ(K)	682	828	666	861	631

Weiser and Eisenreich have described this methodology in a recent review [7]. The NIR spectra of two fuel-rich pyrolants are depicted in Figure 9.40. The actual measured spectrum is depicted as circles, and the best fit is shown as a solid line. The combustion temperature is determined to 1940 and 2180 K, respectively, for MTV and MPP (magnesium/polycarbon monofluoride/polychloroprene). The apparent scatter at $\lambda = 1.17$, 1.25 and 1.52 μm is due to intensive potassium lines that are present as impurities in magnesium metal. The temperatures for several pyrolants based on MgH$_2$, MgB$_2$, Mg$_3$N$_2$ and Mg$_2$Si are given in Table 9.6 [19]. The pyrolant based on magnesium silicide displays the highest combustion temperature being in good agreement with the observation that this fuel yields the highest radiant intensity compared to Mg and the other fuels.

The combustion temperature of MTV increases with increasing pressure as is depicted for oxygen in Figure 9.41. Likewise, the temperature drops for pressures below ambient pressure (Figure 9.42) as has been found both by Griffiths *et al.* [12] and Weiser *et al.* [13].

MTH strands show systematic decrease in temperature with decreasing atmospheric pressure, 0.09−0.03 MPa (Figure 9.32) [13].

The combustion temperature of mesoporous silicon infiltrated with PFE is depicted in Figure 9.43.

9.4.2
Gas-Phase Temperature

Rotationally resolved spectra of molecules present in the flame allow for temperature determination of the gas phase [3]. The emission intensity, I_v, of a rotational-vibrational band of an excited molecule correlates to the temperature of the molecule, T, via the following equation [34]

$$I_n = \frac{C_d v^4}{Q_r}(J' + J'' + 1)e^{-\frac{B'J'(J'+1)hc_0}{kT}} \tag{9.3}$$

where C_d is a constant that depends on the dipole momentum and the total number of molecules in the initial vibrational level and Q_r is the rotational state sum. For a certain vibrational-rotational band at a given temperature, the term $\frac{C_d v^4}{Q_r}$ stays

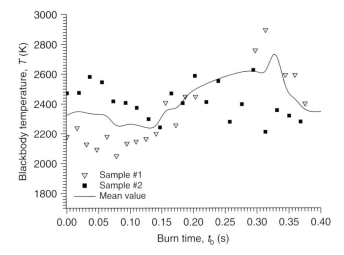

Figure 9.41 Temperature trace of Mg/PTFE (55/45) combustion flame in 0.5 MPa O_2. (Reproduced with kind permission from Volker Weiser and Evelin Roth.)

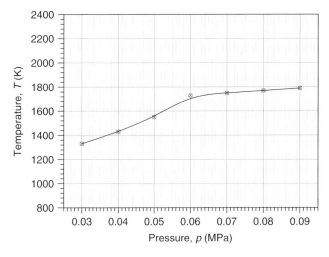

Figure 9.42 Temperature trace of fuel-rich MTH at various pressures in air. (After Ref. [13].)

nearly constant. After rearranging, Eq. (9.3) reads

$$\ln \frac{I_\nu}{J' + J'' + 1} = A - \frac{B' J'(J' + 1)hc_0}{kT} \tag{9.4}$$

where

J', J'': rotational quantum numbers in R and P branch
B': rotational constant of molecule

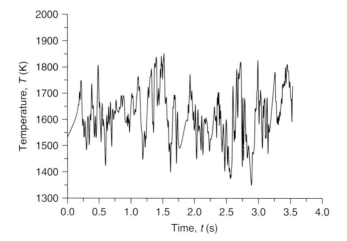

Figure 9.43 Flame temperature of mesoporous silicon infiltrated with perfluoroeicosane. (Reproduced with kind permission from Volker Weiser and Evelin Roth [23].)

h: Planck's constant
k: Boltzmann's constant
c_0: vacuum speed of light

and with the following substitutions

$$y = \ln \frac{I_v}{J' + J'' + 1}, \quad x = J'(J' + 1), \quad a = -\frac{B' h c_0}{kT}, \quad b = A \tag{9.5}$$

A common linear function follows:

$$y = ax + b \tag{9.6}$$

The temperature of the molecule can be determined from the slope of the trace according to

$$T = \frac{B' h c_0}{k|a|} \tag{9.7}$$

Wang *et al.* [34] have determined the rotational-vibrational temperature for gaseous species in a spectral flare composition given below

Wang spectral flare composition [34]
15 wt% aluminium
10 wt% silicon
20 wt% polytetrafluoroethylene
40 wt% ammonium perchlorate
10 wt% manganese dioxide
5 wt% binder

Table 9.7 Temperatures of various molecules [34].

Species	Spectral band (cm^{-1})	Determined temperature (K)
HF	3963–4400	2410
HCl	2865–2400	2610
H_2O	2910–3700	2378
CO	2145–2025	2389

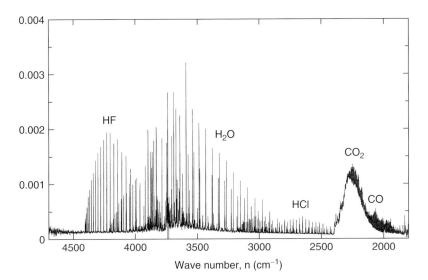

Figure 9.44 FTIR spectrum of Wang's spectral flare formulation. (After Ref. [35].)

They determined the temperatures of HF, HCl, H_2O and CO (Table 9.7).

An FTIR spectral investigation by Franzen and Lissel confirms the determined temperature of HCl (Figure 9.44) [27].

Similarly, temperature determination can be carried out on diatomic molecules in the UV–VIS. Knapp *et al.* recently reported about the gas-phase temperature determination of burning metals in oxygen [36]. For this purpose, it is assumed that the involved atoms and molecules are at thermal equilibrium and excited energy levels are populated according to the Boltzmann statistics. To determine the temperature of the rotationally resolved spectra of diatomic molecules, the intensity distribution has to be calculated. Therefore, the energy levels, transition probabilities and line profiles of all molecular states involved are required (Figure 9.45). The calculation of energy levels and transition probabilities of a diatomic molecule occurs via solving the Schrödinger equation in the Born–Oppenheimer approximation.

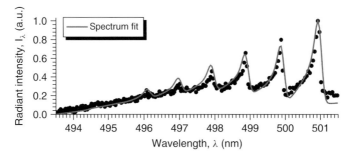

Figure 9.45 MgO $B^1\Sigma - X^1\Sigma$ transition and superimposed calculated spectrum. (Reproduced with kind permission from Sebastian Knapp [13].)

The spectral line intensity is given by

$$I_\nu = N_{nm}A_{nm}hc\nu_{nm}$$

where A_{nm} is the Einstein coefficient describing transition probability and $hc\nu$ the energy of each emitted light quantum.

As the gas-phase molecules are in thermal equilibrium, determination of one species allows assessment of the overall gas-phase temperature of the combustion flame.

References

1. Gaydon, A.G. (1948) *Spectroscopy and Combustion Theory*, Chapman & Hall, London.
2. Gaydon, A.G. and Wolfhard, H.G. (1978) *Flames – Their Structure, Radiation and Temperature*, 4th edn, John Wiley & Sons, Inc., New York.
3. Ludwig, C.B., Malkmus, W., Reardon, J.E. and Thompson, J.A.L. (1973) in *Handbook of Infrared Radiation from Combustion Gases* (eds R. Goulard and J.A.L. Thompson), National Aeronautics and Space Administration, Washington, DC.
4. Pearse, R.W.B. and Gaydon, A.G. (1975) *The Identification of Molecular Spectra*, 4th edn, John Wiley & Sons, Inc., New York.
5. Frerichs, R. (1929) in *Handbuch der Physik*, Licht und Materie, Band **XXI** (eds H. Geiger and K. Scheel), Springer-Verlag, Berlin, p. 273.
6. Lawton, J. and Weinberg, F.J. (1969) *Electrical Aspects of Combustion*, Clarendon Press, Oxford.

7. Weiser, V. and Eisenreich, N. (2005) Fast emission spectroscopy for a better understanding of pyrotechnic combustion behaviour. *Propellants Explos. Pyrotech.*, **30**, 67–78.
8. Workman, G.L. and Duncan, A.B.F. (1970) Electronic spectrum of carbonyl fluoride. *Chem. Phys.*, **52**, 3204–3209.
9. Kubota, N. (2007) *Propellants and Explosives*, 2nd edn, Wiley-VCH Verlag GmbH, Weinheim, p. 491.
10. Markusic, T.E. and Spores, R.A. (1997) Spectroscopic emission measurements of a pulsed plasma thruster plume. 33rd Joint Propulsion Conference, Seattle, WA, July 6–9, AIAA 97-2924.
11. Douglass, C.H., Ladouceur, H.D., Shamamian, V.A. and McDonald, J.R. (1995) Combustion chemistry in premixed C_2F_4-O_2 flames. *Combust. Flame*, **100**, 529–542.
12. Griffiths, V.S., Izod, D.C.A. and O'Sullivan, E. (1970) Some observations of some pyrotechnic compositions.

2nd International Pyrotechnics Seminar, Aspen CO, July 20–24, p. 449.

13. Weiser, V., Roth, E., Knapp, S., Raab, A. and Koch, E.-C. (2011) Combustion behaviour of MTH under reduced atmospheric pressure. 42nd International Annual Conference of ICT, Karlsruhe Germany, 28 June–1 July, p. XX.

14. Weiser, V., Roth, E., Müller, D. and Koch, E.-C. (2006) Investigation of magnesium-flourohydrocarbon-flames using emission spectroscopy. 37th International Annual Conference of ICT, Karlsruhe Germany, June 27–30, p. 161.

15. Koch, E.-C., Weiser, V., Roth, E. and Müller, D. (2006) UV-VIS spectroscopic investigation of magnesium/fluorocarbon pyrolats. 33rd International Pyrotechnics Seminar, Fort Collins CO, July 16–21, p. 763.

16. Farren, R.E., Shortridge, R.G. and Webster, H.A. (1986) Use of chemical equilibrium calculations to simulate the combustion of various pyrotechnic compositions. 11th International Pyrotechnics Seminar, Vail CO, July 7–11, Supplement page 13.

17. Lewis, H.L., Webster, H.A., Burks, L.A. and Rankin, M.D. (1986) Ultraviolet spectra of standard navy colored flare compositions. 11th International Pyrotechnics Seminar, Vail CO, July 7–11, Supplement page 41.

18. Shayesteh, A., Henderson, R.D.E., Le Roy, R.J. and Bernath, P.F. (2007) Ground state potential energy curve and dissociation energy of MgH. *J. Phys. Chem. A*, **111**, 12495–12505.

19. Koch, E.-C., Weiser, V. and Roth, E. (2011) Combustion behaviour of binary pyrolats based on Mg, MgH_2, MgB_2, Mg_3N_2, Mg_2Si and polytetrafluoroethylene. 37th International Pyrotechnics Seminar & EUROPYRO 2011, Reims, France, May 16–19, p. VXX.

20. Cruden, B.A., Rao, M.V.V.S., Sharma, S.P. and Meyyappan, M. (2002) Neutral gas temperature estimates in an inductively coupled CF_4 plasma by fitting diatomic emission spectra. *J. Appl. Phys.*, **91**, 8955–8964.

21. Dagnall, R.M., Fleet, B., Risby, T.H. and Deans, D.R. (1971) Molecular emission

characteristics of various fluorides in a low-temperature hydrogen diffusion flame. *Talanta*, **18**, 155–164.

22. Huczko, A., Osica, M., Rutkowska, A., Bystrzejewski, M., Lange, H. and Cudzilo, S. (2007) A self-assembly SHS approach to form silicon carbide nanofibres. *J. Phys.: Condens. Matter*, **19**, 395022.

23. Weiser, V., Roth, E., Hamberger, P. and Raab, A. (2006) Abbranduntersuchungen an Porösem Silicium, Fraunhofer ICT, 13 Dezember.

24. Son, S.F. and Mason, B.A. (2010) An overview of nanoscale silicon reactive composites applied to microenergetics. 48th AIAA Aerospace Sciences Meeting, Orlando, FL, January 4–7, AIAA 2010-694.

25. Weiser, V., Blanc, A., Deimling, L., Eckl, W., Eisenreich, N., Kelzenberg, S., Koleczko, A., Roth, E. and Walschburger, E. (2006) Verbesserte Wirkmassen für IR-Scheinzielmunition: Spektrale Analyse & Modellierung. Fraunhofer ICT.

26. Nemtchinov, V. and Varanasi, P. (2003) Thermal infrared absorption cross-sections of CF_4 for atmospheric applications. *J. Quant. Spectrosc. Radiat. Transfer.*, **82**, 461–471.

27. Wurfel, B.E., Pugliano, N., Bradforth, S.E., Saykally, R.J. and Pimentel, G.C. (1991) Broadband transient infrared laser spectroscopy of trifluorovinyl radical, $C_2F_3^*$: experiment and ab initio results. *J. Phys. Chem.*, **95**, 2932–2937.

28. Wormhoudt, J.C. (1990) *Development of Laser Spectroscopic Diagnostics to Support Advanced Compound Semiconductor Deposition Techniques*, Aerodyne Research Inc., Billerica, MA.

29. Ladouceur, H.D. (2005) An overview of the known chemical kinetics and transport effects relevant to Mg/PTFE combustion. 2nd Workshop on Pyrotechnic Combustion Mechanisms, Karlsruhe, Germany, June 27, 2005.

30. Spectroscopic Data, *http://webbook. nist.gov/cgi/cbook.cgi?ID=C4605178&Units=SI&Mask=800#Electronic-Spec* (accessed 1 September 2011).

31. Boyer, D.W. (1980) Shock-tube measurements of the band strengths of

HBO$_2$ and OBF in the short wavelength infrared. *J. Quant. Spectrosc. Radiat. Transfer.*, **24**, 269–281.

32. Smit, K.J., Hancox, R.J., Hatt, D.J., Murphy, S.P. and de Yong, L.V. (1997) Infrared emitting species identified in the combustion of boron-based pyrotechnic compositions. *Appl. Spectrosc.*, **51**, 1400–1404.

33. Schulz, O., Weiser, V., Kelzenberg, S., Neutz, J., Raab, A. and Roth, E. (2009) Temperature of burning aluminium particles in pressurized air. 40th International Annual ICT Conference, Karlsruhe, Germany, June 23–26, p. 93.

34. Tianshu, W., Zhu, C., Wang, J., Xu, F., Chen, Z. and Luo, Y. (1995) Study on

spectral characterization of infrared flare material combustion with remote high resolution Fourier transform infrared spectrometry. *Anal. Chim. Acta*, **306**, 249–258.

35. Franzen, G. (2007) Spektralradiometrische Untersuchungen im Infraroten an den Abbränden der Scheinzielwirkmassen W1 und W2. WIWEB-310T, 15 August.

36. Knapp, S., Eckl, W., Kelzenberg, S. and Weiser, V. (2010) Temperature determination analysing emission spectra of di-atomic metal oxides. 41st International Annual ICT Conference, Karlsruhe, Germany, 29 June–2 July, p. 80.

10
Infrared Emitters

10.1
Decoy Flares

Aircrafts are distinct sources of infrared radiation. Whereas the skin of the plane may reflect ambient radiation and undergo aerodynamic heating, the combustion of the jet propellant yields exhaust gases that heat up turbine blades and tail pipes and leave the exhaust nozzle as a hot plume. Figure 10.1 displays the typical peak emission wavelengths and radiant intensity level as a function of the angle of observation. It is because of the intense radiation provided by the turbine blades and tail pipes that infrared-guided missiles have been developed to fight aerial targets [1, 2] (see Chapter 2). The operating mode of infrared-guided missile seekers has been discussed in Ref. [3]. A common cheap photoconductive detector material used in the first-generation infrared-guided missiles is lead sulfide, PbS. It has a peak spectral detectivity, D^*, in the $\lambda = 2.3–3.1\ \mu m$ range, depending on its temperature (see Figure 10.2) [4]. Thus, seekers equipped with either cooled or uncooled PbS detectors are suitable to home on the radiation originating from the hot tail pipes.

To defy infrared-guided missiles, pyrotechnic countermeasure flares have been devised as early as the late 1950s [5]. Infrared-countermeasure (IRCM) flares are pyrotechnic expendables that yield a strong electromagnetic signature that is supposed to interfere with the electro-optical sensor system of the missiles. The flare's signature depends on the duration and intensity of the exothermic reaction of its payload and on the chemical composition of the combustion products [6, 7]. To counter first-generation missiles, flares must yield a blackbody-type signature similar to that of hot tail pipes (Figure 10.3) [8].

The use of IRCM flares is either pre-emptive or reactive. In the pre-emptive use, flares are deployed in a hostile environment in anticipation of a threat. Thus the release serves to reduce the contrast of the scene to impede lock-on and to make tracking of the actual target more difficult. In the reactive mode, the actual target is already tracked by a seeker system and the flare aims to break the lock and to lure the missile away. In view of these necessities, the basic key requirements of decoy flares are the peak intensity, $I_\lambda (W\ sr^{-1})$ and the rise time, dI_λ/dt. A typical time/intensity specification for decoy flare defines rise time profile (dI/dt), peak

Metal-Fluorocarbon Based Energetic Materials, First Edition. Ernst-Christian Koch.
© 2012 Wiley-VCH Verlag GmbH & Co. KGaA. Published 2012 by Wiley-VCH Verlag GmbH & Co. KGaA.

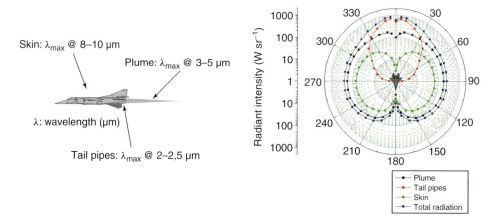

Figure 10.1 Infrared radiation sources on a jet aircraft and their peak emission wavelength together, and the polar diagram of radiant intensity at various aspect angles.

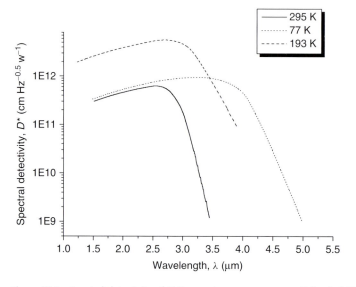

Figure 10.2 Spectral detectivity of PbS at various temperatures. (After Ref. [4].)

intensity level (I) and burn time [9]. Figure 10.4 depicts these parameters together with a schematic design of a typical $1 \times 1 \times 8$ in. flare. Figure 10.6 depicts a US Air Force F-15E Strike Eagle aircraft from the 391st Expeditionary Fighter Squadron pre-emptively deployings MTV flares during a flight over Afghanistan on 12 November 2008.

Figure 10.6 depicts the general scenario on reactive flare release, which is usually affected as early as possible to achieve the maximum miss distance between the missile and aircraft.

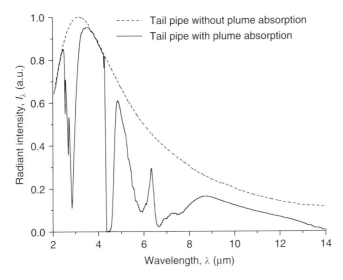

Figure 10.3 Signature of tail pipe with and without absorption by plume. (After Ref. [8].)

The static expulsion of a 36 mm IRCM flare for in-service testing and radiometric measurement is shown in Figure 10.7 from two perspectives.

Figure 10.8 gives a schematic sketch of a magnesium/teflon/viton (MTV) flare travelling in air.

Apart from their widespread use in tactical air-warfare pyrotechnic, infrared decoy flares allegedly serve to increase the survivability of intercontinental ballistic missile warheads (re-entry vehicles) against any ballistic missile defense measures. Practice targets for ballistic missile defense purposes have been developed based on 155 mm parachute illuminant shells. A prototype shell received 2500 g filling of magnesium/Teflon/Hycar®. This material yields burning times between 160 and 190 s and radiant intensities in the 600 W sr⁻¹ range in a particular band at ~18 000 m altitude [11].

10.2
Nonexpendable Flares

Other uses of compositions similar to those used in decoy flares are

- infrared target augmentation flares (ITAF)
- infrared tracking flares.

10.2.1
Target Augmentation

ITAF provides a towed or self-propelled aerial practice target with the appropriate infrared signature to enable lock-on of IR-guided missiles at tactical distances

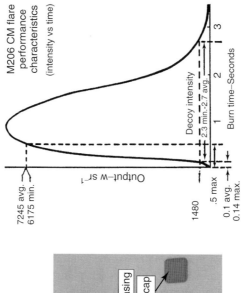

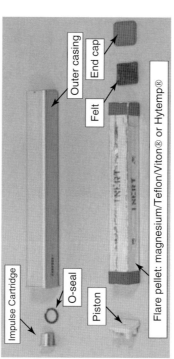

Figure 10.4 Flare intensity specification taken from Ref. [9] and typical design for $1 \times 1 \times 8$ in. flare [10].

Figure 10.5 A US Air Force F-15E Strike Eagle aircraft from the 391st Expeditionary Fighter Squadron pre-emptively deploys MTV flares during a flight over Afghanistan on 12 November 2008. The squadron is deployed to Bagram Air Base. DoD photo by Staff Sgt. Aaron Allmon, US Air Force (released) (accessed at http://www.visualintel.net).

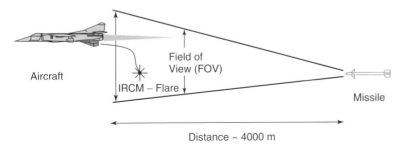

Figure 10.6 General scenario of reactive flare jettonizing. (After Ref. [7].)

in either surface-to-air or air-to-air scenarios. For this purpose, ITAF need long burning times and high radiant intensity. Typical burning times range between 40 and 60 s. Necessary radiant intensity in the 3–5 μm waveband is usually in the $I_\lambda > 600$ W sr^{-1} range. Depending on the purpose, multiple flares may be ignited to provide the target with a greater signature and also to allow target acquisition and lock-on at greater distance. The author has designed a tracking flare for use in towed aerial targets, which is electromagnetically safeguarded while handling and charging (Figure 10.9).

Figure 10.7 Side- and rear-view sequence of static expulsion of 36 mm MTV flare reproduced with kind permission of International Pyrotechnics Society.

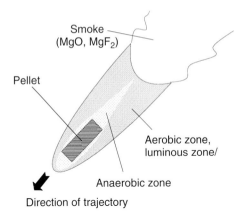

Figure 10.8 Schematic view of dynamic MTV combustion. (After Ref. [6].)

10.2.2
Missile Tracking Flares

Certain surface-to-air missiles use semi-automatic command to line-of-sight (SACLOS) steering. Therefore, these missiles possess a light source at the base of the missiles, such as a solid-state laser or a tracking flare. An example for a short-range missile with tracking flares was the British Blowpipe missile [13]. In addition, tracking flares are used also in larger missiles that travel at greater distances, r. As the missile travels away from the launch site the intensity of the

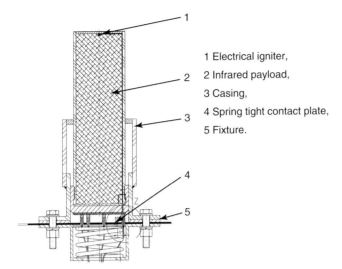

1 Electrical igniter,

2 Infrared payload,

3 Casing,

4 Spring tight contact plate,

5 Fixture.

Figure 10.9 Target augmentation flare design with electromagnetically secure wiring [12].

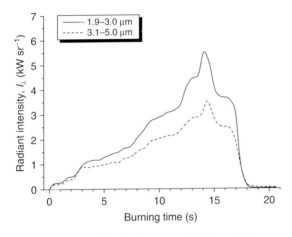

Figure 10.10 Typical static intensity profile for a missile tracking flare at sea level with exponentially rising I_λ.

flare has then to climb up to compensate the $1/r^2$ related drop in intensity detected at the launch site. The challenges in design of these tracking flares include the volume restriction and the peculiarities of the space available between the missile housing and the nozzle, the thermal load from the nozzle and the acceleration forces ranging between $+60$ g and -12 g that ask for high structural integrity of the grain. Finally, after burnout of the propulsion grain, the wake of the missile will bear a reduced pressure proportional to the velocity, which can be of the order of down to 30 kPa, further effecting combustion of the flare pyrolant. A typical intensity curve for an experimental tracking flare developed by the author for medium-range missile guidance is given above (Figure 10.10). It shows the radiant intensity in two bands as a function of time. The payload comprised multiple increments of compositions with increasing burn rate.

10.3
Metal–Fluorocarbon Flare Combustion Flames as Sources of Radiation

Above absolute zero temperature, $T > 0\,K$, all matter emits energy in the form of electromagnetic radiation. Even at thermal equilibrium, matter does emit and absorb radiation. The rate at which energy is emitted by matter – radiant emittance, M ($W\,m^2$) – is proportional to the surface area, A, and the fourth power of the absolute temperature, T.

$$M = A\sigma T^4 \tag{10.1}$$

with $\sigma = $ Stefan–Boltzmann constant, $5.6703 \times 10^{-8}\ W\,m^{-2}\,K^{-4}$.

The amount of energy emitted per wavelength, the spectral radiant emittance, M_λ, is given by Planck's law,

$$M_\lambda = \frac{2 \cdot \pi \cdot c_1}{\lambda^5 \cdot \left(e^{\frac{c_2}{\lambda \cdot T}} - 1\right)} \tag{10.2}$$

with

$c_1 = $ first Planck's radiation constant, $5.955310 \times 10^{-17}\ W \cdot m^2$
and $c_2 = $ second Planck's radiation constant, $1.4388 \times 10^{-2}\ m \cdot K$.

According to this, the wavelength of the maximum radiation intensity, λ_{max}, shifts to the short wavelength region with increasing temperature. This is known as Wien's displacement law.

$$\lambda_{max} = \frac{2897.756}{T}\mu m \tag{10.3}$$

Figure 10.11 displays a series of M_λ graphs for varying temperatures from 1500 to 3000 K. Even though the peak emission is shifted to short wavelengths at high temperature, substantial infrared emission occurs only at high temperatures. Planck's law is an ideal case, and the corresponding radiator is designated blackbody. Real radiators, however, do not emit as much energy as would be expected on basis of this law. To describe the deviation of real radiators from ideal behaviour, the concept of emissivity, ε, has been introduced. The emissivity, ε, is determined as the ratio of the radiant emittance, M', of the real radiator to the radiant emittance, M, of a blackbody at the same temperature. The emissivity is a function of the wavelength as well.

$$\varepsilon_{\lambda,T} = \frac{M'}{M} \tag{10.4}$$

Thus the emissivity may have values between $1 > \varepsilon > 0$. Figure 10.12 shows the radiant emittance of a blackbody ($\varepsilon = 1$) at $T = 3000\,K$ compared to that of a greybody of $\varepsilon = 0.25$ and a fictive selective emitter, both of the same temperature.

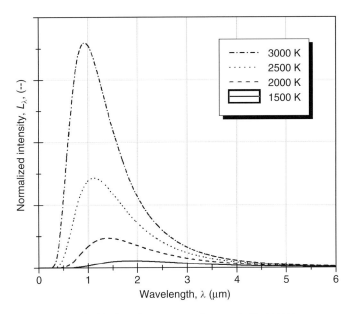

Figure 10.11 Blackbody radiation curves for $T = 1500$–3000 K.

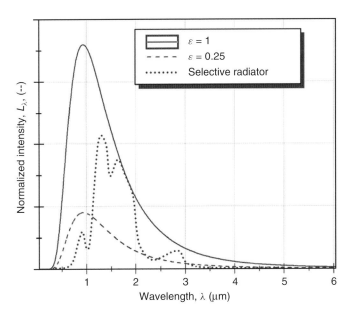

Figure 10.12 Radiant emittance curves for blackbody, $\varepsilon = 1.0$, greybody, $\varepsilon = 0.5$ and selective radiator all of $T = 3000$ K.

The emissivity of a material is a function of both temperature and wavelength. Some materials display invariance of emissivity over a discrete wavelength range and thus may be considered a greybody in this range. However, beyond this particular range, the emissivity may undergo a discontinuous change. These effects are mainly associated with molecular or atomic structure and macroscopic morphology of a material. According to Kirchhoff's law, matter emits as much radiation as it absorbs. Thus, one can write $\varepsilon = \alpha$ with α = absorptivity.

10.3.1
Flame Structure and Morphology

The metal–fluorocarbon flare combustion flame (see MTV plume in Figure 10.13) can be treated as a combustion jet into still air, provided the metallic fuel used is able to undergo vapour phase combustion with both air and fluorine. Typical metals that undergo vapour phase combustion are Li, Ca, Mg, Sm, Eu, Tm, Yb,

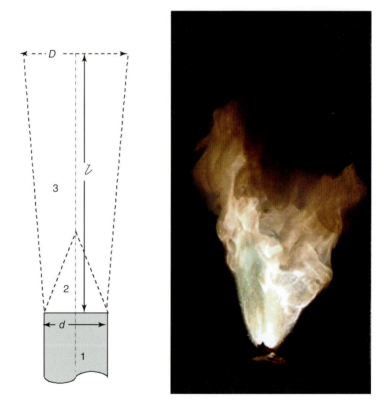

Figure 10.13 Morphology of a flare combustion plume and short exposure photographs (1/8000 s, aperture: 16) of MTV. (Photo taken by V. Weiser, E. Roth Fraunhofer ICT, Pfinztal/Berghausen, Germany, 2006, Ref. [15].)

Zn, (Al) and alloys of it; however, elements such as Ti, Zr, Hf, B, Si and Sb that form volatile fluorides and have vapourization temperatures well above the oxide and/or fluoride sublimation point in accordance with Glassman's criteria for metal combustion will not undergo vapour phase combustion [14]. Hence, the use of these fuels in fluorocarbon-based pyrolants does not yield a voluminous radiating plume but rather a stream of incandescent particles. Figure 10.13 depicts the main morphological characteristics of a static flame of an end-burning cylindrical MTV-flare grain undergoing vapour phase combustion. The aspects of dynamic flare combustion are treated at the end of this chapter.

In Figure 10.13, the combustion of the flare grain (1) yields an anaerobic core zone of conical shape (2) that contains the primary combustion products such as vaporized metal, its fluorides, condensed metal particles, particulate carbon and fluorocarbon species. Owing to momentum transfer mixing of the primary combustion products with the atmosphere and subsequent afterburn, reactions takes place in zone (3) [16]. Thus the aerobic zone typically displays a higher temperature than the core zone [17].

The length, l of the radiating zone is dependent on both atmospheric pressure and stoichiometry of flare. Thus both at low ambient oxygen partial pressure and high fuel surplus, the flame length, l, and terminal diameter will extend. According to Dillehay [18], the plume length (the radiating part of the flare combustion flame) for a magnesium/nitrate-based flare formulation is given by the following relationship:

$$l = \frac{d}{\xi_F} \tag{10.5}$$

with

$d =$ diameter of the flare candle in meters

and $\xi_F =$ the weight fraction of the metallic fuel in the composition that is needed to obtain the maximum adiabatic temperature.

We may assume comparable conditions for fluorocarbon-based pyrolants. Dillehay has described the influence of pressure on the plume length with the following law:

$$l = \frac{d}{\xi_F \cdot \sqrt[3]{\frac{0.1}{p}}} \tag{10.6}$$

with p being the ambient pressure in (MPa).

The terminal diameter of the radiating plume, D, is finally given by the following relationship:

$$D = \sqrt[6]{\frac{0.1}{p}} \cdot d \cdot \sqrt{1 + \frac{1}{\xi_F}} \tag{10.7}$$

Thus the radiating area can be described as the surface area of a truncated cone

$$A_P = \pi \cdot (R)^2 + \pi \cdot m \cdot (r + R) \tag{10.8}$$

with $r = d/2$ and $R = D/2$ and

$$m = \sqrt{(R - r)^2 + l^2}\ [18] \tag{10.9}$$

10.3.2
Radiation of MTV

The radiant flux delivered by a flare plume is influenced by a number of factors such as the mass consumption rate, $\dot{m}$, surface area of the plume, A_P (see above), its temperature, T_P, its emissivity, ε_P and its optical density $\kappa_{\lambda P}(l)$.

Important radiometric parameter are explained below.

Radiometric units

Parameter	Symbol	Unit	Definition
Radiant energy	W	J	–
Radiant intensity	I	$W\ sr^{-1}$	Φ/ω
Radiance	L	$W\ sr^{-1}\ cm^2$	$\Phi/\omega A \cos \Theta$
Spectral efficiency	E	$J\ g^{-1}\ sr^{-1}$	$Q/m\omega$
Solid angle	Ω	sr	A/r^2

All units may be also a function of the wavelength, wavenumber or frequency.

If a flare formulation burns with too low mass consumption rate, excessive thermal losses from the flame prevent formation of a stable plume and thus less energy is radiated than theoretically possible. If on the other hand, the mass consumption rate is too high, the combustion kinetics in the plume may be too slow, and thus, the material could pass the combustion zone without significant oxidation. Therefore, the energy yield would be smaller than theoretically expected. In this context, variations of the spectral efficiency, E_λ in the order of by $\pm 200\%$ have been observed for MTV formulations with the same basic stoichiometry but varying size distributions of both Mg and polytetrafluoroethylene (PTFE) powders to affect the burn rate (for an example, see Table 10.11).

The mass consumption rate will also affect the general particle concentration in the plume and thus directly influence the optical thickness with any given flare candle geometry. The optical thickness is given by

$$\kappa_\lambda(l) = K_\lambda \cdot l \tag{10.10}$$

with l being the plume thickness in meters and K_λ, the mass extinction coefficient in $m^2\ g^{-1}$,

$$K_\lambda = a_\lambda + \sigma_\lambda \tag{10.11}$$

which is the sum of both the absorption coefficient, a_λ, and scattering coefficient, σ_λ of all the plume constituents. Figure 10.14 displays K_λ for soot at various temperatures.

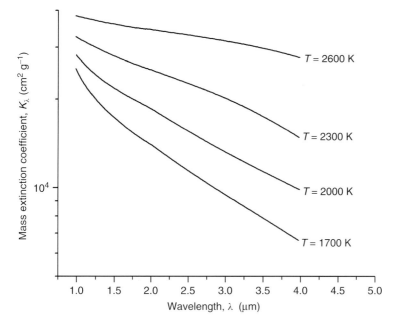

Figure 10.14 Mass absorption coefficient, K_λ, of particulate carbon [19].

If the optical thickness $\kappa_\lambda(l) \ll 1$, the plume is thin and no self-absorption takes place. At $\kappa_\lambda(l) \gg 1$, the plume may be considered thick and the volume elements inside the plume will not contribute to the radiant intensity other than by conductive and convective heat transport.

Particulate carbon is a major reaction product of metal–fluorocarbon flames and mainly determines its spectral characteristics [7]. The influence of soot on the optical properties of rocket plumes has been reviewed in Ref. [19].

According to Mie's theory, the size parameter for scattering is $\left(\frac{\pi\Theta}{\lambda}\right)^4$, whereas the size parameter for absorption is given by $\left(\frac{\pi\Theta}{\lambda}\right)$, with Θ being the particle diameter and λ the wavelength of radiation [20]. As metal–fluorocarbon combustion soot is composed of particles that are very small ($\sim$40 nm) in comparison to the wavelength of consideration (1–14 μm), the scattering contribution can be neglected. Thus the overall spectral emissivity is given by

$$\varepsilon_\lambda = 1 - \exp(-a_\lambda L_m) \tag{10.12}$$

with $-a_\lambda$ = spectral absorption coefficient and L_m = mean path of propagation of radiation in homogeneous volume [20]. The spectral absorption coefficient, a_λ, is given by

$$a_\lambda = Q_\lambda A N \tag{10.13}$$

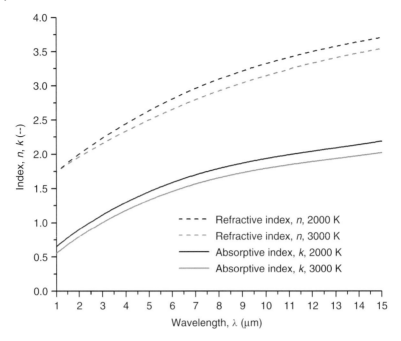

Figure 10.15 Refractive index, *n*, and absorption index, *k*, of particulate carbon at 3000 K [19].

with $Q_\lambda A$ = the spectral absorption cross section and N = the number of particles per volume element.

$$Q_\lambda = \frac{24\pi\Theta}{\lambda} \cdot \frac{n \cdot k}{(n^2 - k^2 + 2)^2 + 4n^2 k^2} \tag{10.14}$$

with n = index of refraction and k = absorption constant, both of which are functions of the wavelength λ. Both indices, n and k, are depicted in Figure 10.15 for carbon at 3000 K.

If we substitute $R(\lambda)$ for the right-hand term in Eq. (10.14), we can write

$$a_\lambda = \frac{24\pi\Theta}{\lambda} \cdot R(\lambda)AN \tag{10.15}$$

With n and k being functions of temperature, we see that the emissivity, ε_λ, of a carbon particle cloud is also a function of both temperature and concentration.

The radiant intensity $I_{\lambda_1 - \lambda_2}$ can be calculated on the basis of a knowledge of both radiating area, A in meters squared

$$I_{\lambda_1 - \lambda_2} = A \cdot \varepsilon_\lambda \int_{\lambda_1}^{\lambda_2} \frac{c_1}{\lambda^5 \cdot \left(\exp\frac{c_2}{\lambda \cdot T} - 1\right)} \tag{10.16}$$

and emissivity, ε_λ, of the carbon particle cloud discussed above.

Radiative calculations of MTV and MT-polyester pyrolants that exclude any contribution of carbon particle radiation have been reported [21–24].

10.4
Infrared Compositions

Flare compositions are characterized by the spectral efficiency, $E_{\lambda_1-\lambda_2}$ (in J g^{-1} sr^{-1}) and their burn rate, u, which at the given density, translates into mass consumption rate, $\dot{m}$ (g s^{-1}). A knowledge of both E_λ and m' allows to calculate intensity, I_λ, in a given band of interest, $\lambda_1 - \lambda_2$, by application of the following relationship:

$$I_{\lambda_1-\lambda_2} = E_{\lambda_1-\lambda_2} \cdot \dot{m} \left(W \ sr^{-1} \right) \tag{10.17}$$

The spectral efficiency of a flare composition is closely related to the combustion enthalpy of the flare composition, the plume temperature and the spectral emissivity of its combustion products as described in Eq. (10.18)

$$E_\lambda = \Delta_c H \cdot \frac{1}{4 \cdot \pi} \cdot F_\lambda \cdot \delta_w \cdot \delta_a \tag{10.18}$$

with

- $\Delta_c H$ = the enthalpy of combustion of the payload (J g^{-1}).
- F_λ = the fraction of energy emitted in a particular band pass
- d_w = the windstream degradation factor
- d_a = the aspect angle factor.

The combustion enthalpy of the flare, $\Delta_c H$, is the sum of both the primary anaerobic reaction enthalpy and the afterburn reaction enthalpy with the atmospheric oxygen. Thus for an Mg/Teflon®/Viton® (60/30/10) payload, this ideally translates into

$$2.4686Mg + 0.3000(C_2F_4)_n + 0.0267(C_{10}F_{13}H_7)_m \xrightarrow{\text{anaerobic}}$$
$$0.7736MgF_{2(s)} + 0.0935H_2 + 1.6950Mg_{(g)} + 0.867C_{(s)} \tag{10.19a}$$

$$0.0935H_2 + 1.6950Mg_{(g)} + 0.867C_{(s)} \xrightarrow{\text{Air}}$$
$$1.6950MgO_{(s)} + 0.0935H_2O_{(l)} + 0.867CO_2; RH = -553 + (-1384) = 1937 \, kJ \tag{10.19b}$$

Thus the enthalpy of combustion of MTV (60/30/10) is 19.37 kJ g^{-1}. However, only a fraction of this energy is transferred into useful radiation in the band of interest. This is described by F_λ. It describes the fraction of the reaction enthalpy that contributes to the radiant energy in the band of interest. Thus it is the ratio of radiation emitted in the band divided by the total integral under curve (Figure 10.16) [20]. For blackbody flares (with known emissivity ε_λ) this can be approximated as follows

$$F_{\lambda_1-\lambda_2} = \frac{\int_{\lambda_1}^{\lambda_2} M_{\lambda s}(\lambda)d\lambda}{\int_0^\infty M_{\lambda s}(\lambda)d\lambda} = \frac{1}{\sigma T^4} \int_{\lambda_1}^{\lambda_2} M_{\lambda s}(\lambda)d\lambda \tag{10.20}$$

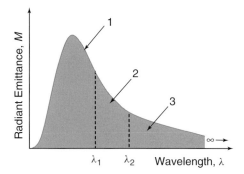

Figure 10.16 Visualization of F_λ by way of example of an ideal blackbody with (1), the intensity trace of the blackbody at given temperature, (2) the area under the trace between λ_1 and λ_2 and (3) the complete integral of the curve.

Table 10.1 Enthalpy of combustion, spectral efficiency and fraction of radiation in band of interest under static conditions [25, 26].

MTV	$\Delta_{comb}H$ (kJ g^{-1})	$E_{1.8-2.5\,\mu m}$ (J g^{-1} sr^{-1})	$E_{3.5-4.8\,\mu m}$ (J g^{-1} sr^{-1})	$F_{1.8-2.5\,\mu m}$	$F_{3.5-4.8\,\mu m}$
57/30/13	19.011	240	110	0.159	0.073
		$E_{2-3\,\mu m}$ (J g^{-1} sr^{-1})	$E_{3-5\,\mu m}$ (J g^{-1} sr^{-1})	$F_{2-3\,\mu m}$	$F_{3-5\,\mu m}$
45/50/5	9.275	155	112	0.210	0.150

with the Stefan–Boltzmann law describing the hemispheric specific exitance of a blackbody in vacuum:

$$M_s = \int_0^\infty M_{\lambda,s}(\lambda)\mathrm{d}\lambda = \sigma T^4 \tag{10.21}$$

Representative F_λ values obtained for MTV are given in Table 10.1.

10.4.1
Inherent Effects

10.4.1.1 Influence of Stoichiometry

Magnesium/Teflon®/Viton® In general, MTV flare formulations are fuel rich in order to take advantage of afterburning reactions with atmospheric oxygen. Afterburning alters the plume temperature and thus affects the radiant intensity, I_λ. The variation of the theoretical explosion enthalpy (anaerobic reaction) and overall combustion enthalpy Eq. 10.19 ($a + b$) as a function of stoichiometry

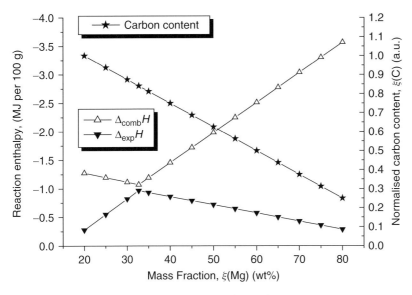

Figure 10.17 Explosion, combustion enthalpy and normalized carbon content of MTV as a function of Mg content.

together with the normalized carbon content of an MTV formulation is depicted in Figure 10.17. According to Eq. (10.18) the spectral efficiency, E_λ, of a formulation could in principal display a similar functional dependence on stoichiometry as the combustion enthalpy. However, the influence of both combustion temperature, T, and spectral emissivity, E_λ, of the flare plume have to be taken into account as well. Thus the decreasing carbon content with increasing Mg content will lead to a decrease in spectral efficiency at some point. In Figure 10.18, the variation of spectral efficiency in two bands as a function of the Mg content for MTV with 5 wt% Viton is depicted. Compositions were prepared from magnesium powder ($<40\,\mu m$), PTFE micropowder ($\sim 5\,\mu m$) and FC-2175 Fluorel [25]. E_λ increases almost linearly between $\xi(Mg)$ 30–65 wt% and inclines around 67 wt% Mg, which is probably due to decreasing carbon content. The slope of E_λ observed is different for each spectral bandpass. It is steeper for ranges close to $\sim\lambda_{max}(T)$ of the corresponding greybody temperature than for longer wavelengths.

The radiance, L_λ, of MTV as a function of Mg content is given in Figure 10.19. Selected data on two MTV compositions are given above in Table 10.1.

Modification of MTV   Modification of MTV has been an important topic since its initial discovery. Modifying agents may fulfil different roles. Thus more powerful oxidizers as well as more energetic fuels have been proposed as well as means to improve the thermal conductivity of the pyrolant grain to further influence the burn rate. Table 10.2 gives a brief synopsis of modifiers and their effect on a fuel-rich MTV formulation.

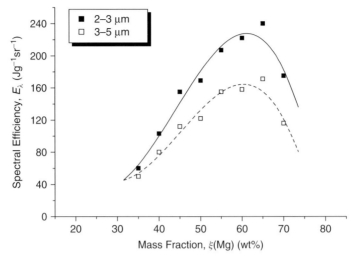

Figure 10.18 Spectral efficiency of MTV in two bands as a function of Mg content [25].

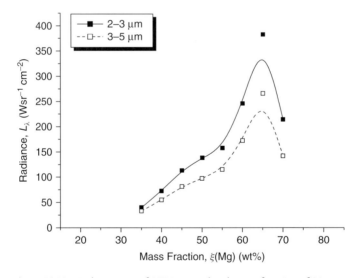

Figure 10.19 Radiance, L_λ, of MTV in two bands as a function of Mg content [25].

Nanometric Aluminium

Nanometric aluminium (27 wt%) added to a standard MTV formulation effects a maximum increase in burn rate of ~46%. At this level, the radiance, L_λ, is increased by ~40%, whereas the spectral efficiency remains essentially the same as that of the baseline composition [27].

Table 10.2 Effect of modifiers on fuel-rich MTV.

Modifier	Burn rate (mm s^{-1})	Spectral efficiency (J g^{-1} sr^{-1})	Radiance (W sr^{-1} cm^{-2})
Nanometric aluminium	Enhanced	Unchanged	Enhanced
Barium stearate	Lowered	Enhanced	Lowered
Carbon nanotubes	Enhanced	Lowered	Enhanced
Graphite	Enhanced	Unchanged	Enhanced
Silicon/ferric oxide	Lowered	Enhanced	Enhanced
Titanium	Enhanced	Lowered	Enhanced
Zirconium	Enhanced	Lowered	Enhanced

Barium Stearate

Long-burning formulations for infrared tracking flares typically comprise burn rate modifiers that extend the burning time. For this purpose, barium stearate has been proposed. Figure 10.20 shows the effect of stearate content on both normalized burn rate and spectral efficiency. Whereas the burn rate drops to levels of 20% of the unaffected composition shown below, the spectral efficiency climbs up to levels of ~140% [28].

Composition according to Villey-Desmeserets [28]

45.0 wt% magnesium (40–80 μm)
30.0 wt% PTFE
 7.5 wt% ferric oxide
 5.0 wt% styrene-based polyester
Variable barium stearate

Graphite and Carbon Nanofibres

Graphite is a good thermal conductor, with thermal diffusivities in the order of 0.6–0.9 cm^2 s^{-1}. As such, it is useful to increase the thermal conductivity of a pyrolant grain [29]. At a 5% weight level, it causes an increase in burn rate but does not affect the spectral efficiency (Table 10.3).

Even better thermal conductors are carbon nanofibres (CNFs) constituted from multiwalled carbon nanotubes. Nielson has proposed to use CNF to enhance the burn rate of MTV and other pyrotechnic compositions [31]. Palaiah has done a parametric study on the influence of CNF on burn rate. Figure 10.21 depicts the effect of CNF on both burn rate and radiant intensity [32].

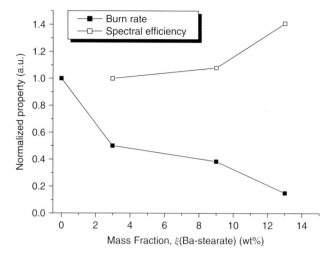

Figure 10.20 Normalized spectral efficiency and burn rate of MT-based formulation modified with barium stearate [28].

Table 10.3 Performance of two MTV compositions modified with MTV [30].

Composition MTV + graphite	ρ (g cm^{-3})	u (mm s^{-1})	$E_{3.5-4.8\,\mu m}$ (J g^{-1} sr^{-1})	$L_{3.5-4.8\,\mu m}$ (W sr^{-1} cm^{-2})
57/32.9/10/0.1	1.76	1.55	122	332
57/28/10/5	1.75	2.04	119	424
60/29.9/10/0.1	1.73	1.96	121	410
60/25/10/5	1.71	2.42	121	500

Even though burn rate is increased, this occurs at the expense of spectral efficiency, as is depicted in Figure 10.22.

Silicon/Ferric Oxide

Additives based on Si/Fe$_2$O$_3$ achieve a boost on spectral efficiency of MTV while slightly reducing its burn rate. This is believed to stem from the increased exothermicity of the afterburn reaction giving rise to higher flame temperatures and larger envelope of aerobic combustion when compared to the MTV baseline system [33]. In addition, as a 'recycled' reaction product, Fe$_2$O$_3$ is known to possess a high emissivity in the IR ($\epsilon_{1-5\mu m} \sim 0.8$) (Tables 10.4 and 10.5).

Miscellaneous Additives

The parametric performance modification of MTV (40/55/5) has been reported [34].

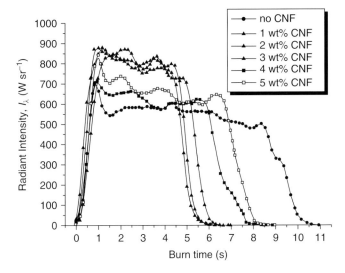

Figure 10.21 Effect of CNF content on burn time and radiant intensity compared to reference MTV composition [32].

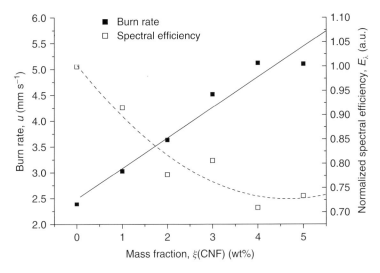

Figure 10.22 Effect of CNF content on burn rate and spectral efficiency [32].

In another study, binary Mg/PTFE (45/55) has been modified with unspecified amounts of various substances leading to higher radiant intensity (Table 10.6) [35].

In a parametric study, the effect of barium peroxide, BaO_2, on binary pyrolant based on Mg/PTFE has been tested for its effect on radiant intensity and burn time [36].

MT Compositions with Binders Other Than Viton   Before Viton was used as the main binder in MT flares, both nitrocellulose [37] and low-molecular-weight Kel-F

Table 10.4 Composition details (wt%) [33].

	1	2
Magnesium	45	45
Polytetrafluoroethylene	50	30
Hexafluoropropene-*co*-vinylidene fluoride polymer	5	5
Silicium	–	13
Ferric oxide	–	7
Experimental density, ρ_{exp} (g cm^{-3})	1.900	1.989

Table 10.5 Performance details [33].

Parameter	1	2
Enthalpy of explosion (kJ g^{-1})	7.16	5.18
Enthalpy of combustion (kJ g^{1})	9.13	13.62
Burn rate, u (mm s^{-1})	3.86	3.26
Spectral efficiency, $E_{2-3\mu m}$ (J g^{-1} sr^{-1})	155	212
$E_{3-5\mu m}$ (J g^{-1} sr^{-1})	112	139
Radiance $B_{2-3\mu m}$ (W sr^{-1} cm^{-2})	113	138
$B_{3-5\mu m}$ (W sr^{-1} cm^{-2})	82	91

Table 10.6 Spectral efficiency and radiance in two band pass ranges.

Composition Mg/ PTFE/Viton®/additive	$E_{3-5\mu m}$ (J g^{-1} sr^{-1})	$L_{3-5\mu m}$ (W sr^{-1} cm^{-2})	$E_{8-14\mu m}$ (J g^{-1} sr^{-1})	$L_{8-14\mu m}$ (W sr^{-1} cm^{-2})
40/55/5/0	103	70	25	12
Charcoal	118	79	24	15
Fe	172	85	34	17
Cu	201	101	39	20
Ti	176	91	34	18
Si	143	89	32	17
Al	179	102	36	21

wax were used as binder. However, both proved unsatisfactory from the standpoint of mechanical strength and sensitivity [38].

Nowadays, beside Viton® a number of non fluorinated binders are used in MT-based formulations. These materials include polyester resins, which are popular in France (MTR) [28], Styrene-butadiene copolymer (Hycar®) (MTH), which was used in the M206 flare payload [39], Polyacrylate binder (Hytemp® 4454) (MTH) and Ethylen-Vinyl-Acetate Elvax® (MTE) [40], which are both in use in the United States. The use of thermoplastic binders such as polystyrene (PS) has been proposed recently (MTTP) [41]. A typical formulation is depicted below.

MTTP formulation [41]

---

66 wt% magnesium spherical
20 wt% Teflon
7 wt% polystyrene
7 wt% dimethyl phthalate

The radiant intensity profile for equal amounts of standard MTV decoy flare mix with the above composition is depicted in Figure 10.23. The spectral efficiency for the latter is about twice ($\sim$2.03) in the 2–2.6 µm band. This is due to the much higher Mg content in MTTP formulation (66 vs 54 wt% Mg). However, despite the high Mg content, the latter formulation shows a very slow rise in intensity, indicating a slower burn rate. Thus this material would require a different grain geometry to meet with operational rise time specifications.

Based on the disclosure [41], a study was conducted to investigate the potential of various thermoplastic binders (PS, polyvinylchloride = PVC, PVC-*co*-PVAc = polyvinylchloride–polyvinylacetate copolymer), softening agents (dioctyladipate = DOA, dimethylphthalate = DMP, glycidylazide polymer = GAP) and burn rate

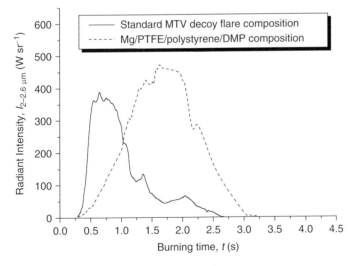

Figure 10.23 Comparison of standard MTV decoy flare composition in 2–2.6 µm band with extruded Mg/PTFE/PS/DMPP composition [41].

Table 10.7 Performance data of various MT pyrolants with alternate binder systems [42].

Binder (wt%)	Softening agent (wt%)	PTFE (wt%)	Mg (wt%)	Modifier (wt%)	Normalized burn rate (mm s^{-1})	Radiant intensity (W sr^{-1})	Spectral efficiency (J g^{-1} sr^{-1})
Viton, 16	–	30	54	–	8.19	315	185
PVC, 8	GAP, 8	30	54	–	0.17	115	316
PVC, 9	DOA, 9	22	60	–	1.80	111	293
PVC, 9	DOA, 9	19	60	Fe_2O_3, 3	6.37	307	232
PS, 8	DMP, 8	20	64	–	5.41	319	278
PVC-*co*-PVAc, 9	DMP, 9	16	60	Fe_2O_3, 6	7.63	325	192

modifiers (iron oxide = Fe_2O_3). Table 10.7 shows the spectral efficiency in a short wave band of a number of MT formulations with these binder systems in comparison to standard MTV (54/30/16).

The formulation reported below was investigated for scale-up in a SERDP Project and had the reported performance data [42]:

Modified MTTP [42]

40.425 wt%	Magnesium, spherical, −100/+200 mesh
17.325 wt%	Magnesium weight, fine spherical, Type III
18.000 wt%	PTFE, Whitcon TL-102
11.250 wt%	Dimethylphthalate
9.000 wt%	PVC-*co*-PVAc, OxyChem 1713
3.000 wt%	Iron oxide, Sicotrans Red L2715 D
1.000 wt%	Graphite
Burn rate:	1.94 mm s^{-1}
Radiant intensity:	110 W sr^{-1}
Spectral efficiency:	256 J g^{-1} sr^{-1}

Figure 10.24 shows a cross-geometry flare grain with applied first fire manufactured from the above-mentioned composition.

Compositions with Oxidizers Other Than Teflon

Magnesium/Polychlorotrifluoroethylene Compositions

Ternary pyrolants containing polychlorotrifluoroethylene (PCTFE) as the main oxidizer show lower radiant intensity than compositions using PTFE (Figures 10.25 and 10.26).

Figure 10.24 Fe$_2$O$_3$.modified extruded MTPP grain with first fire applied from Ref. [42].

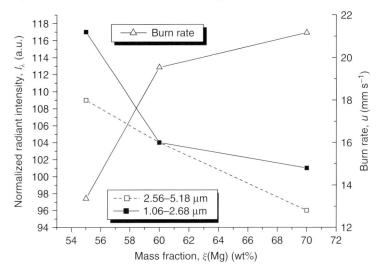

Figure 10.25 Burn rate and normalized radiant intensity of binary Mg/PTFE (Fluon G1) compositions [43].

Magnesium/Graphite Fluoride/Viton Compositions

Substituting graphite fluoride for PTFE in MTV allows for higher performance than with MTV. Table 10.8 depicts the salient features of two graphite fluoride-based compositions [44].

Graphite fluoride is a superior oxidizer compared to PTFE because of

- high reactivity of tertiary C–F bonds and hence high burn rate;
- high emissivity of graphite particles formed on combustion that survive oxidative stress in the aerobic combustion zone and thus display a longer radiative lifetime than carbon soot particulates, which subsequently oxidize in the aerobic combustion zone.
- No melting before exothermic decomposition process (see Figure 3.6 on page 26)

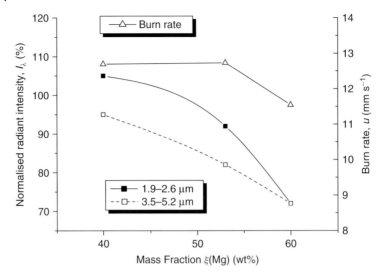

Figure 10.26 Burn rate and normalized radiant intensity of Mg/PCTFE/Viton compositions (5 wt% Viton each) [43].

Table 10.8 Enthalpy of combustion, spectral efficiency and fraction of radiation in band of interest under static conditions.

Composition Mg/PMF/Viton	ρ (g cm^{-3})	u (mm s^{-1})	$E_{2-3\,\mu m}$ (J g^{-1} sr^{-1})	$E_{3-5\,\mu m}$ (J g^{-1} sr^{-1})	$L_{2-3\,\mu m}$ (W sr^{-1} cm^{-2})	$L_{3-5\,\mu m}$ (W sr^{-1} cm^{-2})
35/60/5	1.900	12.5	165	150	393	357
40/55/5	1.876	16.0	170	157	511	472

Thus in advanced flare and first fire compositions, polycarbon monofluoride (PMF) is used either exclusively or in combination with other fluorinated (PTFE, PVDF (polyvinylidene fluoride), PCTFE) or non-fluorinated oxidizers (HC (hexachloroethane), CP (chlorinated paraffine), $KClO_4$ (potassium perchlorate), KNO_3 (potassium nitrate), NH_4ClO_4 (ammonium perchlorate), RDX (hexogen), HMX (octogen), DADNE (1,1-diamino-2,2-dinitroethylene)) to oxidize any fuel or combination of B, Si, Mg, MgAl alloys, Mg_2Si, MgB_2,Ti, TiAl, TiH_x, Zr and both ZrFe and ZrNi alloys.

The use of less fluorinated PMF such as $(C_2F)_n$ allows for even higher burn rates but at the expense of spectral efficiency.

Magnesium/Perfluorotetrazolate/Viton Compositions

Ammonium and guanidinium salts of heptafluoropropyltetrazolates in lieu of PTFE yield superior spectral efficiency compared to MTV and even formulations based on graphite fluoride [45]. This is believed to stem from two effects mainly:

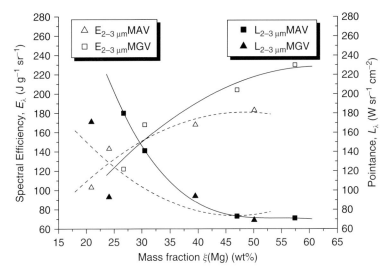

Figure 10.27 Spectral efficiency and pointance of MAV and MGV [45].

- Exothermal decomposition of the tetrazolate moiety
- Release of nitrogen gas that acts as flame expander and thus improves optical density of the plume.

Figure 10.27 depicts both spectral efficiency and pointance in $2-3\,\mu$m band for the ammonium salt-based pyrolant, MAV, and the guanidinium based-salt pyrolant, MGV (with 10.5 wt% Viton each). The exothermal decomposition of the tetrazolate also influences the mass consumption rate, which drops with increasing Mg content. This causes the pointance to decrease with increasing Mg content [45].

The calcium salts of pentafluoroethyl-(CE) and heptafluoropropyl tetrazolate (CP) despite a higher density than the above-mentioned ammonium and guanidinium salts due to high crystal water content prove unsuitable oxidizers and yield both very low spectral efficiencies and low pointance (Figure 10.28) [46]. However, in contrary to the above discussed ammonium and guanidinium salt, these pyrolants (MCEV and MCPV) again show the typical dependency of both burn rate and pointance that do increase with increasing Mg content. This indicates a mechanism that is not influenced by the decomposition of the tetrazolate unit. It can also be speculated that the excessive crystal water content compensates this heat release by forcing a water–gas reaction which depletes the carbon content.

Compositions with Fuels Other than Magnesium

Magnesium–Aluminium Alloy/Fluorocarbon Compositions

A comparison of binary Mg/PTFE and Mg_4Al_3/PTFE formulations is given in Table 10.9 [47]. Although the spectral efficiency is similar to that of Mg-based payloads, the higher burn rate leads to higher pointance.

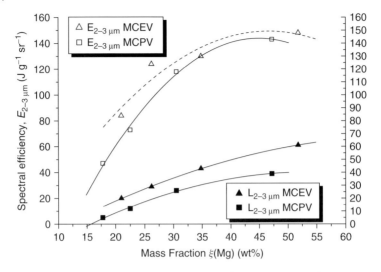

Figure 10.28 Spectral efficiency and pointance of MCMV, MCEV and MCPV [46].

Table 10.9 Enthalpy of combustion, spectral efficiency and fraction of radiation in band of interest under static conditions [47].

Composition metal/PTFE	ρ (g cm^{-3})	u (mm s^{-1})	$E_{3-5\,\mu m}$ (J g^{-1} sr^{-1})	$E_{8-14\,\mu m}$ (J g^{-1} sr^{-1})	$L_{3-5\,\mu m}$ (W sr^{-1} cm^{-2})	$L_{8-14\,\mu m}$ (W sr^{-1} cm^{-2})
50/50 (Mg)	1.40	0.84	139	12	68	6
45/55 (Mg)	1.82	–	–	–	60	20
50/50 (Mg$_4$Al$_3$)	1.84	1.41	135	14	144	15

Mg–Al alloys have been proposed as advantageous fuels in infrared decoy flare payloads and transfer fires as the examples given below [48–50]:

Flare First Fire	Flare Payload	Flare Payload
72 wt% Mg$_4$Al$_3$ 24 wt% PTFE 4 wt% Diphenyl (C$_{12}$H$_{10}$)	65 wt% Mg$_4$Al$_3$ 27 wt% PTFE 8 wt% Diphenyl (C$_{12}$H$_{10}$)	55 wt% Mg$_4$Al$_3$ 10 wt% Mg 30 wt% PTFE 2 wt% Anthracene (C$_{14}$H$_{10}$) 3 wt% Phenanthrene (C$_{14}$H$_{10}$)

Diphenyl, anthracene and phenanthrene are known to yield soot nuclei in the combustion zone and thus add to the blackbody radiation [51].

Titanium and Zirconium/Fluorocarbon Compositions

Titanium has been used as the sole and complementary fuel in RR-81 and RR-82 decoy flares [38].

RR-81 flare composition

21.68 wt% magnesium
13.29 wt% titanium
41.25 wt% PTFE
23.78 wt% Kel-F wax

This composition has an approximate burning rate of 4 mm s^{-1} at ambient pressure [38].

RR-82 flare composition

---

50 wt% magnesium
10 wt% titanium
40 wt% PTFE

The RR-82 mix had an approximate burn rate of 30 mm s^{-1}, which is typical for porous systems [38].

Gongpei tested unspecified Ti particles on MTV (40/55/5) and obtained a significant increase in spectral efficiency E_λ in both 3–5 and 8–14 µm band pass range (Table 10.10) [34].

Zirconium hydride was tested as fuel in tracking flare payload and proved inferior to MTV and even nitrate-based flare systems having a radiance, $L_{0.8-3.5\,\mu m} = 66$ W sr^{-1} cm^{-2} [52]

ZrH$_2$/PTFE flare composition [52]

---

54 wt% ZrH$_2$
46 wt% PTFE

Zirconium does effect a significant increase in burn rate to MTV and thus is able to increase the pointance. However, the spectral efficiency drops to about 80% of the unaffected level (Table 10.11). This is possibly due to the increase in burn rate, which will deteriorate the combustion efficiency of Mg and thus waste energy [26].

Table 10.10 Spectral efficiency and radiance in two band pass ranges.

Composition Mg/ PTFE/Viton®/titanium	$E_{3-5\,\mu m}$ (J g^{-1} sr^{-1})	$L_{3-5\,\mu m}$ (W sr^{-1} cm^{-2})	$E_{8-14\,\mu m}$ (J g^{-1} sr^{-1})	$L_{8-14\,\mu m}$ (W sr^{-1} cm^{-2})
40/55/5/0	103	70	25	12
40/55/5/15	175	91	34	18

Table 10.11 Enthalpy of combustion, spectral efficiency and fraction of radiation in band of interest under static conditions [26]

Composition Mg/PTFE/Zr/Viton	ρ (g cm^{-3})	u (mm s^{-1})	$E_{3-5\,\mu m}$ (J g^{-1} sr^{-1})	$L_{3-5\,\mu m}$ (W sr^{-1} cm^{-2})
65/30/0/5	1.78	12.4	121	267
62/29/4.5/4.5	1.85	16.8	105	326
59/27/7/5	1.88	20.5	104	400
56.5/26/13.5/4.5	1.97	19.7	102	395

Lithium has been proposed as an alternative metallic fuel for IRCM flares [53]. Hahma has developed a series of compositions based on ZrW alloy with both Bi_2O_3 and PTFE as oxidizers. These payloads deliver low specific energy in both A and B band at static conditions and sea level (A. Hahma, personal communication) [54].

10.4.2
Spectral Flare Compositions

Payloads based on boron, PTFE, perchlorates and Viton have been proposed as both flare payloads for spectral flares and spectrally matched igniter compositions [55, 56]. Combustion of these materials yields significant amounts of FBO and CO_2, both of which serve as spectral emitters [57] (see Figures 9.37 and 9.38).

Typical compositions are given below:

Igniter compositions [55]	Spectral flare composition [56]
62 wt% potassium perchlorate	42.075 wt% ammonium perchlorate
14 wt% boron	4.125 wt% boron
12 wt% polytetrafluoroethylene	4.950 wt% polytetrafluoroethylene
12 wt% Viton	14.000 wt% Viton
–	10.150 wt% magnesium
–	9.900 wt% aluminium
–	6.600 wt% potassium nitrate
–	8.250 wt% hexamine
Colour ratio: $\theta_{3.5-4.8/1.8-2.5} = 3.5-4$	Colour ratio: $\theta_{3-5/2-3} = 1.8$

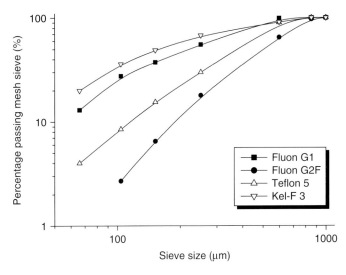

Figure 10.29 Particle size distribution of various PTFEs and PCTFEs [43].

The pressed igniter composition in bulk greater than 50 g upon ignition gives a very fast burn ($\gg$50 mm s^{-1}) that shortly thereafter transitions to deflagration yielding a nice deep-green flash (see Figure 9.29) and a loud bang.

10.4.3
Particle Size Issues

Apart from the influence of chemical and physical modifiers, particle size, surface area and morphology not only play an important role in the rate of reaction of any fuel oxidizer mix but also interfere with combustion efficiency and thus radiant energy release. Figure 10.29 shows the particle size distribution of four different fluoropolymers, including PCTFE. The influence of different origin and particle size distributions of PTFE on radiant intensity is depicted in Figures 10.30 and 10.31.

Table 10.12 shows the variation in spectral efficiency $E_{3-3\mu m}$, for an MTV (45/50/5) as a function of particle size distribution of both Mg and PTFE.

Table 10.12 nicely shows the inverse relationship often encountered with spectral efficiency and burn rate. Even though mixing is far less ideal with *all*-coarse components, slower reaction allows better consumption than with very fast burning mixes.

10.4.4
Geometrical Aspects

From illuminating flares, it is known that performance changes with candle diameter, ϕ, [58]. The same holds true for MTV. The influence of diameter on

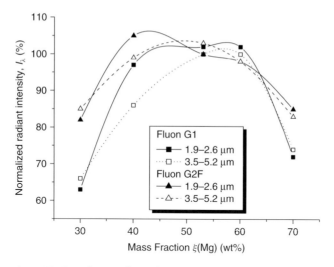

Figure 10.30 Influence of particle size on radiant intensity [43].

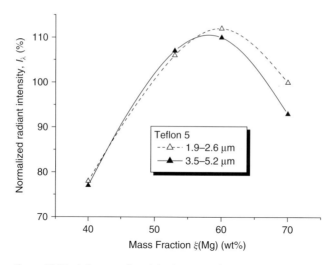

Figure 10.31 Influence of particle size on Radiant Intensity [43].

Table 10.12 Particle size ratio influence on spectral efficiency of MTV (45/50/5) and burn rate.

J g^{-1} sr^{-1}/mm s^{-1}	Mg < 71 μm	Mg < 100 μm
PTFE < 5 μm	112/3.86	153/2.11
PTFE < 600 μm	102/4.01	143/2.56

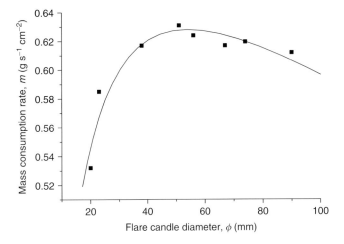

Figure 10.32 Diameter effect on mass consumption rate of stoichiometric MTV pyrolant grain after Ref. [47].

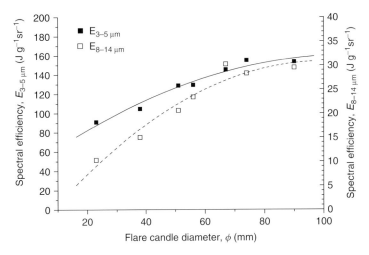

Figure 10.33 Influence of candle diameter on spectral efficiency of stoichiometric MTV grain [47].

a stoichiometric MTV grain at constant density and ambient pressure and static conditions has been studied [47]. It was found that the mass consumption rate will steeply increase from ~20 mm up to 60 mm (Figure 10.32). This effect can be accounted to reduced energy losses from a greater flame that further heat-feedback to the condensed zone and thereby accelerate the combustion. The slight decrease above 60 mm diameter may not be explained in a straightforward manner and could possibly be accounted for statistic scatter. Quite in line with these results are the observations that the spectral efficiency of MTV increases asymptotically from 20 up to 90 mm (Figure 10.33).

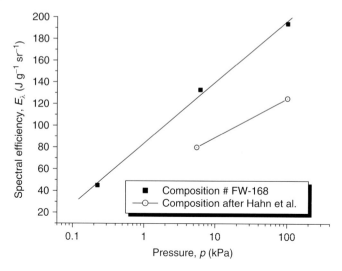

Figure 10.34 Static spectral efficiency of two MT-based formulations as a function of ambient pressure [37, 52].

10.5
Operational Effects

10.5.1
Altitude Effects

At greater altitude, the atmospheric pressure decreases and hence flare plumes expand. This effect could prove beneficial as the radiating surface area increases. A slight increase in intensity has been observed (N. Davies, personal communication); however, this is not the case for greatly diminished pressure because performance of MTV, as we have discussed above, mainly relies on afterburn with atmospheric oxygen (Eq. (10.19)) whose partial pressure decreases with increasing height also. Hence, operational magnesium/PTFE-based payloads show pronounced performance degradation at decreased pressure. Figure 10.34 depicts the performance of two compositions given below at sea level (101.3 kPa) and greater altitude.

IR tracking flare composition [38, 52]	Composition # FW-168 [37]
54 w% magnesium, 60 μm	54 wt% magnesium, 23 μm
23 wt% Teflon, 600 μm	46 wt% Teflon, trimod., 23/52/92 μm
23 wt% Kel-F wax	–
2 wt% nitrocellulose,12.6% N	–
Spectral efficiency, E_λ at sea level	
125 J g^{-1} sr^{-1} @ 1.8–2.7 μm	194 J g^{-1} sr^{-1} @ 1.8–2.8 μm
Burn rate at sea level	
2.3 mm s^{-1}	5.2 mm s^{-1}

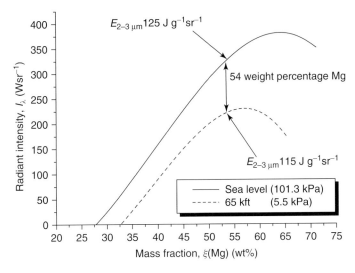

Figure 10.35 Variation of static radiant intensity with stoichiometry at both sea level and 65 kft [52].

Although both compositions feature the same amount of Mg, both burn rate and output at sea level vary significantly. This is due to the different Mg grain sizes (60 versus 23 μm). With smaller grain size, that is, larger surface area, pyrolant formulations usually give off better performance as particle heat-up and combustion time is much smaller than with coarser particles, and hence combustion proceeds faster as with coarser particles. Hence, high-altitude formulations typically feature very fine, atomized magnesium $\ll$40 μm and large particle size, flaky PTFE (approximately few hundred micrometers) to partially overcome this problem [6].

Hahn *et al.* had developed IR tracking flare compositions based on Mg/PTFE and Kel-F [52]. The radiant intensity of a range of compositions in the 2–3 μm range at both sea level and 65 kft (19812 m) is given in Figure 10.35. For the particular stoichiometry, the spectral efficiency can be given. However, at differing stoichiometry, burn rate will vary as well and is not known for the complete range; thus E_λ is not known likewise.

Formulation PL 6328 (see below) i.a. used in Mk 46 decoy flare [38] has been tested under altitude and dynamic conditions [59]. Table 10.13 displays the spectral efficiency in the 1.7–2.95 μm bandpass range.

PL 6328 composition [38]

---

54 wt% magnesium, gran 16
30 wt% Teflon, #7
16 wt% Viton
Density: 1.80 g cm^{-3}, (96% TMD)

Table 10.13 Dynamic spectral efficiency of PL 6328 as determined from Mk 46 Flare configuration at various altitudes released at 308 m s^{-1} (592 kts) [59].

Altitude (m)	1524	4572	9144
Pressure (kPa)	84	58	30
$E_{1.7-2.95\,\mu m}$ (J g^{-1} sr^{-1})	24	21	17
Velocity at burn out (m s^{-1})	45	55	60

10.5.2
Windspeed Effects

Operational use of MTV payloads always occurs under dynamic conditions. That is, payloads do not burn in still air but are projected at velocities varying from 60 to 80 kt (hovering helicopter) to 650 kt (fast jet in afterburn mode). As boot-camp experience teaches us, a little wind will actually aid in burning combustibles, whereas excessive wind will blow out fire. In principle, the same applies to MTV flares.

In practice, it is observed that wind speed levels of up to about 20 m s^{-1} do not affect the radiant intensity level of burning MTV flares [60]. However, above $\sim$20 m s^{-1}, radiant intensity decreases exponentially with wind speed.

To understand windstream effects on combustion, we have to look at the processes occurring with a burning pellet. Steady-state combustion of a pyrolant produces a volume of gas-phase reaction products of temperature T, $V(T)$ (cm^3), that is ejected normal to the surface per unit time, hence we take the first derivative $\dot{V}(T)$. The exit velocity can be calculated from the mole number of the gaseous combustion products and the adiabatic combustion temperature by assuming ideal gas behaviour:

$$\dot{V}(T) = \frac{\dot{n} \cdot R \cdot T_{ad}}{p}$$

(10.22)

The value of $\dot{V}(T)$ equals the exit velocity u_{ex} (cm s^{-1}) of the pyrolant. For fuel-rich MTV, typical values for u_{ex} range between 18 and 25 m s^{-1} (34–49 kts).

At low wind speed levels $\pm$20 m s^{-1}, the pyrolant flame can be considered as a free jet expanding into still air. However, higher wind speeds distort the flare plume and affect faster cooling of the radiating particles by both conductive and convective heat transfer. Hence the flare plume changes in shape and size with increasing wind speed from a rather bulky spherical appearance at still air to a narrow ellipsoidal shape parallel to the wind direction at high wind speeds [60]. Figure 10.36 shows the radiant intensity of an MTV flare arbitrarily normalized to peak intensity at 20 m s^{-1} windspeed, Table 10.14 depicts the differing performance of a flare at various speed and altitude.

Even though wind speed effects faster particle cooling and hence less radiation, it does increase the mass consumption rate of a pyrolant [61] because of increased stagnation pressure at the burning surface facing forward direction as is

Table 10.14 Dynamic spectral efficiency of a fuel-rich binary Mg/PTFE grain determined from Type 400 Mk 2 Flare cartridge configuration at two altitudes at different wind speeds [62].

Altitude (m)	300	12192
Pressure (kPa)	98	19
Speed (m s^{-1})	113	238
$E_{2-3\mu m}$ (J g^{-1} sr^{-1})	58.5	52.3
$E_{3-5\mu m}$ (J g^{-1} sr^{-1})	39.3	29.5
$E_{4-4\mu m}$ (J g^{-1} sr^{-1})	11.2	8.5

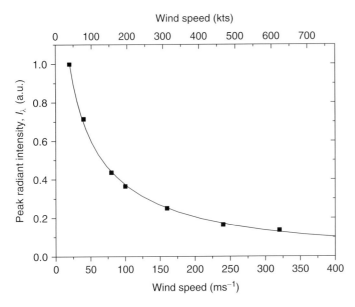

Figure 10.36 Radiant Intensity as a function of wind speed normalized to 20 m s^{-1} peak intensity. (After Ref. [60].)

given below

$$p_{stag} = p_{amb} + \frac{\rho_{air} \cdot v_f}{2} \ [6] \tag{10.23}$$

Figure 10.37 depicts the increasing mass consumption rate of MTV-40 with increasing windspeed.

MTV-40

40 wt% magnesium, 45 μm
55 wt% PTFE, TF 9205
5 wt% fluorel FC 2175
Density: 1.92 g cm^{-3}

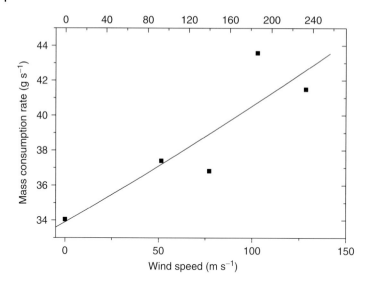

Figure 10.37 Mass consumption rate of MTV-40 as a function of wind speed.

Table 10.15 Spectral efficiency of MTV.

MTV	$E_{2-3\,\mu m}$ $(J\,g^{-1}\,sr^{-1})$	$E_{3-5\,\mu m}$ $(J\,g^{-1}\,sr^{-1})$	$E_{3.7-4.5\,\mu m}$ $(J\,g^{-1}\,sr^{-1})$	$E_{3.5-4.8\,\mu m}$ $(J\,g^{-1}\,sr^{-1})$	$E_{4-4\,\mu m}$ $(J\,g^{-1}\,sr^{-1})$
Static	–	296	–	120	–
150 kts	40	23	10	–	8

At sonic velocities, the radiant intensity of MTV barely reaches 10% of the near-static conditions. Hence the spectral efficiency of MTV at increased windspeed is significantly lower and has to be taken into account when designing decoys. Table 10.15 gives the spectral efficiencies for MTV at 150 kts in four different bands.

The radiant intensity of two MTV flares burned at both static conditions is depicted in Figure 10.38. Besides the dip in intensity, it shows the reduced burn time as discussed above.

The dynamic spectral efficiency of MTV-40 is depicted in Figure 10.39.

Figure 10.40 shows the combustion flame of a 1 × 1 × 8 in. MTV Flare pellet burned in a Windstream at 300 kts [63]. The measurement setup is shown on top (not in scale).

As is evident from the anisotropic flare plume depicted in Figure 10.40, the apparent radiant intensity is a function of the angle of observation as well. Figure 10.41 depicts the polar diagram of an MTV flare burned at 250 kts [64]. The lowest radiant energy level is seen in direction of the wind stream. This is due to the strong convective and conductive cooling and the small hemispherical surface

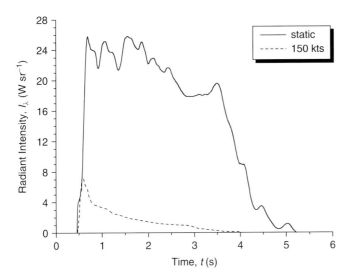

Figure 10.38 Radiant intensity for a large rectangular MTV flare at both static conditions and 150 kts in 3–5 μm band pass range [63].

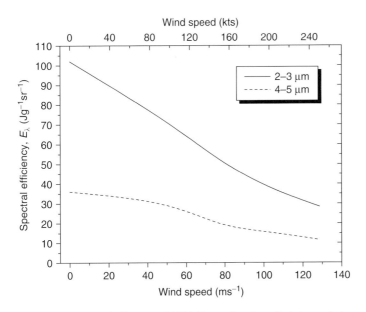

Figure 10.39 Spectral efficiency of MTV 40 as a function of wind speed at sea level.

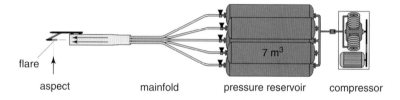

Figure 10.40 Air flow setup for measuring the radiant intensity of flares under dynamic conditions, and flame picture with superimposed traces of air outlet and 118 flare pellet. reproduced with kind permission by Dr. Lukas Deimling, [63].

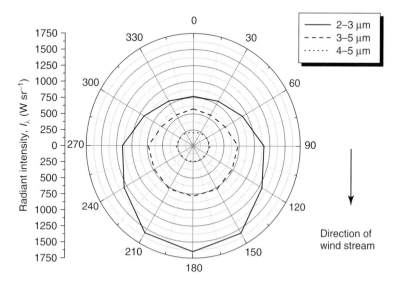

Figure 10.41 Dynamic MTV flare radiant intensity as a function of angle at 250 kts [64].

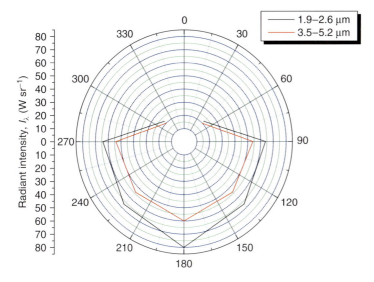

Figure 10.42 Dynamic MTV flare radiant intensity as a function of angle at 529 kts and 15 240 m height [65].

exposed. Almost 50% more radiant energy is projected normal to the wind stream direction, whereas in the wake of the plume, despite the shielding by the plume of the cooled combustion products exposes the largest surface area towards the sensor and thus explains for the maximum radiant energy emitted, which is about 210% of that at $0°$ and 142% at $90°$. The radiation level in band 4 (4–5 μm) is determined by the hot CO_2. It is thus less affected by angle of observation.

Radiant intensity calculations for dynamic flare combustion is treated in Ref. [21].

Figure 10.42 depicts the scenario for a binary fuel-rich Mg/PTFE flare grain at windspeed and high altitude. Even though no data had been collected at $0°$, the same qualitative picture as in Figure 10.41 is obtained [65].

The change of the radiant at intensity two angles with increasing wind speed is depicted in Figure 10.43 [66].

10.6
Outlook

Although spectral threats become increasingly important [57, 67] blackbody-type flares continue to play an important role in the protection of aircraft. However, compositions and corresponding payloads will become more complex to account for rising demands in performance. One such example is the need for high-performance oxidizers that provide flare grains with a greater robustness against windstream effects by release of higher amount of gaseous products to effect a higher exit velocity. The increased need for pre-emptive release measures

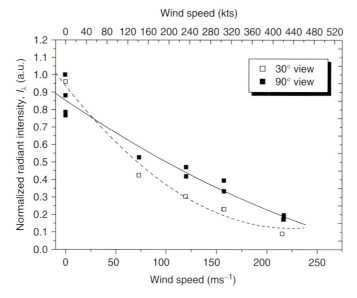

Figure 10.43 Dynamic MTV flare radiant intensity as a function of windspeed at both 30° and 90° [66].

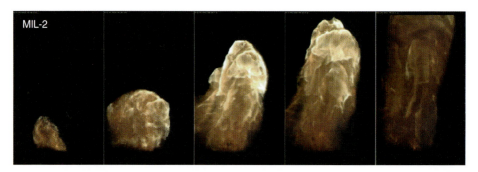

Figure 10.44 MIL Flare combustion plume [71]. (Reproduced with kind permission by Uwe Schaller and Volker Weiser.)

in asymmetric theatres also forces developers to increase volume-specific spectral efficiency by using high-energy fuels such as MgB_2 and Mg_2Si or Li [53] and high-performance oxidizers such as perfluorinated high nitrogen compounds in order to maintain or even increase countermeasure capability and thus survivability. Other means to increase spectral efficiency include the use of easily ionized materials that yield low temperature plasma [68, 69].

Ionic liquid fluorocarbon compounds [70] will come into play as the oxidizers of choice for advanced signature decoy flares that could create structured combustion patterns [71]. Figure 10.44 depicts the combustion plume of Mg/Ionic Liquid (MIL) with 4-amino-1-methyl-1,2,4-triazolium trifluoromethanesulfonate [71].

References

1. Stoll, A. (1973) *Jahrbuch der Wehrtechnik*, Band **7**, Wehr & Wissen Verlagsgesellschaft mbH, Darmstadt, pp. 92–95.
2. Blanchard, D.G. (1996) A brief history of air-intercept missile 9 (Sidewinder). 32nd Joint Propulsion Conference, Lake Buena Vista, FL, July 1–3, AIAA-1996-3154.
3. Hudson, R.D. (1969) *Infrared System Engineering*, John Wiley & Sons, Inc., New York, pp. 235–263.
4. Seyrafi, K. (1985) *Electro-Optical Systems Analysis*, 3rd edn, Electro-Optical Research Company, Los Angeles, p. 306.
5. Moss, T.S., Brown, D.R. and Hawkins, T.D.F. (1957) Infra-Red Decoys. Report No. TN RAD 702, Royal Aircraft Establishment, September 1957, declassified March 14 2001.
6. Brune, N. (1996) in *Countermeasure Systems* (ed. D.H. Pollock), Volume **7** of *The Infrared and Electro-Optical Systems Handbook*, 2nd revised printing (eds J.S. Accetta and D.L. Shumaker), SPIE Optical Engineering Press, Bellingham, WA, p. 287.
7. Koch, E.-C. (2001) Review on pyrotechnic aerial infrared decoys. *Propellants, Explos., Pyrotech.*, **26**, 3–11.
8. Beier, K. (1993) Infrared Radiation Model NIRATAM–NATO Infra Red Air Target Model, Description of Program, DLR Oberpfaffenhofen, Wessling, p. 48.
9. (1986) Military Specification, Flare, Aircraft Countermeasure: M206, MIL-F-63107C(AR), 6 January 1986, USA.
10. Sotzky, L. (2002) Twin screw mixing/extrusion of M206 infrared (IR) decoy flare composition. 33rd International ICT Conference, Karlsruhe, Germany, June 25–28, p. V35.
11. Farnell, P.L. (1994) Development of a high endoatmospheric decoy interceptor flare. 20th International Pyrotechnics Seminar, Colorado Springs, CO, July 25–29, p. 1029.
12. Koch, E.-C. and Friedmann, H. (2000) Elektrisch-mechanische Kupplung für elektrisch initiierbare Munition. DE Patent 19,847,242, Deutschland.
13. Bangash, M.Y.H. (2009) *Shock, Impact and Explosion*, Springer Publisher, pp. 163–166.
14. Glassman, I. (1960) in *Solid Propellant Research*, Progress in Astronautics and Rocketry, Vol. 1 (ed. M. Summerfield), Academic Press, New York, p. 253.
15. Koch, E.-C., Weiser, V., Roth, E., and Müller, D. (2006) UV-VIS Spectroscopic Investigation of Magnesium/Fluorocarbon Pyrolants. 33rd International Pyrotechnics Seminar, pp. 763–768.
16. Dally, B.B., Alwahabi, Z.T., Krishnamoorthy, L.V., Redman, L.D. and Christo, F.C. (2001) Measurements of particles evolution in the near combustion field of MTV formulation using mie scattering technique. 28th International Pyrotechnics Seminar, pp. 219–226.
17. Koch, E.-C. (2010) *Metal Halocarbon Pyrolant Combustion, Handbook of Combustion*, Wiley-VCH Verlag GmbH, pp. 355–402.
18. Dillehay, D.R. (1983) Resonance Line Broadening of Alkali Metals in Pyrotechnic Flames. *Longhorn Army Ammunition Plant, Marshall, TX*.
19. Simmons, F.S. (2000) *Rocket Exhaust Plume Phenomenology*, The Aerospace Press, El Segundo, pp. 173–192.
20. Siegel, R., Howell, J.R. and Lohrengel, J. (1988) *Wärmeübertragung durch Strahlung, Teil 1 Grundlagen und Materialeigenschaften*, Springer-Verlag, Berlin.
21. Alexeev, O.A. (2000) Determination of the parameters of large-sized turbulent-jet vortices using radiometers. *J. Opt. Technol.*, **67** (3), 222–224.
22. Alexeev, O.A. (2004) Model of the radiation of the flares of pyrotechnic thermal decoys. *J. Opt. Technol.*, **71** (2), 94–96.
23. Caliot, C. (2006) Modelisation et simulation de l'emission energetique et spectrale d'un jet reactif compose de gaz et de particules a haute temperature issus de la combustion d'un objet pyrotechnique. Doctoral-Thesis. Institut National Polytechnique de Toulouse, France.

24. Caliot, C., Eymet, V., El Hafic, M., Le Maoult, Y. and Flamant, G. (2008) Parametric study of radiative heat transfer in participating gas–solid flows. *Int. J. Therm. Sci.*, **47**, 1413–1421.

25. Koch, E.-C. and Dochnahl, A. (2000) IR emission behaviour of Magnesium/Teflon/Viton (MTV) compositions. *Propellants, Explos., Pyrotech.*, **25**, 37.

26. Koch, E.-C. (2009) Metal–fluorocarbon pyrolants: IX. Burn rate and radiometric performance of Magnesium/Teflon/Viton (MTV) modified with zirconium. *J. Pyrotech.*, **28**, 16.

27. Shortridge, R.G. and C.K. Wilharm (2004) Improved infrared countermeasures with ultrafine aluminium. 45th AIAA/ASME/ASCE/AHS/ASC Structures, Structural Dynamics & Materials Conference, Palm Springs, California, April 19–22, AIAA 2004-2063.

28. Villey-Desmeserets, F.P.A., Pouliguen, Y.A. and Guillauma, G. (1973) High Luminosity infrared pyrotechnical composition. US Patent 3,770,525, France.

29. Nadler, M.P. (2004) Pyrotechnic pellet decoy method. US Patent 6,675,716, US Navy.

30. Koch, E.-C. (2008) Metal–fluorocarbon pyrolants: VIII. Behavior of burn rate and radiometric performance of two Magnesium/Teflon/Viton (MTV) formulations upon addition of graphite. *J. Pyrotech.*, **27**, 38.

31. Nielson, D.B. and Lester, D.M. (1995) Use of carbon fibrils to enhance burn rate of pyrotechnics and gas generants. US Patent 5,470,408, USA.

32. Palaiah, R.S., Manda, M.N., Shah, N.R., Waghmare, P.L., Danali, S.M. and Raha, K.C. (2011) Effect of carbon nano fibrils on burn rate and IR intensity of Mg/Teflon/Viton infrared flare composition. Book of Abstracts, EUROPYRO-2011, Reims, France, May 16–19, p. 59.

33. Koch, E.-C. (2009) Metal–fluorocarbon pyrolants: X. Influence of ferric oxide/silicon additive on burn rate and radiometric performance of Magnesium/Teflon/Viton® (MTV). *Propellants, Explos., Pyrotech*, **34**, 472.

34. Gongpei, P., Changjiang, Z. and Zhaoqun, W. (1991) A Study on the infrared radiation of pyrotechnic compositions. 17th International Pyrotechnics Seminar, Beijing, China, October 28–31, p. 134.

35. Gongpei, P. (1995) Jamming Performance of Infrared Bait/Chaff, English Translation of the Bulletin of Nanjing University of Science and Engineering No. 7 (77) 1994.

36. Wang, X., Gongpei, P. and Yi, L. (1998) The influence of PTFE effective mass flow rate on 3–5 μm radiant intensity in Mg/PTFE/BaO$_2$ infrared pyrotechnic composition. 24th International Pyrotechnics Seminar, Monterey, July 27–31, pp. 607–612.

37. Knapp, C.A. (1959) New infrared flare and high-altitude igniter compositions. AD Patent 306237, Feltman Research and Engineering Laboratories, Picatinny Arsenal, Dover, NJ.

38. Douda, B.E. (2009) Genesis of Infared Decoy Flares, NSWC/CCR/RDTR-08/63, NSWC-Crane, IN.

39. Vargas. L.M. and Caltagirone, J. (1985) Evaluation of pyrotechnic Fire suppression system for six pyrotechnic compositions. CONTRACTOR REPORT ARLCD-CR85006, ARDC, Dover.

40. Chen, G., Broad, R., Valentine, R.W. and Mannix, G.S. (2001) Magnesium-fueled pyrotechnic compositions and processes based on elvax-cyclohexane coating technology, US Patent 6,174,391, USA.

41. Nielson, D.B. and Lester, D.M. (2002) Extrudable black body decoy compositions, US Patent 6,432,231, USA.

42. Campbell, C. (2006) Twin screw Extruder Production of MTTP Decoy Flares, SERDP WP-1240, Final Report, Thiokol Propulsion, Brigham, UT.

43. Jackson, D. and Dadley, D.A. (eds) (1969) Investigation into the Manufacture and Properties of Magnesium/Fluorocarbon Compositions for Pyrotechnic Applications, Part I, RARDE Memo 31/69, declassified now available at the National Archives at DEFE 15/2191.

44. Koch, E.-C. (2005) Metal fluorocarbon pyrolants: VI. Combustion behaviour and radiation properties of Magnesium/Poly(carbon Monofluoride)

Pyrolant. *Propellants, Explos., Pyrotech.*, **30**, 209.

45. Koch, E.-C., Hahma, A., Klapötke, T.M. and Radies, H. (2009) Metal–fluorocarbon pyrolants: XI. radiometric performance of pyrolants based on magnesium, perfluorinated tetrazolates, and v iton A. *Propellants, Explos., Pyrotech.*, **35**, 248.

46. Koch, E.-C., Klapötke, T.M., Radies, H., Lux, K. and Scheutzow, S. (2011) Metal–fluorocarbon pyrolants: XIII. Calciumsalze der 5-perfluoroalkylierten Tetrazolate–Synthese, Charakterisierung und Leistungsbewertung als Oxidationsmittel in ternären Mischungen mit Magnesium und Viton^{TM}. *Z. Naturforsch. B*, **66b**, 378–368.

47. Gongpei, P., Changjiang, Z., Xue, W. and Yi, L. (1999) The influencing factor of radiation intensity of IR pyrotechnic composition. 3rd International Autumn Seminar on Propellants, Explosives and Pyrotechnics, IASPEP-99, Chengdu, China, October 5–8, 1999, p. 198.

48. Deluga, Z., Plachta, E., Rembiszewski, W., Florczak, B., Daniluk, E., Koch, M., Gilewicz, M. and Maziejuk M. (1995) Mieszania pirotechniczna do wytwarzania dymow maskujacych zwlaszcza w podczerwieni. PL Patent 175,254, Poland.

49. Deluga, Z., Plachta, E., Rembiszewski, W., Florczak, B., Szymczak, J. and Olejniczak E. (1995) Mieszania pirotechniczna do wytwarzania pozornich celow termicznych. PL Patent 175,249, Poland.

50. Deluga, Z., Plachta, E., Rembiszewski, W., Florczak, B., Szymczak, J. and Slabik, P. (1999) Mieszania pirotechniczna do wytwarzania pozornich celow termicznych. PL 187,030, Poland.

51. Koch, E.-C. (2007) Pyrotechnic countermeasures. III: the influence of oxygen balance of an aromatic fuel on the color ratio of spectral flare compositions. *Propellants, Explos., Pyrotech.*, **32**, 365–370.

52. Hahn, G.T., Rivette, P.G. and Weldon, R.G. (1997) Infrared tracking flare. US Patent 5,679,921, USA.

53. Jone, T. (1993) Flare arrangement, GB Patent 2,266,944, Great Britain.

54. Hahma, A. (2011) Flugkörper mit einem pyrotechnischen Satz, EP Patent 2,295,927, Germany.

55. Posson, P.L. and Bagget, A.J. Jr (2002) Pyrotechnic composition and uses thereof. US Patent 6,427,599, USA.

56. Herbage, D.W. and Salvesen, S.L. (1995) Spectrally balanced infrared flare pyrotechnic composition. US Patent 5,472,533, USA.

57. Koch, E.-C. (2006) Pyrotechnic countermeasures: II. Advanced aerial infrared countermeasures. *Propellants, Explos., Pyrotech.*, **31**, 3–19.

58. Dillehay, D.R. (2004) Illuminants and illuminant research. *J. Pyrotech.*, **19**, 53–60.

59. Toothman, H. and Loughmiller, C. (1971) F-4B and F-8 Flare Effectiveness Against the ATOLL Missile (AA-2), Naval Research Laboratory, Declassified, 22 August 2002.

60. Towning, J.N. (1989) Pyrotechnic flares: radiant intensity and radiance calculations. 14th International Pyrotechnics Seminar, Jersey, UK, September 18–22, p. 537.

61. Weiser, V., Blanc, A., Deimling, L., Eckl, W., Eisenreich, N., Kelzenberg, S., Neutz, J. and Roth, E. (2007) Pyroorganic flares–a new approach for aircraft protection. Europyro 2007, Beaune, France, p. 725.

62. Hughes, N.D.P. and Jenkins, S. (1972) Emission from infrared decoy flares. Technical Report 72016, Royal Aircraft Establishment, Declassified 1 January 2003.

63. Deimling, L., Blanc, A., Billeb, G. and Klahn, T. (2010) Air flow test facility for investigation of pyrotechnical decoy. 41st International Annual Conference of ICT, Karlsruhe, Germany, June 29–July 2, p. 127.

64. Davies, N. and Holley, D. (1996) Factors affecting the radiometric output of infrared flares. 22nd International Pyrotechnics Seminar, Fort Collins, July, p. 469.

65. Brown, D.R. and Heywood, D.P. (1961) An Interim Report on the Infra-Red Emission From Teflon-Magnesium Flares, Technical Memorandum No.

Rad 581, Royal Aircraft Establishment, Declassified 26 October 2006.

66. Magalhaes, L.B. and Alves, F.D.P. (2010) Estimation of radiant intensity and average emissivity of Magnesium/Teflon/Viton (MTV) flares. *Proc. SPIE*, **7662**, 766218–766211.

67. Koch, E.-C. (2009) 2006–2008 annual review on aerial infrared decoy flares. *Propellants, Explos., Pyrotech.*, **34**, 6–12.

68. Guo, Y.-X. and Li, X.-X. (2007) Study of thermal decomposition of cesium nitrate. *Initiators Pyrotech.*, 24.

69. Li, X.-X. (2008) Study on thermo-decomposition kinetics of $CsNO_3$ employed in composite decoy. *Initiators Pyrotech.*, 29.

70. Kirsch, P. (2004) *Modern Fluoroorganic Chemistry: Synthesis, Reactivity, Applications*, Wiley-VCH Verlag GmbH, Weinheim.

71. Schaller, U. (2011) Triazolium based energetic ionic liquids. 8th Workshop on Pyrotechnic Combustion, Reims, France, May 14.

11
Obscurants

11.1
Introduction

Unlike clouds, fog or haze, obscurants are man-made aerosols. There purpose is to scatter, absorb and emit radiation in the visible, infrared or millimetric wavelength range and thereby to interrupt the line of sight between an object and observer. Applying an obscurant serves camouflage, concealment and deception [1]. Obscurants may be disseminated mechanically but are mainly generated *in situ* by way of a pyrotechnic reaction [2].

The yield factor, Y_f, of a pyrotechnic obscurant composition describes the aerosol mass m_s available from a unit mass of pyrotechnic payload m_p:

$$Y_f = \frac{m_s}{m_p}$$

The typical wavelength ranges and mechanisms of interaction with particles are shown in Figure 11.1.

Figure 11.2 shows an idealised scenario in which the line of sight between an object and observer is interrupted by an obscurant cloud. The radiation originating from the object, I_0, is scattered, I_{scat}, and absorbed, I_{abs}, by the obscurant. The ratio of transmitted radiation $I_t = I_0 - I_{scat} - I_{abs}$ to the incident radiation is called *transmission* T_{obs}, which is a dimensionless number. If we consider the obscurant cloud to have a defined pathlength, l, and to possess a uniform particle concentration c, then the mass extinction coefficient, for the wavelength range of consideration, α_λ, can be written as

$$\alpha_\lambda = \frac{\ln T_{obs}}{cl} \, (m^2 \, kg^{-1}) \tag{11.1}$$

which is Lambert Beer's law.

Absorption and scattering phenomena of radiation by small particles are treated in great detail in two monographs [3, 4].

Depending on the chemical and physical properties of the primary particle, the mass extinction coefficient, α_λ, can be a function of the relative humidity as well.

Metal-Fluorocarbon Based Energetic Materials, First Edition. Ernst-Christian Koch.
© 2012 Wiley-VCH Verlag GmbH & Co. KGaA. Published 2012 by Wiley-VCH Verlag GmbH & Co. KGaA.

Frequency:	3 – 7	7 – 4.3	4.3–1.9×10¹⁴	10–6×10¹³	3.8–2.1×10¹³ Hz	300–30 GHz	30–3 GHz
Wavelength:	100–420	420–700	700–1600 nm	3–5 µm	8–14 µm	1–10 mm	1–10 cm
	UV	VIS	NIR	MIR	FIR	MMW	CMW

| Mode of Interaction : | Scattering, absorption, emission | | | | Scattering, absorption | |

| Sensor: | Missile seeker | Eye Video | Night vision Nd-YAG laser | FLIR Missile seeker | Thermal Imager CO₂-laser | Tracking radar | Battlefield radar |

Figure 11.1 Wavelength ranges, sensor systems and possible mode of interaction of electromagnetic radiation with matter.

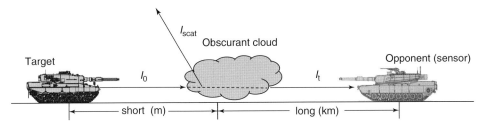

Figure 11.2 Obscuring cloud interrupting the line of sight between an observer and object.

(a)   (b)

Figure 11.3   (a) Test rack of 76 mm vehicle smoke discharge launchers and (b) corresponding 76 mm grenade with metal/halocarbon payload.

Typically, the mass extinction coefficient is multiplied with the yield factor to give the figure of merit:

$$\text{FoM} = \alpha_\lambda \times Y_F$$

The chemistry and performance of metal/chlorocarbon obscurants have been described by the author recently [5]. A typical vehicle smoke discharge launcher and the corresponding grenade are shown in Figure 11.3a,b, respectively.

11.2
Metal–Fluorocarbon Reactions in Aerosol Generation

The use of metal–fluorocarbon pyrolants in obscurant formulations may fulfil two purposes. The basic reaction yields carbonaceous material and metal fluorides, which in some cases are volatile species. These reaction products may be considered the desired aerosol. In other cases, the highly exothermic metal–fluorocarbon reaction supplies the necessary energy to facilitate decomposition and/or vaporisation of certain materials to yield the actual aerosol or its precursor. Both categories of aerosol generation are discussed in the following.

11.2.1
Metal–Fluorocarbon Reactions as an Exclusive Aerosol Source

Zinc hexachloroethane (HC) mixtures have been successfully applied since World War I as a source of hygroscopic $ZnCl_2$-based aerosol in the so-called HC smokes [5]. However, on hydrolysis, $ZnCl_2$ yields highly corrosive solutions that destroy material and cause severe lesions to living tissue and are held responsible for a numerous deaths among soldiers on inhalation of HC-based smokes [6].

An obscurant based on zinc and polytetrafluoroethylene (PTFE) has been proposed by Rozner and Helms [7]. They advise to use Zn/PTFE with 10–15 wt% auxiliary magnesium fuel and energetic additives based on either B/KNO_3 or Ni/Al. The author has investigated the potential of a number of metal/fluorocarbon mixtures to yield multi-spectral screening smoke effective in the VIS, IR and MMW range [8]. In the course of this study, a number of systems were tested including the aforementioned Zn/PTFE system. Although systems based on magnesium, titanium, zinc or zirconium with PTFE or polycarbon monofluoride (PMF) yield very good mass extinction coefficients in both the visible and infrared ranges, the attenuation behaviour in the 94 GHz range was acceptable only with stoichiometric Mg/PMF and Ti/PMF. These mixtures were found to generate fine particulate carbon including multi-walled carbon nanotubes (MWCNTs) exhibiting non-classical attenuation behaviour for electromagnetic radiation patented earlier by Koch and Dochnahl [9]. Figure 11.4a,b shows video stills taken from the combustion of both Mg/PMF and Ti/PMF in test tunnel experiments. The extinction coefficients are listed in Table 11.1.

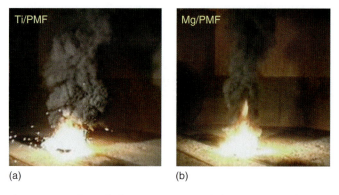

(a) (b)

Figure 11.4 Video stills from (a) Ti/PMF and (b) Mg/PMF combustion.

Table 11.1 The mass extinction coefficient and yield factor
of metal/graphite fluoride aerosols.

System (weight proportions; wt%/wt%)	Mass extinction coefficient at 94 GHz/m^2 g^{-1}	Yield factor, $Y_f{}^a$	Burn rate ratio, $u/u_{Mg/MPF}$
Ti/PMF (28/72)	4.1	~1.2	2–3
Mg/PMF (28/72)	5.5	~1	1

^{a}Not experimentally determined but calculated on the basis of the reactions.

Table 11.2 Main resonance wavelengths of metal fluorides
and state at ambient temperature [10].

	MgF$_2$	TiF$_4$	ZnF$_2$	AlF$_3$	BF$_3$	SiF$_4$
λ(μm)	11.9	12.6	13.3	11.0	6.6	9.7
State	Solid	Solid	Solid	Solid	Gas	Gas

The formed metal fluorides do not significantly alter the attenuation behaviour in the MMW range; however, some metal fluorides show strong absorption in the FIR range as listed in Table 11.2.

11.2.2
Metal–Fluorocarbon Reactions to Trigger Aerosol Release

As discussed in Chapter 1 (Introduction to Pyrolants), metal–fluorocarbon pyrolants possess the highest volumetric combustion enthalpies among energetic materials. In addition, the combustion speed of metal–fluorocarbon pyrolants can be easily adjusted as well. Thus, these materials may be applied *in situ* as a heat source to volatilise or decompose other materials to generate aerosols.

11.2.2.1 Metal–Fluorocarbon Reactions to Trigger Soot Formation

It is known from combustion chemistry that the thermal decomposition of aromatic compounds via cyclodehydrogenation yields soot [11]. Following Berger's invention of zinc/HC smokes [12], Gowdy shortly after that modified the composition by replacing zinc with magnesium and adding anthracene to induce soot formation and to obtain a dense black smoke [13]. This concept was consequently adapted to metal–fluorocarbon-based compositions. Espagnacq *et al.* proposed to use formulations similar to the one discussed in [13] modified with 10–20 wt% polyvinylidene fluoride (PVDF) [14, 15]. The infrared mass extinction coefficients for these formulations given below are common for carbon-based materials [16]. Espagnacq *et al.* refined these formulations to give FIR-784, which is reportedly in use with 81 mm vehicle protection discharge smoke grenade 'Galix 13' [17, 18].

IR opaque obscurant [14, 15]	Composition FIR-784 [17]
16.7 wt% Magnesium	15.0 wt% Magnesium
66.7 wt% Hexachlorobenzene	50.0 wt% Chlorocarbon compound
8.3 wt% Naphthalene	25.0 wt% Aromatic compound
8.3 wt% Polyvinylidene fluoride	5.0 wt% Polyvinylidene fluoride
	5.0 wt% Unknown binder
Burn rate	
0.57 mm s^{-1}	Unknown
Mass extinction coefficient	
$\alpha_{0.3-6\,\mu m} = 0.95\,\mathrm{m^2\,g^{-1}}$	$\alpha_{3-5\,\mu m} = 1.00\,\mathrm{m^2\,g^{-1}}$
	$\alpha_{8-12\,\mu m} = 0.36\,\mathrm{m^2\,g^{-1}}$

Similar formulations comprising only three ingredients, magnesium, chloro-paraffin and PVDF, were proposed by Vega and Morand [19]. Christofi *et al.* investigated the yield factor and mass extinction coefficient in the 0.4–12 μm range for a series of formulations based on 10% (Magnesium/Teflon/Viton (MTV)) [20]. MTV with a fixed ratio was modified with both bitumen and anthracene. For anthracene, the amount was increased from an unknown initial amount x wt% to $x + 10$ wt%, $x + 20$ wt% and $x + 30$ wt%. For bitumen, only two concentrations were tested y and $y + 10$ wt%. The variation of mass extinction coefficient as a function of relative anthracene content is shown in Figure 11.5.

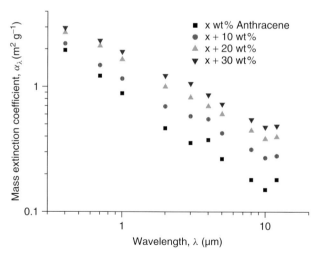

Figure 11.5 The mass extinction coefficient of anthracene-modified MTV [20].

DERA IR-obscurant formulations [20]

- Magnesium
- Polytetrafluoroethylene
- Viton
- Bitumen or anthracene

With decreasing magnesium content, the burn rate decreases nearly linearly from 5.25 to 1.95 mm s^{-1}.

Cudziło investigated the combustion behaviour and obscuration performance of a number of Mg/PTFE-based obscurant compositions [21]. Table 11.3 lists the tested compositions. Table 11.4 lists both mass extinction coefficients and chemical equilibrium compositions of the major products.

Mixtures containing polystyrene or diphenyl yield a lower performance than formulations based on naphthalene. An even stronger influence of the carbon framework is observed with on the co-oxidizing chlorocarbon. Thus, HC-based formulations show a significantly lower performance compared to those containing hexachlorobenzene. Büsel and Schneider proposed a non-toxic IR-opaque screening smoke based on Mg/PVDF/anthraquinone as given below (J. Schneider, personal communication) [22]. The mass extinction coefficient in an unspecified infrared band is reported to be greater than 1 m^2 g^{-1}.

IR-opaque obscurant (J. Schneider, personal communication) [22]

---

20 wt% Magnesium
27 wt% Polyvinylidene fluoride
48 wt% Anthraquinone
 5 wt% Chlorinated paraffin

From mass spectroscopy, it is known that anthraquinone on thermal load undergoes de-carbonylation yielding highly reactive benzene, C$_6$H$_4$ [23]. This reactive highly

Table 11.3 The composition of Mg/PTFE-based obscurants [21].

	1	2	3	4	5	6
Magnesium	25	25	25	25	25	25
Polytetrafluoroethene, $(C_2F_4)_n$	10	10	10	10	10	10
Hexachloroethane, C_2Cl_6	–	41	–	41	–	41
Hexachlorobenzene, C_6Cl_6	50	–	50	–	50	–
Polystyrene, $(C_8H_8)_n$	15	24	–	–	–	–
Naphthalene, $C_{10}H_8$	–	–	15	24	–	–
Diphenyl, $(C_6H_5)_2$	–	–	–	–	15	24

Table 11.4 The mass extinction coefficient and chemical equilibrium composition [21].

Parameters	1	2	3	4	5	6
$\alpha_{1.06\,\mu m}(m^2\,g^{-1})$	2.18	1.54	2.27	1.85	2.18	1.59
$\alpha_{10.6\,\mu m}(m^2\,g^{-1})$	0.22	0.15	0.20	0.15	0.24	0.15

species rapidly polymerises to give long helical structures capable of absorbing and scattering infrared radiation. A similar reaction behaviour is anticipated with analogous molecules such as acridine, $(C_{13}H_9N)$ or phthalic acid anhydride $(C_8H_4O_3)$.

11.2.2.2 Metal–Fluorocarbon Reactions to Trigger Phosphorus Vaporisation

Red phosphorus (RP)-based obscurant formulations require the vaporisation of the phosphorus to yield gaseous $P_4(g)$. This is the actual aerosol precursor formed in any RP-based ammunition. $P_4(g)$ spontaneously combusts with air to give P_4O_{10} and subsequently reacts in a strongly exothermic manner with moisture to yield phosphoric acid. The chemistry of RP-based screening smokes has been discussed by the author recently [24, 25]. To provide sufficient supply of $P_4(g)$ that is necessary to sustain a high aerosol concentration, RP is usually blended with energetic materials that have a high specific enthalpy of the reaction. Typical examples are

- metal/oxide (e.g. Mg/CuO)
- metal/sulphate (e.g. B/CaSO$_4$)
- metal/nitrate (e.g. Zr/KNO$_3$)
- metal/fluorocarbon

The required heat for the volatilisation of RP from a composition can be calculated from its heat capacity, $c_p(RP)$ and its enthalpy of vaporisation $\Delta_{vap}H(RP)$ [26]:

$$c_p(RP) = 25\ J\ mol^{-1}\ K^{-1}$$
$$\Delta_{vap}H(RP) = 51.9\ kJ\ mol^{-1}$$

The vaporisation temperature of RP is 431 °C.

Heating 1 mol of RP from ambient temperature to this point and vaporising it requires 62.05 kJ. The rate at which the heat is released directly influences the burn rate and thus the P_4 release wave.

In 1968, a number of RP-based formulations with alternative non-oxygen–based oxidisers were investigated at Crane Naval Ammunition Depot/USA [27]. Both Mg/HC and Mg/PTFE were tested as energetic materials to vaporise RP. The compositional data and burn rates of consolidate strands at atmospheric pressure are given below.

Mg/PTFE/RP obscurant [27]	Mg/HCB/RP obscurant [27]
16 wt% Magnesium	16.8 wt% Magnesium
26 wt% Polytetrafluoroethylene	25.2 wt% Hexachlorobenzene
50 wt% Red phosphorus	50.0 wt% Red phosphorus
8 wt% Viton	8.0 wt% Viton
Burn rate	
0.47 mm s^{-1}	0.977 mm s^{-1}

Table 11.5 The composition of Mg/PTFE-based obscurants [28].

	1	2	3
Magnesium	25.3	17.8	12.0
Polychlorotrifluoroethene, $(C_2ClF_3)_n$	36.7	40.6	40.1
Zinc oxide	1.9	1.8	1.8
Red phosphorus, P_n	36.1	39.8	46.1

HCB offers a higher combustion rate than PTFE because of lower thermal stability of the aromatic C–Cl bond compared to that of the aliphatic C–F bond ($\Delta H_{BDE} \sim$ 100 kJ).

According to the report, the composition based on Mg/PTFE deteriorated after some time and a 'phosgene-like smell'[1] was detected [27]. Following this work, experiments were conducted with polymerised chlorotrifluoroethylene (CTFE) as the only binder and oxidiser in RP-based smoke compositions [28]. For this purpose, RP, magnesium powder, a small amount of ZnO and gaseous CTFE were filled in a can under such pressure that CTFE would liquefy. On exposure of the filled can to ^{60}Co-γ photons, the liquid eventually polymerises to a solid. Three stoichiometries were tested and are listed in Table 11.5. The evolved smoke was off-white for composition 1 and appeared a little greyish for compositions 2 and 3 owing to higher CTFE/Mg ratio that causes higher release of carbon.

In 1980s, a Swedish company studied obscurant formulations based on Mg/PTFE/RP- and chloroprene-type binders [29].

Figure 11.6 shows the burn rate that is a linear function of the magnesium content at a fixed PTFE content of 17 wt% (Table 11.6).

1) It is assumed that this should actually read 'phosphine'.

Table 11.6 The burn rate and colour of smoke [28].

Parameters	1	2	3
u (mm s^{-1})	0.61	0.53	0.35
Smoke colour	White	Light grey	Light grey

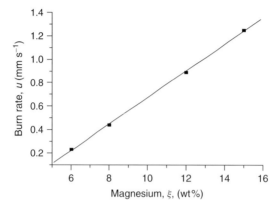

Figure 11.6 The burn rate of Mg/PTFE/RP obscurant as a function of magnesium content at the PTFE level (17 wt%) [29].

On the basis of findings of the study, the following composition was selected as an IR-screening smoke payload.

IR-opaque obscurant Z 97 [29]

67.5 wt% Red phosphorus
9.5 wt% Magnesium
17.0 wt% Polytetrafluoroethylene
6.0 wt% Polychloroprene

Cudziło *et al.* patented a similar mixture in 1996 [30]. Their composition is given below.

IR-opaque obscurant based on disclosure by Cudziło *et al.* [30]

50.00 wt% Red phosphorus
22.50 wt% Magnesium
20.25 wt% Polytetrafluoroethylene
3.25 wt% Polyvinylchloride
3.25 wt% Polystyrene

Table 11.7 Performance parameters of smoke composition.

Enthalpy of combustion $(kJ\,kg^{-1})$	u $(mm\,s^{-1})$	Y_f	Mass extinction coefficient $(m^2\,g^{-1})$							
3100	1.05	0.8	Wavelength(μm)	0.54	0.92	1.06	1.55	3.5	4.8	10.6
				5.50	2.76	1.95	1.52	0.85	0.59	0.40

The mass extinction coefficient and other important performance parameters are listed in Table 11.7.

Cudziło reported about magnalium, an Al_3Mg_4-based composition similar to the one discussed above [31]. In addition to the spectral mass extinction coefficients, he investigated the temperature distribution of the aerosol cloud to assess the emissive effects of the obscurants.

IR-opaque obscurant [31]

---

20 wt% Magnesium–aluminium alloy, Al_3Mg_4
20 wt% Polytetrafluoroethylene
51 wt% Red phosphorus
 5 wt% Polystyrene
 5 wt% Chlorinated polyvinyl chloride

Burn rate
$0.52\ mm\,s^{-1}$

Yield factor
0.70

Mass extinction coefficient
$\alpha_{1.06\,\mu m} = 1.44\ m^2\ g^{-1}$

In another study, Cudziło showed that the above obscurant in comparison to other pyrotechnic smoke composition yields aerosols with in comparison high temperatures useful for the protection of targets with hot spots such as smoke stack on ships or exhaust pipes on tanks [32]. In yet another study, Cudziło presented the effect of varying the RP ratio to fixed set Al_4Mg_3/PTFE levels. The mass consumption rate of these formulations decreases nearly linearly with increasing RP, that is, decreasing Mg content [33]. Fayed *et al.* reported the modification of RP/Fe_2O_3/Mg with variable amounts (5–15 wt%) of Kel-F wax [34].

In general, compositions of magnesium or magnalium and RP are capable of undergoing side reactions to yield the corresponding phosphides. Thus, a high metal content will both enhance thermal conductivity of pyrolant and offer supplementary reaction energy released by the reaction with phosphorus, which reads [33]

$$3 Al_3Mg_4 + 17 P \longrightarrow 9 AlP + 4 Mg_3P_2, \quad \Delta H = -3336 \text{ kJ}, 3.11 \text{ g}^{-1}$$

However, these side reactions are problematic for two reasons. First, they reduce the available phosphorus content and thus lower the actual yield factor. Secondly, the phosphides of both magnesium and aluminium react with atmospheric moisture to give toxic and self-inflammable phosphanes [24]. To avoid the formation of phosphane, zirconium may be applied as fuel, which yields very stable ZrP that does not react with moisture or dilute acids.

References

1. Hook D.W. Jr and Sutherland, R.A. (1996) in *Countermeasure Systems*, (ed. D.H. Pollock), Volume 7 of *The Infrared and Electro-Optical Systems Handbook* (eds J.S. Accetta, D.L. Shumaker), SPIE Optical Engineering Press, Bellingham, p. 359.

2. McLain, J.H. (1980) *Pyrotechnics*, The Franklin Institute Press, Philadelphia, pp. 63–79.

3. van de Hulst, H.C. (1981) *Light Scattering by Small Particles*, Dover Publications Inc., New York.

4. Bohren, C.F. and Huffman, D.R. (1998) *Absorption and Scattering of Light by Small Particles*, John Wiley & Sons, Inc., New York.

5. Koch, E.-C. (2010) in *Handbook of Combustion, New Technologies*, Vol. 5 (eds M. Lackner, F. Winter and A.K. Agarwal), Wiley-VCH Verlag GmbH, Weinheim, p. 388.

6. Subcommittee on Military Smokes and Obscurants, Committee on Toxicology, Board on Environmental Studies and Toxicology, Commission on Life Sciences, National Research Council, (1997), *Toxicity of Military Smokes and Obscurants, Volume 1*, National Academy Press, Washington DC.

7. Rozner, A.G. and Helms, H.H. (1972) Smoke generating compositions and methods of use. US Patent 3,634,283, USA.

8. Koch, E.-C. (2005) Recent developments in pyrotechnic obscurant technology. 36th International Annual Conference of ICT & 36th International Pyrotechnics Seminar, Karlsruhe, Germany, June 28–July 1, p. V–1.

9. Koch, E.-C. and Dochnahl, A. (2002) Düppel. DE Patent 10,102,599, Germany.

10. Nakamoto, K. (1997) *Infrared and Raman Spectra of Inorganic and Coordination Compounds Part A Theory and Applications in Inorganic Chemistry*, 5th edn, John Wiley & Sons, Inc., New York.

11. Warnatz, J., Maas, U. and Dibble, R.W. (2001) *Verbrennung*, Springer-Verlag, 3rd edn, pp. 284–286.

12. Berger, E.-E.-F. (1916) Nouveau procede d'obtention de fumees par combustion de melanges. FR Patent 501,836, France.

13. Gowdy, R.C. (1919) Signal rocket. US Patent 1,326,494, USA.

14. Espagnacq, A. and Sauvestre, G.D. (1988) Pyrotechnical composition which generates smoke that is opaque to infrared radiance and smoke ammunition as obtained. US Patent 4,724,018, France.

15. Espagnacq, A. and Sauvestre, G.D. (1987) Method for opaquing visible and infrared radiance and smoke producing ammunition which implements this method. US Patent 4,697,521, France.

16. Appleyard, P.G. and Davies, N. (2002) The estimation of extinction properties of obscurant clouds by modelling the fundamental electromagnetic scattering properties of individual particles. 29th International Pyrotechnics Seminar,

Westminster, Colorado, July 14–19, p. 509.

17. Vaullerin, M., Morand, P. and Espagnaqc, A. (2001) Optimization of a smoke producer composition by experiment design. *Propellants Explos. Pyrotech.*, **26**, 229.

18. Vaullerin, M. and Espagnacq, A. (1999) Optimisation de la composition fumigene FIR 784 par plans d'experiences. Europyro 99, Brest, France, June 7–11.

19. Vega, J.F. and Morand, P.C. (1987) Castable smoke generating compounds effective against infrared. US Patent 4,609,108, France.

20. Christofi, J.M.B., Collins, P.J.D. and Edwards H. (1995) The development and assessment of pyrotechnic compositions as sources for multispectral screening smokes. 21st International Pyrotechnics Seminar, Moscow, Russia, September 11–15, p. 139.

21. Cudziło, S. and Papliński, A. (1999) An influence of the chemical structure of smoke-generating mixtures on laser radiation attenuation at 1.06 μm and 10.6 μm wavelengths. *Propellants Explos. Pyrotech.*, **24**, 242.

22. Schneider, J. and Büsel, H. (1995) Composition generating an IR-opaque smoke. US Patent 5,389,308, Germany.

23. a) MassSpectroscopy of Anthraquinone, NIST Chemistry Webbook, *http://webbook.nist.gov/cgi/cbook.cgi?ID=C84651&Units=SI&Mask=200#MassSpec*; (accessed 1 September 2011); (b) Hesse, M., Meier, H. and Zeeh, B. (1987) *Spektroskopische Methoden in der organischen Chemie*, Georg Thieme Verlag, Stuttgart, p. 219, 251.

24. Koch, E.-C. (2005) Special materials in pyrotechnics: IV the chemistry of phosphorus and its compounds. *J. Pyrotech.*, **21**, 39.

25. Koch, E.-C. (2008) Special materials in pyrotechnics: V. Military applications of phosphorus and its compounds. *Propellants Explos. Pyrotech.*, **33**, 165.

26. Blachnik, R. (ed.) (1998) *D'Ans Lax–Taschenbuch für Chemiker und Physiker*, Elemente, anorganische Verbindungen und Materialien, Minerale, Band **III**, 4th neubearbeitete und revidierte Auflage, Springer-Verlag, Berlin, p. 150.

27. Biggs, W.T. (1968) NOSC–Red Phosphorus Study Part I–Smoke. RDTR-No. 135, US Naval Ammunition Depot Crane, Indiana, 3 December.

28. Webster, H.A. III and Johnson, D.M. (1976) New Potentials in Red Phosphorus Compositions, Naval Weapons Support Center Applied Sciences Department Crane, Indiana.

29. Norman, L.E. (1991) Final Qualification of RP Smoke Composition. Report FGT R91:125, Swedish Ordnance.

30. Cudziło, S., Papliński, A. and Włodarczyk, E. (1996) Mieszaniny aerozolotwórcze oparte na czerwonym fosforze, ich granulat I sposób jego wytwarzania. PL Patent 183,882, Poland.

31. Cudziło, S. and Trzcinski, W.A. (1999) Comparison investigations of camouflage capability of different pyrotechnic smoke compositions in IR region. Europyro 99, Brest, France, June 7–11.

32. Cudziło, S. (2001) Studies of IR-screening smoke clouds. *Propellants Explos. Pyrotech.*, **26**, 12.

33. Cudziło, S. (2002) Combustion characteristics and screening capability of red phosphorus-based mixtures. 32nd International Annual Conference of ICT, Karlsruhe, Germany, July 3–6, p. 42.

34. Fayed, M.S., Soliman, M.G. and Mansour, A. (1996) *Investigation of Thermal Screening Characteristics of Selected Smoke Producing Mixtures that Contain Halogenated Polymers, Theory and Practice of Energetic Materials*, Beijing Institute Technology Press, Beijing, p. 246.

12
Igniters

Igniters are self-contained energetic devices that serve the ignition of bulk energetic material in rocket motors, propellant charges and gas generators. Typical pyrolants used in igniters comprise black powder [1], boron/potassium nitrate [1], aluminium/potassium perchlorate [2] and Magnesium/Teflon/Viton (MTV) [3]. Igniters based on metal/fluorocarbon pyrolants yield mainly hot incandescent particles, continuum infrared radiation and provide a low permanent gas content that is desirable for low brisance and soft ignition applications. The condensable reaction products from MTV combustion have threefold effect on ignition [4]:

- heat release by phase change upon impingement on the propellant;
- any deposit after its conductive heat transfer will act as thermal barrier;
- gas flow and hence convective heat transfer proportionally decreases with increasing mass fraction of condensed matter.

The equilibrium composition for a MTV (54/30/16) mixture as a function of temperature is depicted in Figure 12.1 [5].

A number of magnesium/fluorocarbon-based igniter materials are specified in Refs. [6, 7]. Magnesium/PTFE (polytetrafluoroethylene)-based material were and are used in both strategic and tactical missile applications as shown below in Table 12.1.

The characteristics of a rocket motor igniter are shown in Figure 12.2 taken from Ref. [9].

The igniter ignition delay describes the sensitivity of the pyrolant material itself towards ignition. The period $t_2 - t_1$ characterizes the pressure rise rate of the pyrolant grain. t_2 depicts the time to reach the initial peak pressure. t_3 is the propellant ignition delay.

The propellant ignition delay at $-53\,°C$ for double base rocket propellant (NC/NG: 50/33) ignited by magnesium/PTFE/PCTFE (polychlorotrifluoroethylene) pyrolants as function of Mg content is shown in Figure 12.3. Thus the ignition delay is short at the stoichiometric ratio, and increases at both fuel-lean and fuel-rich stoichiometries [9].

The propellant ignition delay at $-53\,°C$ for various other metal/fluorocarbon combinations is given in Table 12.2. The rocket propellant ignition delay, t_d has

Metal-Fluorocarbon Based Energetic Materials, First Edition. Ernst-Christian Koch.
© 2012 Wiley-VCH Verlag GmbH & Co. KGaA. Published 2012 by Wiley-VCH Verlag GmbH & Co. KGaA.

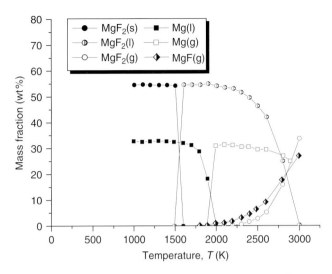

Figure 12.1 Condensed phase products at 2.5 MPa. (After Ref. [5].)

Table 12.1 Strategical and tactical missile igniter formulations.

Composition	UGM-27 polaris	AIM-54 phoenix	AIM-9 sidewinder
Magnesium	60.0	32.5	54.0
Polytetrafluoroethylene	40.0	67.5	30.0
Graphite	1.0	–	–
Viton®	–	–	16.0
Additives	–	2.0	–

From Ref. [8].

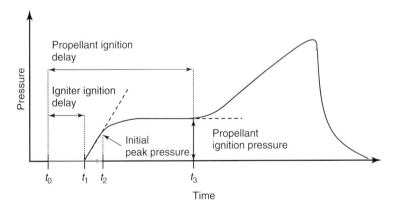

Figure 12.2 Typical time pressure characteristics of a rocket motor with pyrotechnic igniter. (After Ref. [9].)

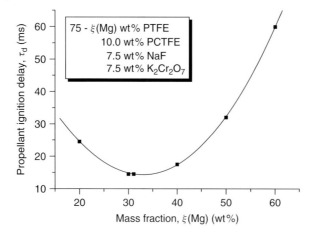

Figure 12.3 Double base (NC/NG: 50/33) propellant ignition delay at −53 °C as function of Mg content. (After Ref. [9].)

Table 12.2 Rocket propellant ignition delay for various metal/PTFE/PCTFE pyrolants at −53 °C.

Composition	1	2	3	4	5
Magnesium	25	–	–	–	–
Boron	25	–	–	–	–
Thorium	–	66.4	–	–	–
Zirconium	–	–	33.3	–	–
Titanium	–	–	–	21.53	–
Molybdenum	–	–	–	–	27.31
Aluminium	–	–	9	9	9
PTFE	25	28.6	36.61	46.82	42.69
PCTFE	10	–	–	–	–
$K_2Cr_2O_7$	7.5	2.5	3	3	3
NaF	7.5	2.5	–	–	–
PbF_2	–	–	9	9	9
$LiClO_4$	–	–	9	9	9
Propellant ignition delay (ms)	26.9	70.7	31.4	42.5	28.8

After Ref. [9].

been found to correlate with the mass flow rate of the igniter pyrolant, m' [4].

$$t_d \approx m'^{-1.5}$$

In a comparative study, it was found that MTV (SR 886B) (mass ratio: 55/40/5) has a lower rocket propellant ignition delay than typical AP-based propellant formulations used in pyrogen igniters [4].

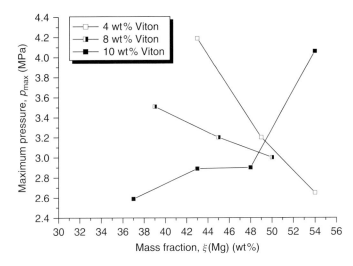

Figure 12.4 Maximum pressure as function of stoichiometry for various MTV compositions. (After Ref. [10].)

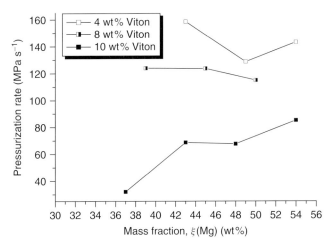

Figure 12.5 Pressure rise rate as function of stoichiometry for various MTV compositions. (After Ref. [10].)

The pressurisation rate of various MTV igniter materials as function of both Mg and Viton® content has been investigated by Özkar *et al.* [10]. At low Viton content (4–8 wt%), the highest pressure is obtained with rather fuel-lean compositions. However, at 10 wt% Viton the maximum pressure is obtained for $\xi(Mg) > 48$ wt% (Figures 12.4 and 12.5).

The pressure rise rate is highest for compositions with 4–8 wt% Viton and significantly lower with 10 wt% Viton.

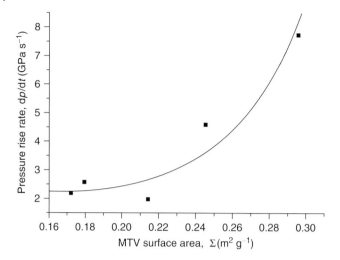

Figure 12.6 Influence of surface area on pressure rise rate. (After Ref. [11].)

The influence of surface area of the pyrolant grain on pressure rise rate is depicted in Figure 12.6. In an igniter composition, the PTFE surface area was adjusted to achieve a range of final grain surface area. It is observed that with increasing surface area the pressure rise rate increases exponentially.

The investigation of MTV and MTV modified with either ammonium perchlorate or boron has been reported in Refs. [12, 13]. The use of titanium/Teflon®/Viton® (TTV) pyrolants in igniters with safety partitions has been described in Ref. [14]. Taylor has investigated the behaviour of confined igniter materials including MTV (SR 886) and noted a significant extension of MTV afterburn reactions in space though MTV plume moves slower than Black powder, Al/CuO or B/KNO$_3$ [15].

References

1. Klingenberg, G. (1984) Experimental study on the performance of pyrotechnic igniters. *Propellants Explos. Pyrotech.*, **9**, 91–107.
2. Goddard, T.P. (1980) High Deflagration Rate (HIDEF) Igniter Technology Applications. Final Report BDM/M-003-80, BDM Corporation, Montery, CA.
3. Crosby, R.E., Swenson, I.C. and Mullenix, G.C. (1972) Design and development of a hot particle igniter. AIAA/SAE 8th Joint Propulsion Specialist Conference, New Orleans, LA, 29 November – 1 December, pp. 72–1196.
4. Hall, A.R., Southern, G.R. and Sutton, D. (1979) Some measurements of ignition delay and heat transfer with

pyrogen igniters. Solid Rocket Motor Technology – Meeting of the 53rd AGARD Propulsion and Energetics Panel, Oslo, Norway, April 2–5, CP-259.
5. Valenta, F.J. (1988) MTV as a pyrotechnic composition for solid propellant ignition. 13th International Pyrotechnics Seminar, Grand Junction, CO, USA, July 11–15, p. 811.
6. N.N. (1988) Ignition pellets, Magnesium-Fluorocarbon, MIL-PRF-82736A(OS), Naval Sea Systems Command, Indian Head, USA.
7. N.N. (1974) Ignition Compound, MIL-I-81970 (AS), Naval Air Systems

Command, Department of the Navy, USA.

8. N.N. (1971) Solid Rocket Motor Igniters, NASA SP-8051, National Aeronautics and Space Administration, Lewis Research Center, Cleveland, OH.

9. Julian, E.C., Crescenzo, F.G. and Meyers, R.C. (1973) Igniter Composition. US Patent 3,753,811, USA.

10. Özkar, S., Göçmez, A., Yilmaz, G.A. and Pekel, F. (1999) Development of MTV compositions as igniter for HTPB/AP based composite propellants. *Propellants Explos. Pyrotech.*, **24**, 65–69.

11. Caulder, S.M., Leon, J.R. and Luense, L. (2006) Compositional factors affecting the ballistic properties of Magnesium/Fluorocarbon (MTV) igniters. 33rd International Pyrotechnics Seminar, Fort Collins, CO, July 16–21, p. 53.

12. Mul, J., Meulenbrugge, J. and de Valk, G. (1990) Development of an MTV composition as igniter for rocket propellants. 15th International Pyrotechnics Seminar, Boulder, CO, July 9–13, p. 721.

13. Leenders, J. and Meulenbrugge, J. (1994) Adjusted closed vessel experiments for testing new lova igniter compositions. 20th International Pyrotechnics Seminar, Colorado Springs, CO, July 25–29, p. 615.

14. Kuwahara, T., Tohura, C., Kohno, T. and Shibamoto, H. (2005) Ignition characteristics of Ti/Teflon pyrolants: safety igniter with partition. *Propellants Explos. Pyrotech.*, **30**, 425–429.

15. Taylor, M.J. and Gransden, J.I. (2007) Study of confined pyrotechnic compositions for Medium/Large calibre gun igniter applications. *Propellants Explos. Pyrotech.*, **32**, 435–446.

13
Incendiaries, Agent Defeat, Reactive Fragments and Detonation Phenomena

Metal–fluorocarbon pyrolants are highly effective infrared emitters. Besides the exploitation of this effect for both decoy flare and igniter applications, it can also be used for destructive purposes. Thus, metal/fluorocarbon-based energetics finds use in incendiaries, document destruction devices, agent defeat payloads and reactive fragments.

13.1
Incendiaries

The purpose of an incendiary is to heat a target substrate above its ignition temperature. The ignition temperature, T_{ig}, is the minimum temperature necessary to maintain self-combustion [1]. It determines the time, t_{ig}, necessary to set an item to fire with thermal conductivity, k, specific heat, c_p, ambient temperature, T_0, and density, ρ, at a given level of radiant heat, q'':

$$t_{ig} = \left(\frac{\pi}{4}\right) \cdot k\rho c_p \cdot \left[\frac{T_{ig} - T_0}{q''}\right]^2 \tag{13.1}$$

Ignition temperatures for common materials are listed in Table 13.1.

There are two categories of incendiaries:

- compositions for directed use, which provide a high radiant intensity level to a relatively small combustion zone
- compositions for scattered use, which provide a low-to-intermediate radiant intensity level to large target areas.

The latter group often comprises compositions that rely completely on combustion with atmospheric oxygen such as Napalm, gasoline and similar materials. The former group most often comprises pyrotechnic compositions. Within the context of this chapter, we will consider only pyrotechnic incendiaries.

In order to be applicable as incendiaries, pyrotechnic compositions must fulfil a number of thermo-chemical requirements such as high enthalpy of reaction, low activation energy, suitable rate of combustion, high radiant efficiency and large radiating combustion plume and/or highly conductive slag. In addition, these

Metal-Fluorocarbon Based Energetic Materials, First Edition. Ernst-Christian Koch.
© 2012 Wiley-VCH Verlag GmbH & Co. KGaA. Published 2012 by Wiley-VCH Verlag GmbH & Co. KGaA.

Table 13.1 Ignition temperatures for common materials.

Material	Ignition temperature, t_{ig} (°C)
Acetylene, C_2H_2	305
Asphaltum	400
Benzene, C_6H_6	555
Ethanol, C_2H_5OH	425
Hydrogen, H_2	560
Fuel oil	220
Wood	280–340
Paper – newspaper type	175
Paper – this book page	360
Polyethylene	488
Polyvinylchloride	507
Match head	80
Cotton cloth	450
Straw	280–330

materials should not be susceptible to high humidity, temperature changes and when considered for scatter applications they should exhibit a good adhesion to a target substrate. Finally, from a military point of view, these materials should be safe in both storage and handling and they should be difficult to extinguish [2, 3]. Pyrotechnic incendiaries very often contain metallic fuels. As such, the Glasmann criterion applied to incendiaries reads as follows.

Incendiaries based on volatile metal fuels yield large diffusion flames and thus lead to high levels of radiant intensity. In contrast, incendiaries based on non-volatile fuels undergo solid-phase combustion and predominantly transfer heat by conduction and convection. Pyrotechnic incendiaries can be either fuel rich or provide a surplus of oxidizer to facilitate combustion of a target substrate.

13.2
Curable Fluorocarbon Resin–Based Compositions

Compositions based on glycidyl methacrylate and 1,1,7-trihydrododecafluoro methacrylate as a binder and an oxidizer and various metals such as Zr, Ta, U and cerium mischmetal have been proposed [4].

Curable Incendiary Composition
81.269 wt% Tantalum
11.734 wt% 1,1,7-Trihydrododecafluoroheptyl methacrylate
4.563 wt% Glycidyl methacrylate
2.303 wt% 2-Ethylhexyl methacrylate
0.130 wt% Butylene dimethacrylate

The aforementioned composition was cured with 0.435 part benzoyl peroxide. After polymerization, the composition would burn slowly. After a delay of ~30 min, the material undergoes another exothermic reaction that brought the material to incandescence.

A system proposed as Napalm substitute incendiary is given below [5].

Scatter composition

45–56 wt% Polytetrafluoroethylene
22–28 wt% Magnesium
12–26 wt% Iron
1.5–2.5 wt% Nitrocellulose
0–8 wt% Glass powder

Incendiaries against metallic targets have been proposed to comprise the following system [6]:

Directed composition

---

32 wt% Fluoroalkyl phosphate ester (Zonyl S-13)
12 wt% Ferric oxide
28 wt% Magnesium
12 wt% Silicon
16 wt% Potassium perchlorate

Imparting silicon in a metal–fluorocarbon composition offers the possibility to form low-melting silicofluorides of the type $M[SiF_6]$, which are known to improve the flowing behaviour of pyrotechnic slags [7]. Both zirconium and hafnium sponge powder pressed with 2–15 wt% polytetrafluoroethylene (PTFE) or other fluoropolymer binders have been proposed as incendiaries [8]. However, it can be argued that volatile ZrF_4 and HfF_4 that evaporate under the reaction conditions reduce the slag temperature and, thus, reduce the effectiveness as an incendiary material.

13.3
Document Destruction

Quick, safe and reliable destruction of bank notes, classified printed documents and digital storage media may become necessary in a number of situations encountered in civilian as well as military environment. For these purposes, solutions based on energetic materials have the advantage to meet with the aforementioned requirements and to be independent of any electrical power energy supply as, for example document shredder.

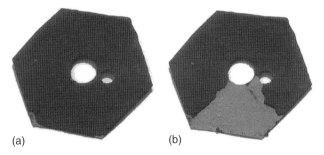

(a) (b)

Figure 13.1 Swiss made document destruction sheet with pyrolant coating (a) and partially removed exposing the NC substrate (b).

Figure 13.2 Combustion of incendiary sheets and remaining slag.

Simple document destruction devices, *Brandplatten*, comprising low-nitrogen-content nitrocellulose sheets covered with chlorate-based pyrolant compositions are in use in Switzerland for some decades (J. Schneider, personal communication). Modifications of these devices comprise slow-burning Magnesium/Teflon/Viton (MTV). Figure 13.1 shows a typical hexagonal sheet (70 mm diameter) with pyrolant coating and after removal of the coating exposing the low-nitrogen NC substrate. Figure 13.2 depicts the combustion sequence and the residual slag.

The destruction of compact discs by applying a pyrolant coating on the top side has been investigated by Karametaxas et al. [9]. They looked into a variety of pyrolant systems to be applied on either caddy or CD top layer. Figure 13.3 shows the hexagonal evaluation matrix. Figure 13.4 shows the results obtained with three different compositions **A** being stoichiometric $CaSi_2/MnO_2$, B = Enerfoil® (see also Chapter 16) coating based on stoichiometric Mg/PTFE and the finally selected system **C** based on complex pyrolant comprising (Al, Zr, Fe, Mn_2O_3, MnO_2, 13/21/20/30/16). All pyrolants except Enerfoil were compounded with binder B-14 a styrene copolymer with modified colophony resin.

Vapour-deposited pyrolant layers have been proposed for fast and safe eradication of microcircuits [10]. Figure 13.5 depicts a proposed setup.

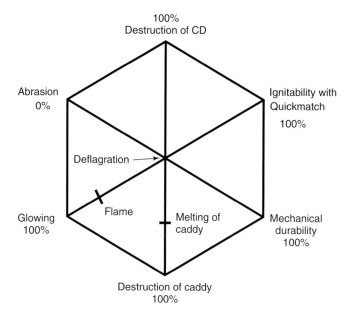

Figure 13.3 Hexagonal evaluation matrix for CD + Caddy destruction. (After Ref. [9].)

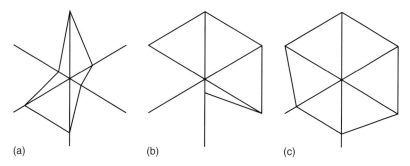

(a) (b) (c)

Figure 13.4 Results obtained with three different pyrolant coatings: (a) (CaSi$_2$/MnO$_2$), (b) (Enerfoil$^®$) and (c) (Zr/Al/Fe/Fe$_2$O$_3$/MnO$_2$). (After Ref. [9].)

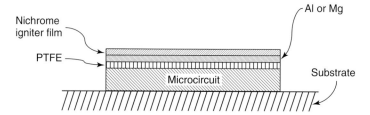

Figure 13.5 Micro-circuit with vapour-deposited pyrolant cover for safe eradication. (After Ref. [10].)

13.4
Agent Defeat

Agent defeat warheads are designed to neutralize stockpiled biological and chemical warfare agents in soft and hardened above ground and both shallow and deeply buried shelters. It is imperative in fighting non-nuclear weapons of mass desctruction (WMD) storage shelters that as minimum as possible warfare agent is released. Thus, new weapon tactics with respect to penetration and fusing have evolved [11].

The requirements for agent defeat payloads are not as straightforwardly derived as for incendiary compositions. This stems from the different concepts envisaged for defeating both biological and chemical stockpiles and possibly for a good part also in the lack in operational experiences with life chemical and biological stockpiles.

Agent defeat payloads are supposed to yield high temperatures to facilitate the decomposition of chemical warfare agents and pyrolysis of biological material. Chemical agents require pyrolysis at significantly higher temperatures to avoid *de novo* formation upon quenching. Although thermal decomposition of nerve agent Sarin is easily accomplished at comparatively low temperatures $\sim$1000 K [12], the pyrolysis of blister agent Mustard (HD) requires minimum 2000 K to achieve satisfactory destruction levels [13] (Figure 13.6).

Figure 13.7 shows the decomposition of nerve agent Sarin (GB) [13] as a function of time at $T = 1000$ k [12] Figure 13.8 depicts the survival of Anthrax protein as a function of time and temperature after ref. [12].

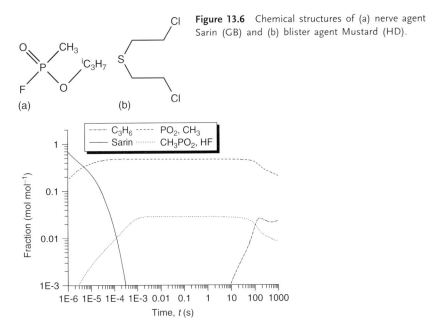

Figure 13.6 Chemical structures of (a) nerve agent Sarin (GB) and (b) blister agent Mustard (HD).

Figure 13.7 Thermal decomposition of Sarin (GB) nerve agent at 1000 K [12].

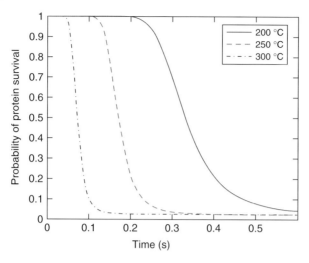

Figure 13.8 Survival of Anthrax in thermal environment.

With biological material, it may also be aimed at destruction by chemical means via release of halogens, halogen compounds, and photocatalyst such as TiO_2 [12] that would aid in detoxification by sunlight and antiviral material such as I_2 and Ag [14].

A two-stage high-temperature incendiary (HTI) releasing halogens and hydrogen halides for fighting chemical and biological payloads is known as *HTI-J-1000* [15, 16].

HTI-J-1000

First stage
14 wt% Boron
60 wt% Titanium
20 wt% Polytetrafluoroethylene

Second stage
70–85 wt% Alkali metal perchlorate
15–30 wt% Viton®

Combustion of the first payload is mainly an intermetallic reaction yielding eutectic TiB_2, TiC accompanied by reaction of excess Ti with PTFE:

$$1.29\,B + 0.645\,Ti \longrightarrow 0.645\,TiB_2$$
$$0.2\,C_2F_4 + 0.605\,Ti \longrightarrow 0.400\,TiC + 0.200\,TiF_4 + 0.005\,Ti$$

The reaction products then come into contact with oxidizer pellets comprised of an alkali metal perchlorate and possibly a hydrogen rich fluoroelastomer such as

Viton:

$$0.7\,LiClO_4 + 0.07\,(C_{10}F_{13}H_7)_n \longrightarrow 0.7\,CO_2 + 7\,HF + 3\,F_2 + 0.7\,O_2 + 0.7\,LiCl$$
$$2.5\,O_2 + TiB_2 \longrightarrow TiO_2 + B_2O_3$$
$$2\,O_2 + TiC \longrightarrow TiO_2 + CO_2$$

In a ratio of 1 : 2, these payloads give an overall temperature of >3000 K and yield ~20 mol% Cl and F and 5 mol% HCl and HF. In view of the aimed production of these highly toxic materials, one may argue that the cure causes more harm than good.

13.5
Reactive Fragments

When metal fragments hit a metallic target structure at velocities higher than the sonic velocity of the materials involved on top of mechanical and hydraulic effects, blast effects take place that are due to shock-wave-induced vapourization of either or both target and/or fragment material and subsequent oxidation of the vapour with the oxygen of the air [17]. This effect is known as *vapourific effect*. In addition to fragments, a vapourific effect also occurs with shaped charge liners of metals that are readily oxidized such as aluminium, magnesium, zinc and alloys thereof [18].

Research into reactive fragments that show increased effects started in the 1980s [19]. Reactive fragments are composites made from two or more materials that are able to undergo an exothermal reaction. Upon high-velocity impact on a target structure, these composites are shock initiated and deflagrate after a short delay. Figure 13.9 shows the different effects of inert and reactive fragments on a missile body and its guidance section [20].

The shock ignition of pyrotechnics was first studied by Hardt [21–23]. The compression of a condensed solid pyrolant mixture along a Hugoniot curve – which describes the dynamic compressibility of a given material – yields an increase in

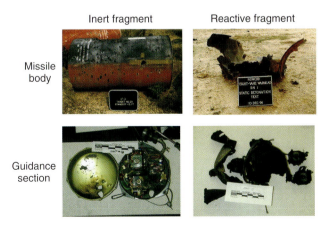

Figure 13.9 Comparison between inert and reactive fragment effects. (From Ref. [20].)

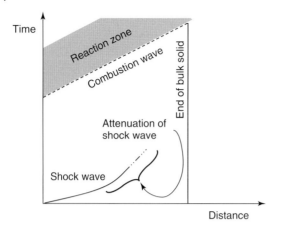

Figure 13.10 Shock initiation of a pyrotechnic solid by a short pulsed shockwave.

temperature similar to the adiabatic compression of gas. However, unlike gaseous detonation products, the decrease in pressure behind the shock wave does not result in a drop in temperature but causes attenuation of the shock while the shocked solid stays hot. Thus, unlike homogeneous high explosives, the shock wave does not induce simultaneous bond breaking, at least not to a significant extent, but after some delay leads to thermal ignition of condensed solids.

Figure 13.10 schematically depicts the processes in a pyrotechnic solid shocked by a short pulse of sufficient energy E_i. The shock wave travels at speed U_s with pressure P, through the material with initial density ρ_0 and affects a compression. The shock front is not immediately sustained by a chemical reaction and thus eventually decelerates and dies out. With some delay, the combustion wave starts in the heated material and proceeds through the solid, however, at a lower rate than the shock wave [24]. Because of the thermal nature of the ignition process in the shocked material, Hardt could show that shock initiation energy, E_i, is equivalent to the auto-ignition enthalpy, H_i, and thus is much easily obtained from thermo-chemical measurements rather than cumbersome shock initiation experiments:

$$E_i = 0.5 \cdot \frac{P^2}{U_s^2 \cdot \rho_0^2} \tag{13.2}$$

$$H_i = C_p \cdot (T_i - T_0) \tag{13.3}$$

Low-velocity impact initiation of metal/metal/PTFE mixtures was first reported by Davis in 1995 [25]. It was then observed that the addition of PTFE to Ti/Si would decrease the time-to-emission peak by orders of magnitude from few micro-seconds to few ten micro-seconds. This increase in reactivity was understood as a consequence of the shock-compression of PTFE yielding fluorine-rich decomposition products that react with both Ti and Si and, thus, trigger the alloying reaction. In parallel, it was also observed for the first time that Ti/PTFE would fiercely react in a mass ratio (80/20) ($\rho = 2.05\ \mathrm{g\,cm^{-3}} = 55.4\%\mathrm{TMD}$) when exposed to a low-velocity impact (13 m s^{-1}). It was further elucidated that the Ti/PTFE ignition delay in this velocity regime would decrease with density being indicative of a hot spot ignition mechanism [26, 27]. Later, it was found that for high-density Ti/PTFE (80/20) (89.2% TMD) shock initiation (>10 GPa) performed with a gas gun yields a stable detonation with CJ pressure of 8.8 GPa and $V_{det} = 3.39$ km s^{-1} [28].

High-pressure (2 GPa) decomposition reaction of aluminium with fluorocarbon polymers was studied by Ladouceur [29, 30].

A known reactive material 'RM4' was patented by Joshi in 2003. RM4 is prepared by sintering (375–385 °C) a mixture of pre-compressed (55–97 MPa) Al/PTFE [31].

RM4

26.5 wt% Aluminium, H-5 (mean diameter 5 μm)

73.5 wt% Polytetrafluoroethylene, Teflon® 7A (DuPont) (mean diameter 35 μm)

The material obtained this way exhibits an increased tensile strength of over 400% and an elongation of over 300% over the non-sintered material. Its TMD is 2.39 g cm^{-3} and the longitudinal speed of sound at 10 MHz is $c_{sl} = 1570$ m s^{-1}. The response of RM4 to mechanical stress has been reported in [32]. Compositions similar to RM4 have been investigated for their sensitivity parameters. 1943-77A is the so-called nano-RM4 and uses nano-metric Al with the same mass fractions Al and PTFE. In STR 22037 and STR 22080, the mass proportions are slightly different than that in RM4 (28.3 wt% Al/71.7wt% PTFE) and use either H-5 or H-95 aluminium grade, and finally 22235 is a fuel-rich formulation with 44.2 wt% H-5 (Table 13.2).

The formulations are all friction insensitive and all display auto-ignition onset >260 °C. The electrostatic sensitivity varies significantly. This is due to different mass fractions and different grades of aluminium used in the formulation and expectedly are highest for the nano-RM4 and lowest for the formulation using H-95 aluminium [33]. The performance of a composite charge of nitromethane with suspended RM4 has been calculated with CHEETAH in [34]. The influence of Al

Table 13.2 Sensitivity data for metal–fluorocarbon-based reactive materials, STR 22037 and 1943-77-A.

Formulation	ABL friction (lbs at fps)	SBAT (°C)	TC-ESD (J)	TC impact (cm)	ABL impact (cm)
STR 22037	800/8	>260	6.75	114.3	21
STR 22080	800/8	>260	>8	>116.8	80
1943-77-A	800/8	>260	<0.05	>116.8	n.t.
STR 22235	800/8	>260	4.5	>116.8	21

Taken from Ref. [33].

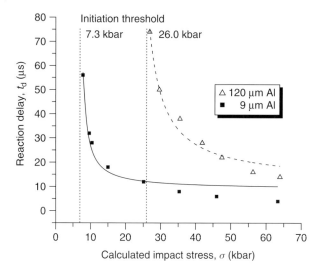

Figure 13.11 Reaction delay as a function of impact stress with two different Al particle sizes in vacuum [36].

particle size, surrounding atmosphere and impact stress on the reaction delay of RM4 has been investigated in Refs. [35, 36]. The reaction delay measured as the time between impact and observation of first light is depicted in Figure 13.11 for two different Al powder types. In general, the light output in air is greater than that in vacuum. Coarse Al ignites with more difficultly in vacuum than fine Al; however, in air, the initiation stress for 120 μm is similar for both 9 and 120 μm Al, points out the influence of the after-burn reaction.

The calculated stress relate to impact velocities between 172 (7.8 kbar) and 969 m s^{-1} (63.9 kbar). Spectroscopic investigation of the impact combustion plume shows signals evident of Al, AlO and C_2 species. Even though AlF could not be detected, the presence of C_2 is indicative of polymer decomposition and release of fluorine-containing species that must have reacted with Al [37]. The logistic advantage of RM4 is its relative inertness and thus its UN hazard classification as 4.1 material (solid flammable) opposed to 1.3 G that is assigned to MTV. In view of this, Mock et al. proposed to use Al/PTFE as a low-sensitivity initiator material for projectiles [38].

As RM4, its above-mentioned derivatives and other metal–fluorocarbon-based reactive materials are not able to undergo sustained combustion and their reaction efficiency largely depends on the degree of impact initiation. RM4, Zr/THV (52/48), Ta/THV (74/26) and Hf/THV (69/31) were tested for these effects [39]. The sensitivity data of similar formulations are listed in Table 13.3.

The reaction efficiency of reactive materials highly depends on the impact stress the material is subjected to. Hence, the determination of the reaction enthalpy is carried out in a vented combustion calorimeter. In this setup, the fragments penetrate a cylindrical combustion chamber that is confined by a thin skin of aluminium on one of the terminal sides. Inside the chamber, an impact anvil of

Table 13.3 Sensitivity data for metal–fluorocarbon-based reactive materials.

Formulation and mass ratio	TMD (g cm^{-3})	ABL friction (lbs at fps)	SBAT (°C)	DSC (°C)	TC-ESD (J)	TC impact (cm)	ABL impact (cm)
Ta/THV 220 (71.7/28.3)	5.31	800/8	260	360/435	>8	>116.8	–
Zr/THV 220 (65/35)	3.58	800/8	>260	–	<0.05	111.8	13
Hf/THV 220 (90/10)	8.41	800/8	>260	–	<0.05	>116.8	21

Taken from Ref. [33].

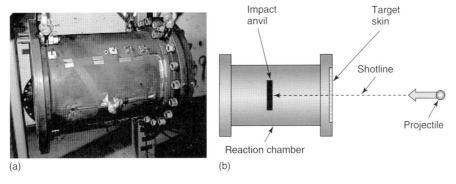

(a)　　　　　　　　　　　　　(b)

Figure 13.12 (a) Side view photograph of test chamber and (b) schematic setup. (From Ref. [39].)

hardened steel is provided to catch and interact with the pre-stressed fragment. It has been observed that blast performance of pre-stressed fragments is superior to those not exposed to any mechanical barrier before impact. Figure 13.12 shows a photograph of test chamber used in [39] together with a schematic setup.

Table 13.4 lists the theoretical explosion, after-burn and phase-change enthalpies for a series of fluorocarbon-based reactive materials [39]. Although the gravimetric reaction enthalpy of the aluminium-based formulation is superior to the other materials, the volumetric reaction enthalpy is about equal for all of them with a slight advantage for both Hf- and Ta-based formulation.

Now the vented combustion calorimetry (VCC) method shows that the experimental reaction enthalpy increases about asymptotically with increasing velocity. At low impact speed, the hafnium-based formulation outperforms any other composition and already reaches an efficiency of 52%, whereas the other formulations barely deliver a quarter of the reaction enthalpy. The tantalum-based composition is exceptionally poor as only about 3% of the theoretically stored energy is released. The situation changes at higher speeds (1800–2400 ms^{-1}) and shows the gravimetric superiority of Al/PTFE over the other formulations (Figure 13.13).

Table 13.4 Theoretical explosion enthalpy of various metal–fluorocarbon-reactive fragment materials according to Ref. [39].

System	Mass ratio	Volume ratio	TMD (g cm⁻³)	$\Delta_R H$ (kJ g⁻¹)	$\Delta_R H$ (kJ cm⁻³)	$\Delta_{AR} H$ (kJ g⁻¹)	$\Delta_{AR} H$ (kJ cm⁻³)	$\Delta_{PT} H$ (kJ g⁻¹)	$\Delta_R H + \Delta_{PT} H$ $\Delta_{AR} H +$ (kJ g⁻¹)	$\Delta_R H + \Delta_{PT} H + \Delta_{AR} H$ (kJ cm⁻³)
Al/PTFE	26.5/73.5	24/76	2.394	5.44	13.02	5.78	13.84	2.93	14.15	33.87
Zr/THV[a]	52/48	25/75	3.127	3.22	10.07	5.48	17.14	1.09	9.79	30.62
Hf/THV[a]	69/31	25/75	4.839	3.20	15.49	3.77	18.24	0.13	7.10	34.36
Ta/THV[a]	74/26	26/74	5.637	2.05	11.56	4.29	24.18	–	6.34	35.74

[a]THV-500 (see Chapter 3.8 for details).

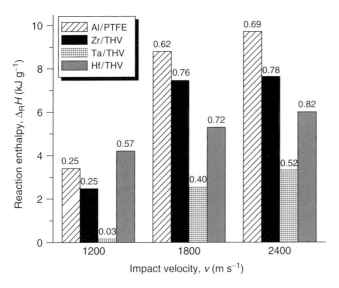

Figure 13.13 Reaction enthalpy and efficiency (superimposed) of metal/fluorocarbon-based reactive materials as a function of impact speed [39].

Compositions based on either THV 220 or 500 fluoroelastomer are chemically equivalent but show different compression yield strength and melting points: THV 220 (5 MPa, 124 °C) versus THV 500 (12 MPa, 165 °C). An interesting finding is that the material strength of the formulations inversely correlates with the available reaction enthalpy. Thus, the formulations with a weaker binder THV 220 yield higher reaction efficiencies than those using a tougher binder THV 500. A possible explanation for this is that a stronger material will effect a better distribution of the mechanical energy in the composite upon impact. Hence, the formation of hot spots will be delayed and thus subsequent ignition (Figure 13.14).

13.6
Shockwave Loading of Metal–Fluorocarbons and Detonation-Like Phenomena

A number of metal/fluorocarbon materials have been studied with respect to their response on exposure to a shock wave. As mentioned earlier, Davis et al. found indications for a sustained shock wave in compressed Ti/PTFE (80/20) (3.39 g cm^{-3}, 89.2% TMD) initiated by a steel flyer [28].

Dolgoborodov *et al.* have shown that mechanochemically activated mixtures of aluminium or magnesium with PTFE can be shock initiated to undergo fast reactions [40–47]. Figure 13.15 shows the reaction velocity as a function of stoichiometry. For two stoichiometries, the influence of density is also indicated, clearly showing that Al/PTFE rather behaves like an ammonium perchlorate-type explosive with respect to its density [48].

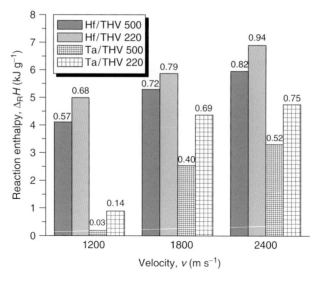

Figure 13.14 The influence of the binder on the reaction enthalpy and efficiency (superimposed) at various impact velocities [39].

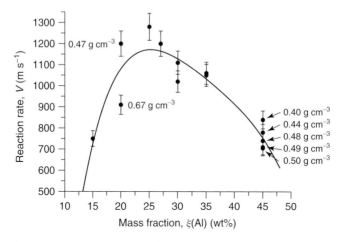

Figure 13.15 Reaction velocity of shock initiated Al/PTFE.

Mechanochemically activated Mg/PTFE (35/65) yields reaction rates that are dependent on the activation dose. Thus, up to activation doses of 9 kJ kg, the reaction rates increases but soon after decreases as is shown in Figure 13.16. Both Al/PTFE and Mg/PTFE reactions yield cellular structures as imprints on the witness plates. This indicates that the propagation process is accompanied by the formation of jets of product particles. These jets may spread the reaction from one centre to another.

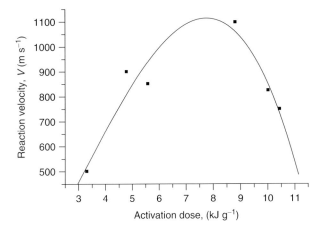

Figure 13.16 Influence of mechanochemical activation dose on Mg/PTFE (35/65) on the reaction velocity.

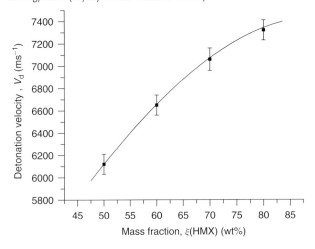

Figure 13.17 Detonation velocity of HMX/(Mg/PTFE: 50/50). (After Refs. [50, 51].)

Tulis investigated Al/PTFE (26.5/73.5) ($\rho = 0.833$ g cm^{-3}) in a 2.65 cm diameter, 1 m long steel pipe but did not achieve a detonation of the material [49]. However, blends of this pyrolant with PETN resulted in detonation. A mixture Al/PTFE/PETN (18.5/51.5/30.0) ($\rho = 0.845$ g cm^{-3}) detonates at 2848 m s^{-1}, which is about the level for the low-velocity detonation of PETN. For a pyrolant Al/PTFE/PETN (22.5/62.5/15) ($r = 0.808$ g cm^{-3}), a lower detonation velocity (1662 m s^{-1}) was noted due to the lower PETN content. However, a second detonation front was also detected in this experiment, which is probably due to detonation of shocked Al/PTFE. In the first experiment, the shock front from the higher percentage PETN is probably sufficient to immediately induce a reaction. Gogulya et al. investigated the effect of HMX admixtures on Al/PTFE. A pyrolant (Al/PTFE/HMX: 38.25/46.75/15.00) after-shock initiation yields a steady-state detonation velocity of $4.8\,km\,s^{-1}$ at charge diameters greater or equal to 40 mm. A pyrolant with even lower HMX content (Al/PTFE/HMX: 40.50/49.50/10.00) is initiated, but the detonation dies out after a few centimeters at 30 mm diameter charge [52]. Cudziło and Trzcinzski investigated charges from HMX mixed with equal amounts of Mg/PTFE. The detonation velocity as a function of HMX content is shown in Figure 13.17. In the range of 50–90 wt% HMX, a steady-state detonation is obtained [50, 51]. In the presence of Mg/PTFE, about 30–50 times longer infrared radiant intensity is measured in comparison to pure HMX, which results in a narrow 1–2 ms peak release, indicating only after-burn effects of the Mg/PTFE pyrolant.

References

1. Fisher, G.J.B. (1946) *Incendiary Warfare*, McGraw-Hill Book Company, Inc., New York, London.
2. Koch, H.W. and Licht, H.H. (1974) Brandstoffe, Brandmunition, Brandwirkung. Bericht CO 34/74, Deutsch-Französisches Forschungsinstitut, Saint Louis, Frankreich.
3. Headquarters US Army Materiel Command (1967) *Engineering Design Handbook*, Military Pyrotechnics Series Part One, Theory and Application, 5–4 Incendiaries, AMCP 706-185, Headquarters US Army Materiel Command, Washington, DC.
4. Waite, H.R. (1971) Incendiary composition containing a metallic fuel and a solid fluorocarbon polymer. US Patent 3,565,706, USA.
5. Beckert, W.F. and Dengel, H.O. (1976) Fast-burning compositions of fluorinated polymers and metal powders. US Patent 4,000,022, USA.
6. Shaw, G.C. and Reed, R. Jr (1977) Incendiary compositions of magnesium and fluoroalkyl phosphate esters. US Patent 4,013,491, USA.
7. Faber G., Florin H., Grommes, P.-J. and Röh, P. (1993) Verzögerungssätze mit langen Verzögerungszeiten. EP Patent 332,986, Dynamit Nobel AG, Germany.
8. Lucy, C.F. (1978) Metallic sponge incendiary compositions. US Patent 4,131,498, USA.
9. Karametaxas, G., Kutzli, J., Schweizer, P. and Tobler, M. (1998) Kontrollierte Vernichtung von Datenträgern duch dünne pyrotechnische Schichten als Bestandteil des Datenschutzes in Sicherheitskritischen Bereichen. 29th International Annual Conference of ICT, Karlsruhe, Germany, June 30–July 3, p. 41.
10. Keister, F.Z. and Rust, J.B. (1973) Pyrotechnic eradication of microcircuits. US Patent 3,725,671, USA.
11. Younger, S. (2004) Global strike technology strategy. Precision Strike Association Winter Roundtable, Washington, January 21.
12. Weiser, V., Roth, E., Neutz, J. and Kelzenberg, S. (2005) Untersuchungen zur Bekämpfung freigesetzter B- und C-Kampfstoffe. ICT-Symposium, Karlsruhe.
13. Glaude, P.A., Melius, C., Pitz, W.J. and Westenbrook, C.K. (2002) Detailed chemical kinetic reaction mechanism for incineration of organophosphorus and fluoroorganophosphorus compounds. *Proc. Combust. Inst.*, **29**, 2469–2476.
14. Bless, S. and Pantoya, M. (2008) *Advanced Energetic Materials for Agent Defeat: Impact-Driven Reactions in Biocidal Reactive Materials for WMD Applications*, Institute for Advanced Technology, The University of Texas, Texas, IAT R 0553, September 2008.
15. Baker, J.J., Gotzmer, C., Gill, R., Kim, S.L. and Blachek, M. (2009) Agent defeat bomb. US Patent 7,568,432, USA.
16. Vizard, F. (2003) *Extinguishing the Threat U.S. Special Weapons May Target Iraqi Chemical and Biological Threats*, American Scientific, February 17, *http://www.globalsecurity.org/org/news/2003/030218-hpm01.htm* (accessed February 7 2011).

17. Cook, M.A. (1958) *The Science of High Explosives*, Robert E. Krieger Publishing Company, Malabar, FL, pp. 259–264.

18. Kennedy, D.R. (1993) The application of reactive metal technologies to non-nuclear warheads. 14th International Symposium on Ballistics, Quebec, Canada, September 26–29, p. 131.

19. Held, M. (1985) Untersuchungen mit Korruskativstoffen. 16th International ICT Annual Conference Combined with 10th International Pyrotechnics Seminar, Karlsruhe, Germany, July 2–5, p. 40.

20. National Research Council (2004) *Advanced Energetic Materials*, The National Academies Press, Washington, DC, p. 21.

21. Hardt, A.P. and Martison, R.H. (1974) Initiation of pyrotechnic mixtures by shock. 8th Symposium on Explosives and Pyrotechnics, The Franklin Research Center, Philadelphia, PA, p. 53.

22. Hardt, A.P., McHugh, S.L. and Weinland, S.L. (1986) Chemistry and shock initiation of intermetallic reactions. 11th International Pyrotechnics Seminar, Vail, CO, July 7–11, p. 255.

23. Hardt, A.P. (1988) Shock initiation of thermite. 13th International Pyrotechnics Seminar, Grand Junction, CO, July 11–15, p. 425.

24. Sheffield, S.A. and Schwarz, A.C. (1982) Shock wave response of titanium subhydride-potassium perchlorate. 8th International Pyrotechnics Seminar, Steamboat Springs, CO, July 12–16, p. 972.

25. Woody, D.L., Davis, J.J. and Miller, P.J. (1995) in *Decomposition, Combustion, and Detonation Chemistry of Energetic Materials*, Symposium Proceedings of Materials Research Society, vol. 418 (eds T.B. Brill, T.P. Russell, W.C. Tao and R.B. Wardle), Materials Research Society, Pittsburgh, PA, pp. 445–449.

26. Davis, J.J. and Miller, P.J. (1997) Reaction of pyrotechnic material under deformation: experiments and modelling. JANNAF Propulsion Systems Hazards Subcommittee Meeting, USA.

27. Woody, D.L., Davis, J.J. and Deiter, J.S. (1997) Recovery studies of impact-induced Metal/Polymer reactions in titanium based composites. Shock Compression of Condensed Matter, 1997, pp. 667–669.

28. Davis, J.J., Lindfors, A.J., Miller, P.J., Finnegan, S. and Woody, D.L. (1998) Detonation like phenomena in metal-polymer and Metal/Metal oxide polymer mixtures. 11th Symposium on Detonation, Snowmass Village, CO, August 31–September 4, p. 1007.

29. Parker, L.J., Ladouceur, H.D. and Russell, T.P. (1999) Teflon and Teflon/Al (Nanocrystalline) decomposition chemistry at high pressures. Shock Compression of Condensed Matter, 1999, p. 941.

30. Parker, L.J., Ladouceur, H.D. and Russel, T.P. (1999) Teflon and Teflon/Al decomposition chemistry at high pressures. 30th International Annual Conference of ICT, Karlsruhe, Germany, June 29–July 2, p. 78.

31. Joshi, V.S. (2003) Process for making polytetrafluoroethylene-aluminum composite and product made. US Patent 6,547,993, USA.

32. Casem, D. T. (2008) Mechanical Response of an Al-PTFE Composite to Uniaxial Compression over a Range of Strain Rates and Temperatures. ARL-TR-4560, Army Research Laboratory, Aberdeen Proving Ground, Aberdeen, MD.

33. Nielson, D.B., Ashcroft, B.N. and Doll, D.W. (2005) Reactive material enhanced munition compositions and projectiles containing same. US Patent 2005/0,199,323, USA.

34. Ashcroft, B.N., Nielson, D.B. and Doll, D.W. (2010) Reactive compositions including metal. US Patent 2010/0,276,042, USA.

35. Mock, W. and Holt, W. (2005) Impact initiation of rods of pressed Polytetrafluoroethylene (PTFE) and aluminum powders. Shock Compression of Condensed Matter – 2005, pp. 1097–1100.

36. Mock, W. and Drotar, J.T. (2007) Effect of aluminum particle size on the impact initiation of pressed PTFE/Al composite rods. Shock Compression of Condensed Matter – 2007, pp. 971–974.

37. Lee, R.J., Mock, W., Carney, J.R., Holt, W.H., Pangilinan, G.I. and Gamache, R. (2005) Shock Compression of Condensed Matter – 2005, p. 169.

38. Mock, W., Hanna, B.L. and Holt, W.H. (2009) Reactive material initiator for explosive filled munitions. US Patent 7,587,978, USA.

39. Ames, R.G. and Waggener, S.S. (2005) Reaction efficiencies for impact initiated materials. ICT Annual Conference, p. 2.

40. Gogulya, M.F., Makhov, M.N., Brazhnikov, M.A. and Dolgobrodov, A.Y. (2004) Investigation of shock-induced processes in Teflon/Al – based mixtures. 35th International Annual Conference of ICT, Karlsruhe, Germany, June 29–July 2, p. P79.

41. Dolgobrodov, A.Y., Makhov, M.N., Streletskii, A.N., Kolbanev, I.V., Gogulya, M.F., Brazhnikov, M.A. and Fortov, V.E. (2004) Detonation-like phenomena in non-explosive oxidizer-metal mixtures. 31st International Pyrotechnics Seminar, Fort Collins, CO, July 11–16, p. 569.

42. Dolgoborodov, A.Y., Makhov, M.N., Kolbanev, I.V., Streletskii, A.N. and Fortov, V.E. (2005) Detonation in an Aluminum-Teflon mixture. *JETP Lett.*, **81**, 311–314.

43. Dolgoborodov, A.Y., Makhov, M.N., Kolbanev, I.V., Streletskii, A.N. and Fortov, V.E. (2006) Detonation in Metal-Teflon mechanoactivated composites. 13th International Symposium on Detonation, USA.

44. Dolgoborodov, A.Y., Kolbanev, I.V., Makhov, M.N., Streletskii, A.N. and Fortov, V.E. (2006) Detonation of Teflon-Based mechanoactivated energetic composites. 33rd International Pyrotechnics Seminar, Fort Collins, CO, July 16–21, p. 535.

45. Dolgoborodov, A.Y., Streletskii, A.N., Makhov, M.N., Kolbanev, I.V. and Fortov, V.E. (2007) Explosive compositions based on the mechanoactivated metal-oxidizer mixtures. *Russ. J. Phys. Chem. B*, **1**, 606–611.

46. Streletskii, A.N., Dolgoborodov, A.Y., Kolbanev, I.V., Makhov, M.N., Lomaeva, S.F., Borunova, A.B. and Fortov, V.E. (2009) Structure of mechanically activated high-energy Al+polytetrafluoroethylene nanocomposites. *Colloid J.*, **71**, 852–860.

47. Dolgoborodov, A.Y., Streletskii, A.N., Kolbanev, I.V. and Makhov, M.N. (2010) Explosive compositions on the basis of mechanoactivated nanocomposites of metals and solid oxidizers. 14th International Symposium on Detonation, Coeur d'Alene Resort, ID, 11–16 April.

48. Price, D. (1981) Critical Parameters for Detonation Propagation and Initiation of Solid Explosives. NSWC TR 80-339, Research and Technology Department, Naval Surface Weapons Center, Dahlgren.

49. Tulis, A.J. (1989) On the detonation of pyrotechnics. 4e Congres International de Pyrotechnie du groupe de Travail de Pyrotechnie, La Grande Motte, France, June 5–9, p. 73.

50. Cudzilo, S., Trczinski, W.A. and Maranda, A. (1996) Detonation behaviour of HMX based explosives containing magnesium and polytetrafluoroethylen. 27th International Annual Conference of ICT, Karlsruhe, Germany, June 25–28, p. 71.

51. Cudzilo, S. and Trzcinski, W.A. (2003) Studies of HMX-based explosives containing magnesium and polytetrafluoroethylene. *Chimiczeskaya Fizika*, **22**, 82–89.

52. Gogulya, M.F., Makhov, M.N., Brazhnikov, M.A. and Dolgoborodov, A.Y. (2006) Detonation-like processes in Teflon/Al-Based explosive mixtures. 13th International Symposium on Detonation, USA.

Further Reading

Gordopov, Y.A., Batsanov, S.S. and Trofimov, V.S. (2009) *Shock Induced Solid-Solid Reactions and Detonations*, in Shock Wave Science and Technology Reference Library, Heterogeneous Detonation (ed. F. Zhang) Springer, New York, **4**, 287–314.

14
Miscellaneous Applications

14.1
Submerged Applications

14.1.1
Underwater Explosives

The combustion of metals in water is of practical importance in underwater propulsion [1, 2], hydrogen gas generation [3, 4] underwater explosives with increased shock and bubble energy [5, 6]. The influence of Al/O stoichiometric ratio on both shock and bubble energy of an explosive is shown in Figure 14.1 taken from Ref. [6] and other applications [7].

Pantoya et al. investigated the deflagration performance of Al/PTFE (polytetrafluoroethylene) (ξ(Al) = 0.30–0.45) using nanometric aluminium with low densities in the range 10–40% theoretical maximum density (TMD) [8, 9]. For the low-density deflagrations, <20% TMD the bubble energy, E_b [8], has been calculated from the bubble radius (Figure 14.2) and pressure for three stoichiometries and is shown in Figure 14.3. The bubble energies are largest for fuel-rich stoichiometries. In addition to stoichiometry, the density of the payload influences the deflagration behaviour. Dense materials are less likely to undergo volumetric combustion and, thus, deliver lower bubble size as shown in Figure 14.3:

$$E_b = \frac{4/3 \cdot \pi \cdot P_{hydro} \cdot R_{max}^3}{M_{sample}} \tag{14.1}$$

Unlike the experiments with nanometric aluminium, mixtures containing micrometric Al (99 wt% Al) (2–3 μm) extinguish immediately after submerged ignition.

14.1.2
Underwater Flares

PTFE has been considered both as hydrophobic co-oxodizer and as exclusive oxidizer in underwater illuminating flares. A patent from 1959 devises the use of Mg/Al/Fe$_3$O$_4$/Ba(NO$_3$)$_2$/PTFE as payload for underwater torches [10] with the PTFE content ranging from 17 to 33 wt%.

Metal-Fluorocarbon Based Energetic Materials, First Edition. Ernst-Christian Koch.
© 2012 Wiley-VCH Verlag GmbH & Co. KGaA. Published 2012 by Wiley-VCH Verlag GmbH & Co. KGaA.

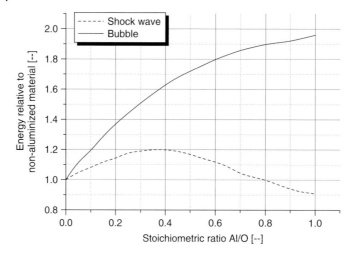

Figure 14.1   Influence of aluminium on both shock and bubble energy of explosive [6].

Underwater thermite flare [10]

34.8−27.6 wt% Magnesium
21.7−17.2 wt% Aluminium
17.4−13.8 wt% Barium nitrate
8.7−6.9 wt% Ferrous oxide
17.4−34.5 wt% Polytetrafluoroethylene

Douda refers to stoichiometric (Mg/F = 1:2) underwater flares [11]. One particular composition is shown below:

Rolle underwater flare [11]	NOTS PL 6246 underwater flare [12]
34.8−27.6 wt% Magnesium	30.5 wt% Magnesium gran 16
21.7−17.2 wt% Aluminium	49.7 wt% Teflon® #7
17.4−13.8 wt% Barium nitrate	16.3 wt% Viton® A
17.4−35.4 wt% Polytetrafluoroethylene	1.0 wt% Aluminium
8.7−6.9 wt% Ferrous oxide	1.0 wt% Boron
	1.5 wt% Sodium chloride

14.1.3
Underwater Cutting Torch

Compositions for use in underwater cutting torch [12] have been developed by Helms and Rozner [13]. The original composition is given below:

Pyronol [14]

31.0 wt% Nickel, 80–325 mesh (40% >150 mesh)
27.4 wt% Aluminium, 200 mesh
34.4 wt% Ferric oxide, 325 mesh
7.0 wt% Polytetrafluoroethylene, 35 μm

Figure 14.2 Morphology of submerged plume at various densities starting left: 40%, 35% and <20% TMD [7] reproduced with kind permission by Prof. Michelle Pantoya.

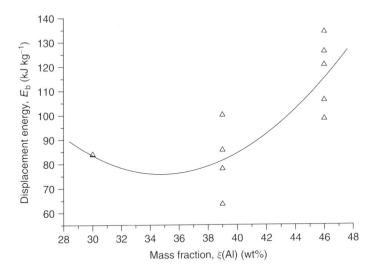

Figure 14.3 Bubble energy for three stoichiometries.

Pyronol is typically pressed to annular tablets of densities of 3.45 g cm^{-3}. These tablets are assembled in a steel tube closed at one end by a multi-hole graphite nozzle to allow for an either linear or centrosymmetric array of liquid metal jets

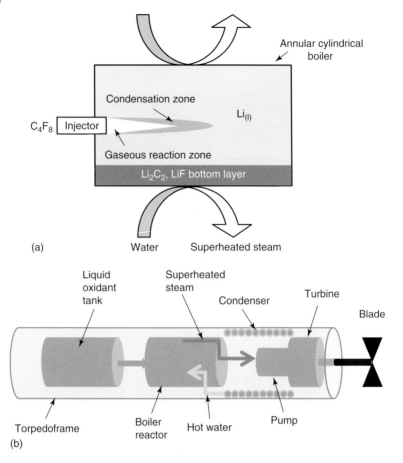

Figure 14.4 Schematic setup of (a) SCEPS-boiler reactor and (b) Torpedo propulsion unit.

to increase the cutting performance. The PTFE partly decomposes and reacts with aluminium to yield gaseous products, C_2F_4, AlF_3 (sublimation at 1275 °C), which help to project the mainly condensed material to create some jet of molten NiAl alloy. The perforation performance of the pyronol torch is related to three principal effects on the target substrate, which are

• a strong single compression pulse with a duration of ~1 µs
• a weaker compressive load by the jet flowing outwards
• shear loads affected by the liquid moving over the surface [14].

14.2
Mine-Disposal Torch

Disposal of landmines typically requires the use of either plastic high explosives or shaped charges to induce sympathetic detonation. Alternatively, mines and

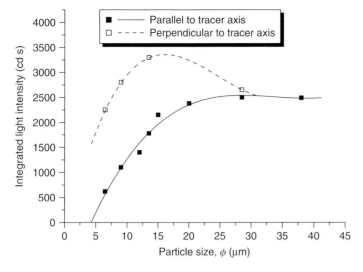

Figure 14.5 Light intensities integrated over time versus particle size, observed parallel and perpendicular to tracer axis.

unexploded ordnance may undergo a low-order fast cook reaction when subjected to localized thermal stimuli. For this purpose, a variety of the so-called mine-disposal torches have been developed [15]. A mine-disposal torch designated propellant torch system (PTS) (also described in the review [15]) has been developed by Tulis et al. [16]. The composition is given below:

PTS mine-disposal torch

26.1 wt% Aluminium
56.5 wt% Potassium perchlorate
 8.7 wt% Polytetrafluoroethylene
 8.7 wt% Zinc

The use of zinc has been proposed because of its high density ($\rho = 7.13\,\mathrm{g\,cm^{-3}}$), which at given jet velocity (v_{jet}) improves the pressure, p, of a jet of the molten material, with c_{jet} the sound velocity of the liquid jet, on a target surface of given density ρ_{target} and c_{target} the sound velocity of the target by [14]

$$p = v_{jet} \cdot \frac{\left(\rho_{jet} \cdot c_{jet}\right) \cdot \left(\rho_{target} \cdot c_{target}\right)}{\left(\rho_{jet} \cdot c_{jet}\right) + \left(\rho_{target} \cdot c_{target}\right)} \tag{14.2}$$

In addition, because of its low enthalpy of vapourization, Zn allows good heat transfer by condensation and finally Zn has a hypergolic effect on the molten TNT [16].

Table 14.1 Li-based SCEPS storage data.

Oxidizer	m_r (g mol^{-1})	B_p (°C)	ρ (kg m^{-3})	$\Delta_f H°$ (kJ mol^{-1})	$\Delta_R H°$ (kJ g^{-1})	$\Delta_R H°$ (kJ cm^{-3})	Ox/fuel (wt%/wt%)
C_4F_8	200.031	−5.9	1499.28	−1528	12.52	11.37	2.40
SF_6	146.05	−64	1339.10	−1222	14.56	13.26	2.63

14.3
Stored Chemical Energy

14.3.1
Heating Device

The exothermal reaction of fuel-rich mixtures from Li and PTFE has been proposed as payload for one-shot heating [17]. The device comprises a thin-walled steel tube containing a lithium tube with a core filling of coarse and an electrical igniter. To reduce the ignition time, the system may be pressurized with either sulphur hexafluoride, SF_6, or Freon®.

14.3.2
Stored Chemical Energy Propulsion

Similar to the application discussed above, the production of heat can be used in a more complex design to generate mechanical energy in the so-called stored chemical energy propulsion systems (SCEPS). SCEPS are typically used for underwater propulsion of torpedoes (see Figure 14.4b) and use the reaction between molten lithium and any gaseous fluorine compound (see Figure 14.4a) such as SF_6 [18] or fluorocarbons such as Freon [19].

Table 14.1 compares key energy storage data of both SF_6 and C_4F_8 taken from [20]. The ideal reactions are

$$12\ Li_{(l)} + C_4F_{8(g)} \longrightarrow 2\ Li_2C_{2(s)} + 8\ LiF_{(s)} \tag{14.3}$$

$$8\ Li_{(l)} + SF_{6(g)} \longrightarrow Li_2S_{(s)} + 6\ LiF_{(s)} \tag{14.4}$$

14.4
Tracers

Pyrotechnic tracers serve as an optical guide to follow the path of a bullet or shell. Because of the release of gaseous combustion products, tracers also positively affect the wake drag of a bullet and, hence, are able to effect higher ranges. Upon firing, the wake of a bullet experiences very high pressure in the gun barrel and pressures lower than ambient level after leaving the barrel. This pressure drop affects most burning pyrotechnics. As has been discussed, the burn rate of Magnesium/Teflon/Viton (MTV) using fine Mg grain sizes is not strongly affected by pressure changes. Thus, tracer compositions based on MTV have been proposed.

Visible tracers based on MTV and modifications with strontium nitrate have been proposed by Ramnarace [21]. A typical pink burning composition is given below:

Pink tracer [21]

---

45 wt% Magnesium, Gran 16
23 wt% Teflon #7
14 wt% Viton A
17 wt% Strontium nitrate (100–230 mesh)

Diewald et al. proposed to use MTV compositions for tracers in small arms ammunition [20]. Flückiger investigated the effects of different Mg particle sizes on both static burn rate and static integrated light intensity. He found that the light intensity significantly varies with Mg particle size for both observation parallel and perpendicular to the tracer axis (Figure 14.5).

Flückiger also described the use of MTV tracers in 35 mm APDS-T ammunition [22].

Small arms tracer [23]

---

57 wt% Magnesium, Gran 16
39 wt% Teflon #7
 4 wt% Viton B

14.5
Propellants

Solid rocket propellants typically yield hot combustion flames with a major amount of gaseous combustion products having as low possible as molecular mass. The specific impulse, I_{sp}, of a propellant is proportional to the square root of the ratio of combustion temperature, T_c, to mean molecular weight of the combustion products, m_r [24, 25] (see Figure 14.6):

$$I_{sp} \approx \sqrt{\frac{c_{p1} T_1 - c_{p2} T_2}{m_r}} \tag{14.5}$$

Metal/fluorocarbon pyrolants, though inherently rich in condensed reactions products with some metals, have been proposed as modifiers for perchlorate-based propellant compositions. This serves primarily to increase the density and thus the density-related impulse [26]. As both boron and silicon give gaseous fluorides at ambient temperature and pressure, these are ideal target reaction products, and thus, both elements are preferred fuels in some propellant compositions with lithium perchlorate as a preferred co-oxidizer to burn carbon and carbides (SiC, B_4C) to carbon monoxide [27].

Propellant [27]

5.50 wt% Boron   8.00 wt% Silicon
30.60 wt% Polytetrafluoroethylene   6.00 wt% Aluminium
27.25 wt% Fluoroelastomer   40.75 wt% Polytetrafluoroethylene
36.65 wt% Lithium perchlorate   14.50 wt% Fluoroelastomer
                                                       32.20 wt% Lithium perchlorate

Burn rate at 4.1 MPa
10.8 mm s^{-1} 10.0 mm s^{-1}
Burn rate at 6.9 MPa
19.5 mm s^{-1} 14.0 mm s^{-1}

Another type of metal/fluorocarbon-modified propellant composition has been proposed using relying on potassium perchlorate and aluminium [28].

Propellant [28]
20.0 wt% Aluminium
25.0 wt% Polytetrafluoroethylene
15.0 wt% Viton
35.0 wt% Potassium perchlorate
2.5 wt% Sodium fluoride
2.5 wt% Potassium dichromate

The combustion behaviour of AP-based propellants (44 wt%) having a highly fluorinated binder (62 wt% fluorine) and fuels selected from the group of boron, silicon and zirconium have been investigated in [29] and are listed in Table 14.2.

Table 14.2 Burning rates and pressure exponent for metal/fluorocarbon/AP propellant at various pressures.

Fluorocarbon binder (wt%)	NH_4ClO_4, 200 μm (wt%)	Fuel (wt%), particle size (μm)	3.4 MPa	6.9 MPa	10.3 MPa	N
31	44	Zr, 25, 30	4.826	6.35	–	–
31	44	B, 15, 2Al, 10, 5	5.842	–	8.382	0.33
31	44	Si, 25	4.064	5.588	6.604	0.45
40	42	B, 18, 2	5.080	6.096	7.874	0.40

Air-breathing propulsion systems for use in the atmosphere require a pyrolant to provide a constant stream of combustible gases and hot particles [25]. Peretz investigated the applicability of metal/fluorocarbon reactions to provide combustible products for fuel purpose in both ducted rockets (Figure 14.7) and solid fuel ramjet (SFRJ) propulsion [30] (Figure 14.8). In a ducted rocket design, a pyrolant generates combustible gases that enter the ramburner and mix and burn with the air from air intakes. The hot combustion gases leave the ramburner through the nozzle.

In an SFRJ design, the pyrolant material is placed as a tubular insert in the ramburner itself. In place of the gas generator is a pyrotechnic igniter that will facilitate ignition of the air–gas mixture.

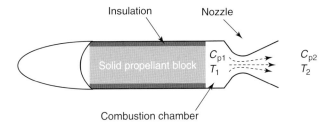

Figure 14.6 Principal sketch of a solid propellant rocket motor.

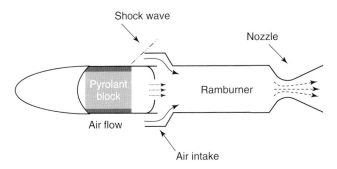

Figure 14.7 The structure of a ducted rocket.

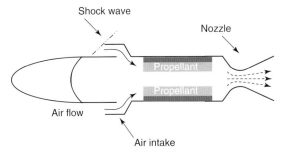

Figure 14.8 Structure of a solid fuel ramjet (SFRJ).

Some tactical surface-to-air missiles use an air-breathing propulsion unit with a metal–fluorocarbon-based gas generator. A typical gas generator for surface-to-air missiles comprise

---

Missile gas generator

Magnesium
Polytetrafluoroethylene
Calcium carbonate

Graphite fluoride has been proposed as combustion modifiers for boron-based rockets and RAM propellants [31]. This stems from the removal of the oxide coating by volatile CF_2 formed upon decomposition of graphite fluoride the following reaction (Eq. (14.6)):

$$B_2O_{3\,(s)} + 3\,CF_{2(g)} \longrightarrow 2\,BF_3 + 3\,CO + 1785\,kJ \qquad (14.6)$$

Propellant grains as given below were tested for the influence of fluorine content, agglomerate size and AP particle size ratio on the burn rate.

RAM propellant

---

37 wt% Boron/graphite fluoride agglomerate (CF_x/B: 1 : 14 or 1 : 7)
28 wt% Ammonium perchlorate
27 wt% HTPB
 8 wt% Mg_4Al_3

Propellants with 1 : 7 CF_x/B ratio display the highest burn rate and a pressure exponent of $n = 0.25$. In comparison at 1 : 14 CF_x/B, the highest burn rate is obtained with large agglomerates of the order of 1180 μm diameter, whereas lower burn rates are obtained with smaller agglomerates of 300 μm diameter [31]. A US patent teaches the application of graphite fluoride as a preferred oxidizer in propellants with Al and carboxy-terminated-polybutadiene (CTPB) in air-augmented ramjet engines (Figure 14.8) [32].

The repeated ignition and extinction of metal/fluorocarbon propellants has been reported for Al/PTFE with ξ(Al) < 30 wt% by application and interruption of a DC arc discharge [33].

References

1. Miller, T.F. and Herr, J.D. (2004) Green rocket propulsion by reaction of Al and Mg powders and water. 40th AIAA/ASME/SAE/ASEE Joint

Propulsion Conference and Exhibit, Fort Lauderdale, FL, July 11–14, AIAA 2004-4037.

2. Foote, J.P., Thompson, B.R. and Lineberry, J.T. (2002) in *Advances in Chemical Propulsion* (ed. G.D. Roy), CRC Press, Boca Raton, FL, pp. 133–146.

3. Risha, G.A., Son, S.F., Yetter, R.A., Yang, V. and Tappan, B.C. (2007) Combustion of nano-aluminum and liquid water. *Proc. Combust. Inst.*, **31**, 2029–2036.

4. Connel, T.L. Jr, Risha, G.A., Yetter, R.A., Young, G., Sundaram, D.S. and Yang, V. (2010) Combustion of alane and aluminium with water for hydrogen and thermal energy generation. *Proc. Combust. Inst.*, doi: 10.1016/j.proci.2010.07.088.

5. Bates, L.R. (2004) Underwater explosive behaviour of compositions containing nanometric aluminium powder. *Mater. Res. Soc. Symp. Proc.*, **800**, 257–263.

6. Swisdak, M.M. Jr (1978) Explosion Effects and Properties: Part II – Explosion Effects in Water. Research and Technology Department, Naval Surface Warfare Center, Dahlgren, 22 February.

7. Wilson, M.A. (2001) The Development of an Underwater Flare for the Olypic Torch Relay. Sydney 2000, 28th International Pyrotechnics Seminar, 4–9 November, Adelaide, Australia, p. 759.

8. Shawn, S.C. (2008) Hydrodynamical analysis of nanometric aluminum/Teflon deflagrations. Ph D thesis. Texas Tech University.

9. Shawn, S.C., Pantoya, M., Prentice, D.J., Steffler, E.D. and Daniels, M.A. (2009) Nanocomposites for underwater deflagration. *Adv. Mater. Processes*, 33–35.

10. Rolle, E.W. (1963) Thermit reaction underwater flare. US Patent 3,107,614, USA.

11. Douda, B.E. (2009) Genesis of Infrared Decoy Flares, The Early Years From 1950 into the 1970s. Report NSWCCR/RDTR-08/63, NSWC, Crane, IN.

12. Helms, H.H., Rozner, A.G. and Spencer, D.E. (1973) Incendiary cutting torch for underwater use. US Patent 3,713,636, USA.

13. Helms, H.H. and Rozner, A.G. (1972) Pyrotechnic composition. US Patent 3,695,951, USA.

14. Rozner, A.G. and Helms, H.H. (1976) Pyronol torch – a non-explosive underwater cutting tool. 8th Annual Offshore Technology Conference, Houston, TX, May 3–56, OTC 2705.

15. N.N. (2005) Operational Evaluation Test of Mine Neutralization Systems. Institute for Defense Analyses, 4850 Mark Center Drive, Alexandria, VA.

16. Tulis, A.J., Dillehay, D.R., Frolov, Y.V., Patel, D.L. and Dillon, J. (2001) Chemical and physical aspects of a pyrotechnic torch in penetrating and non-detonatively neutralizing all UXO. 28th International Pyrotechnics Seminar, Adelaide.

17. Olson, D.R. (1974) Chemical heater tube. US Patent 3,811,422, USA.

18. Paulinkonis, R.S. (1962) Fuel system comprising sulfur hexafluoride and lithium containing fuel. US Patent 3,325,318, USA.

19. Zheng, H. and Bu, J. (1996) The submerged jet reaction process of sulfur hexafluoride into molten lithium. *J. Chem. Ind. Eng.*, **47**, 565.

20. Diewald, G. and Bickel, H. (1978) DE-Patent 2,17,642, Germany.

21. Ramnarace, J. (1978) Tracer and composition. US Patent 4,094,711, USA.

22. Flückiger, R. (1990) Tracer testing in the test center Ochsenboden. 15th International Pyrotechnics Seminar, Boulder, CO, July 9–13, p. 311.

23. Flückiger, R. (1989) Magnesium/polytetrafluoroethylene tracer compositions. 14th International Pyrotechnics Seminar, Jersey, Channel Islands, UK, p. 81.

24. Kubota, N. (2004) Propellant chemistry, in *Pyrotechnic Chemistry*, Chapter 11 (ed. K. Kosanke), Journal of Pyrotechnics, Whitewater, CO, pp. 11-1–11-25.

25. Kubota, N. (2007) *Propellants and Explosives – Thermochemical Aspects of Combustion*, Wiley-VCH Verlag GmbH, Weinheim.

26. Reed, R., Zentner, B.A., Marrs, C.D. and Mason, B.E. (1990) Energetics of the fluorine metal reaction in energetic

materials. 15th International Pyrotechnics Seminar, Boulder, CO, July 9–13, p. 815.

27. Burnside, C.H. (1970) Composite Solid Propellants.

28. Eldridge, J.B., Julian, E.C., Dow, R.L. and Rice, G.B. (1975) Fluorocarbon solid propellant with burning rate modifier. US Patent 3876,477, USA.

29. Reed, R., Chan, M.L., Zurn, D.E. and Atwood, A.I. (1982) A new class of fluorocarbon propellants. JANNAF Combustion Meeting, CPIA Publication 366, Vol. 1, p. 1.

30. Peretz, A. (1987) Some theoretical considerations of metal-fluorocarbon compositions for ramjet fuels. ISABE, 87-7027.

31. Liu, T.-K., Shyu, I.-M. and Hsia, Y.-S. (1996) Effect of fluorinated graphite on combustion of boron and boron-based fuel-rich propellants. *J. Propul. Power*, **12**, 26–33.

32. Fields, J.L. and Martin, J.D. (2004) Combustible compositions for air-augmented rocket engines. US Patent 6,736,912, USA.

33. Tachibana, T. and Kimura, I. (1988) DC arc discharge ignition and combustion control of solid propellants. *J. Propul.*, **4**, 41–46.

15
Self-Propagating High-Temperature Synthesis

15.1
Introduction

Combustion synthesis also termed self-propagating high-temperature synthesis (SHS) is a strong exothermic chemical reaction that leads to the formation of new thermodynamically stable materials with structures often not obtained under different reaction conditions. In some way or the other, both Goldschmidt and Matignon (Chapter 2) can be considered to be the first to have applied this concept successfully to synthesize materials. However, the term *SHS* in today's perception is linked mainly to the work of Russian Chemist Alexander G. Merzhanov (*1931) who has been working on this concept since the 1960s [1, 2].

In the SHS, the reaction occurs in a narrow zone (designated the combustion wave) of high temperature that travels at a rate ranging from a few millimetres per second to several ten metres per second.

Now metal/fluorocarbon reactions beside their use in flares and igniters constitute a versatile toolbox for the synthesis of new materials. Depending on the metallic substrate, reactions of fluorocarbons with metals and their alloys and compounds yield materials such as

- fluorides
- carbon and
- carbides.

In accordance with the general simplified reactions, where w describes the maximum valency of the metal and "omega" describes the maximum valency of the element X.

$$M^w + C_n F_m \xrightarrow{\Delta} MF_w + C \tag{15.1}$$

$$M^w + C_n F_m \xrightarrow{\Delta} MF_w + MC + (C) \tag{15.2}$$

$$M^w X + C_n F_m \xrightarrow{\Delta} MF_w + X_2 \uparrow + (C) \tag{15.3}$$

$$M^w X^\omega + C_n F_m \xrightarrow{\Delta} MF_w + XF_\omega + (C) \tag{15.4}$$

$$M^w X^\omega + C_n F_m \xrightarrow{\Delta} MF_w + XC + XF_\omega + (C) \tag{15.5}$$

Metal-Fluorocarbon Based Energetic Materials, First Edition. Ernst-Christian Koch.
© 2012 Wiley-VCH Verlag GmbH & Co. KGaA. Published 2012 by Wiley-VCH Verlag GmbH & Co. KGaA.

Because of the high exo-thermicity of metal–fluorocarbon reactions, high reaction temperatures are obtained that facilitate the volatilization of product species and thus help to obtain high-purity products. In addition, high thermal gradients, for example high cooling rates give rise to the formation and quenching of meta-stable product phases often not obtained by other means. Thus, combustion synthesis via metal–fluorocarbon reactions have become an important tool in the synthesis of new inorganic materials.

While exploring advanced oxidizers for infrared countermeasure (IRCM) applications in spring 1999, the author tested graphite fluoride as an oxidizer in mixtures with magnesium and other metallic fuels [3]. After testing flare grains in a test tunnel, it was noted that there would always remain a black dust layer that would not form with polytetrafluoroethylene (PTFE)-based compositions. Subsequent investigation of this material by transmission electron micrograph (TEM), scanning electron micrograph (SEM), energy dispersive X-ray spectrsocopy (EDX) and X-ray diffraction (XRD) revealed the formation of magnesium fluoride, soot and exfoliated graphite [4, 5] (Figure 15.1).

The TEM picture revealed the presence of exfoliated graphite, also confirmed by SEM (Figure 15.2).

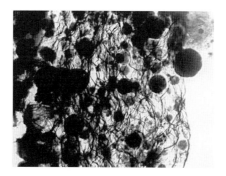

Figure 15.1 Transmission electron micrograph of aerobic combustion residue from magnesium/polycarbon monofluoride/Viton® (MPV) flare grain.

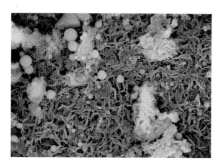

Figure 15.2 Scanning electron micrograph of MPV flare grain combustion residue.

Then just a little earlier, Manning had found exfoliated graphite in poly(carbon monofluoride) treated in a 5000 K hot atmospheric Ar plasma [6]. Manning also found indications for the formation of nanotubular structures under same conditions [7]. As combustion conditions encountered were basically similar to Manning's setup, the author aimed at investigating the possibility to produce carbon nanotubes and other carbon-based materials by the same concept. Diehl Group at Nuremberg/Germany provided the funding for preliminary research (2001–2004) on both magnesium/PTFE and magnesium/PMF (poly(kohlenstoffmonofluorid)) reactions as possible sources for the formation of carbon nanotubes and other nanoallotropes of carbon.

At about the same time, Ric Kaner et al. reported, for the first time, about the formation of multi-walled carbon nanotubes by the reaction of PTFE with lithium acetylide, Li–C≡C–Li, in a solid-state combustion reaction (R.B. Kaner, personal communication) [8, 9].

15.2
Magnesium

The XRD analysis of combustion residues from Mg/PTFE (32/68) under 2 MPa argon atmosphere displays only reflexes for MgF_2 but no crystalline carbon (Figure 15.3). The HRTEM analysis of the combustion residues reveals the presence of MgF_2 spheres (Figure 15.4a) and amorphous carbon (Figure. 15.4b).

In contrast, the combustion residues from Mg/PMF (27/73) at 2 MPa in XRD display both reflexes for magnesium fluoride and graphite indicative of a reductive de-fluorination of the parent compound carbon framework (Figure 15.5).

HREM of the above-mentioned sample reveals MgF_2 spheres and expanded graphite (Figure 15.6). A polished micro-section of a MgF_2 sphere reveals spheroidal constituents of ~200 nm diameter. This indicates condensation of MgF_2 from the gas phase and coalescence of particles to give larger agglomerates.

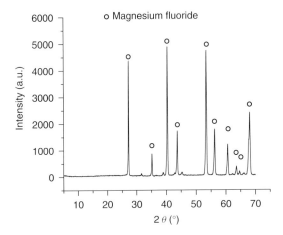

Figure 15.3 (a) MgF_2 spheres and amorphous carbon and (b) amorphous carbon.

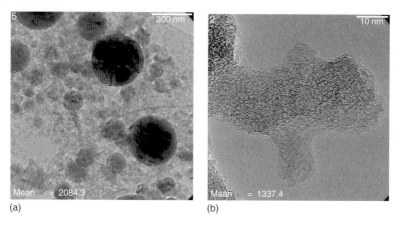

Figure 15.4 XRD plot of Mg/PTFE combustion residues in 2 MPa argon.

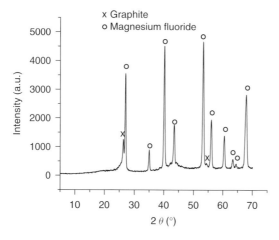

Figure 15.5 XRD plot of Mg/PMF combustion residues in 2 MPa argon.

Analysis of residues from Mg/PMF (27/73) combustion at 0.02 MPa Ar quenched in a calorimetric bomb at liquid nitrogen temperature shows the same XRD pattern but exhibits a secondary carbon layer on the expanded graphite as can be seen in the next picture (Fig. 15.7). This is strongly indicative of carbon transport through the gas phase which is a necessary prerequisite for the formation of nanocarbon particles. The picture on the right-hand side shows the carbon structures deposited on a MgF$_2$ sphere (Figure 15.7).

Raman spectroscopy of these samples shows the characteristic radial breathing mode (RBM) of single-walled carbon nanotube (SWCNT) [10] at 283 and 267 cm^{-1} together with the D and G bands at both 1592 and 1283 cm^{-1} (Figure 15.8). On the basis of a general relationship,

$$\omega_{RBM} \propto (1/d_t) \tag{15.6}$$

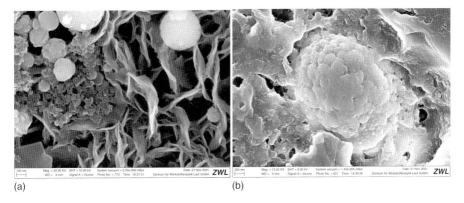

(a)   (b)

Figure 15.6   HREM of combustion residues of Mg/PTFE in 2 MPa argon (a) expanded graphite and magnesium fluoride and (b) polished microsection graph of MgF_2 sphere.

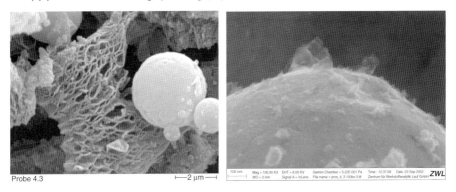

Figure 15.7   HREM of Mg/PMF combustion residues in 0.2 kPa argon.

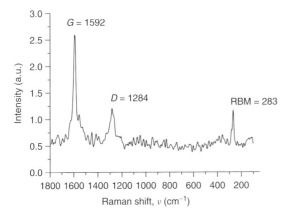

Figure 15.8   Raman spectrum of Mg/PMF combustion residues under 0.02 MPa argon, indicating the radial breathing mode characteristic for single walled carbon nanotubes (SWCNTs).

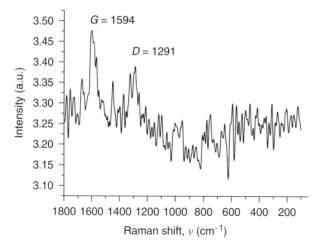

Figure 15.9 Raman spectrum of Mg/PMF combustion residues under 0.1 MPa argon, indicating just weak signals for both D and G band but no RBM for SWCNT.

with ω_{RBM} the Raman frequency of the RBM lines (cm^{-1}) and d_t the diameter of an SWCNT (nm), the diameter of the SWCNTs can be estimated to be in the order of <1 nm. Figure 15.9 shows the Raman spectrum of the reaction products obtained at 2 MPa argon pressure. Obviously the pressure influences the product distribution. Thus, at low pressures (0.02 MPa Ar), carbon transport occurs, whereas at higher pressures (2 MPa) the carbon transport is insufficient for the formation of SWCNTs.

Although no true combustion reaction, Chen et al. observed the formation of micron-sized carbon spheres on reaction of PTFE with liquid Mg in supercritical CO$_2$ at $T = 650\,^\circ$C [11]. The high exo-thermicity of the Mg/PTFE has been used to boost other low enthalpy solid-state reactions [12].

15.3
Silicon and Silicides

Cudziło et al. and Huczko et al. noted that silicon and many silicides such as CaSi$_2$ would yield nanofibrous mono-crystalline SiC on reaction with PTFE [13, 14]. Figure 15.10a,b shows the bulk of these fibres as obtained after purification steps. Figure 15.11a,b shows HRTEM of single SiC fibre with either stack or herring bone structure and amorphous SiO coating.

Cudziło reasoned that the growth of the SiC fibres proceeds via a transport mechanism using volatile Si and C species – probably SiF$_4$ and CF$_2$ and their transport via thermo-phoresis to a cooler decomposition and growing site to yield SiC and another volatile specie [15]. This is confirmed by the negative influence of increasing pressure on the yield reaction. In addition, it was also found that the yield of SiC nanofibres, though present in all Si containing combustion residues, is highest in CaSi$_2$/PTFE. This is understood to result from the superior heating rate

Figure 15.10 REM of SiC nanofibres obtained from CaSi$_2$/PTFE combustion. (Reproduced with kind permission by Prof. S. Cudziło.)

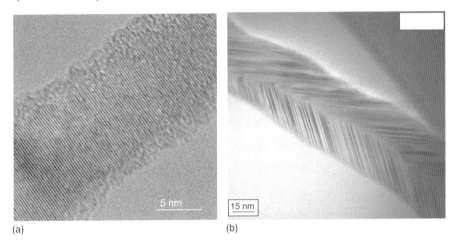

Figure 15.11 HRTEM of SiC nanofibres obtained from CaSi$_2$/PTFE combustion. (a) stack structure and (b) herring-bone structure. (Reproduced with kind permission by Prof. S. Cudziło.)

of these pyrolants that are in the order of 20 000 K s^{-1} compared to Si/PTFE that at best is only 3000 K s^{-1}. Thus, fast heating facilitates expulsion of gaseous species and serves kinetic rather than thermodynamic control [16]. Detailed analysis of the formation and structure of these fibres is reported in Refs. [17–23].

The use of poly-carbon monofluoride as an oxidizer with silicon and silicides yields SiC again, however, in significantly lower yields than that with PTFE [24, 25].

At higher Ar pressures, the author obtained spherical products, which according to EDX are constituted from a triple-phase material CaSi$_x$C$_y$. Figure 15.12 shows both open and fractured spheres and a close-in on the shell showing it being constituted from small spheres of 100–400 nm diameter.

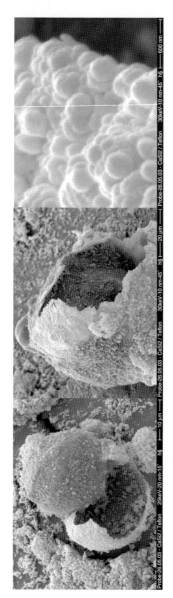

Figure 15.12 Scanning electron micrograph of CaSi$_2$/PTFE combustion residue with SiC broken hollow spheres.

Ultra-fine β-silicon carbide powders have been synthesized from mechano-activated ternary mixtures comprising Si, C and 5–6 wt% PTFE [26, 27] (Figure 15.13).

Cudziło and co-workers continue to do extensive investigations on the identity and morphology of combustion products of a great variety of metal/halocarbon reactants. Table 15.1 lists the systems investigated so far and the identified products.

 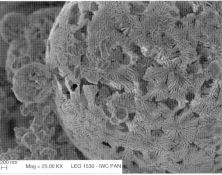

Figure 15.13 HREM of TiSi$_2$/PTFE combustion residues. (Reproduced with kind permission by Prof. S. Cudziło.)

Table 15.1 Pyrolants used for SHS purposes and obtained products.

Pyrolant	Crystalline products	References
Zn/PTFE	ZnF$_2$	[13]
Al$_4$Mg$_3$/PTFE	MgF$_2$, Al, AlF$_3$, Al$_4$C$_3$, C	[13, 28]
ZrTi/PTFE	ZrF$_4$, TiF$_3$	[13]
Cr/PTFE	CrF$_2$, CrF$_3$	[13]
B/PTFE	B$_4$C, B$_{13}$C$_2$, C	[13, 28]
Si/PTFE	Si, SiC, C	[13, 28]
FeSi/PTFE	Si, Fe$_3$Si, Fe$_5$Si, SiC, C	[13, 28]
CaSi$_2$/PTFE	Si, CaF$_2$, SiC, C	[13, 28]
TiSi/PTFE	TiC, Si	[13]
Mg/PMF	MgF$_2$, C	[5]
NaN$_3$/PMF	C, NaF	[25]
B/PMF	B$_4$C, C	[25]
Ti/PMF	TiC, TiF$_3$, C	[25]
ZrTi/PMF	TiC, ZrF$_4$, TiF$_3$, C	[25]
Si/PMF	Si, SiC, C	[25]
AlSi/PMF	SiC, AlF$_3$, Si,	[25]
CaSi$_2$/PMF	CaF$_2$, SiC, Si	[25]
TiSi/PMF	TiC, SiC, Si	[25]
MoSi$_2$/PMF	Mo$_3$C$_2$, Si, Mo$_{4.8}$Si$_3$C$_{0.8}$	[25]
Li$_2$C$_2$/PTFE	Multi-walled carbon nanotube (MWCNT)	[8] (R.B. Kaner, personal communication)

Possibly by a similar mechanism, the reaction of $TiSi_2$ with PTFE yields dendritic TiC spheres indicative of rapid growth of TiC via the deposition of a volatile specie such as TiF_4 or TiF_3. Again, the carbide precipitates at temperatures below $3100\,^{\circ}C$ and, thus, may also provide a means of shifting the reaction towards TiC instead to the thermodynamically more stable TiF_4.

SHS using fluorocarbons as either primary reactants or promoters will continue to play an important role in the synthesis of new carbon-based materials.

References

1. Merzhanov, A.G. (1994) Pyrotechnical aspects of self-propagating high-temperature synthesis. 20th International Pyrotechnics Seminar, Colorado Springs, CO, July 25–29, PL-1.
2. Borisov, A.A., DeLuca, L. and Merzhanov, A.G. (2002) *Self-Propagating High Temperature Synthesis of Materials*, Taylor and Francis, New York.
3. Koch, E.-C. (1999) Pyrotechnischer Satz zur Erzeugung von IR-Strahlung. DE Patent 19,964,172, Germany.
4. Koch, E.-C. (2000) Poly(kohlenstoffmonofluorid) (PMF) ein extrem energiereiches Oxidationsmittel. 10. Vortragstagung Fachgruppe Festkörperchemie und Materialforschung – Anorganische Funktionsmaterialien, Münster, Germany, September 26–29, p. A62.
5. Koch, E.-C. (2001) Fluoreliminierung aus Graphitfluorid mit Magnesium. *Z. Naturforsch.*, **56b**, 512–516.
6. Manning, T.J., Mitchell, M., Stach, J. and Vickers, T. (1999) Synthesis of exfoliated graphite from fluorinated graphite using an atmospheric-pressure argon plasma. *Carbon*, **37**, 1159–1164.
7. Manning, T.J., Noel, A., Mitchell, M., Miller, A., Grow, W., Gaddy, G., Riddle, K., Taylor, K. and Stach, J. (2000) in *Science and Application of Nanotubes* (eds D. Tománek and R.J. Enbody), Kluwer Academic/Plenum Publishers, New York, pp. 205–213.
8. O'Loughlin, J.L., Kiang, C.-H., Wallace, C.H., Reynolds, T.K. and Kaner, R.B. (2001) Rapid synthesis of carbon nanotubes by solid-state metathesis reactions. *J. Phys. Chem. B*, **105**, 1921–1924.
9. Mack, J.J., Tari, S. and Kaner, R.B. (2006) Enhanced solid-state metathesis routes to carbon nanotubes. *Inorg. Chem.*, **45**, 4243–4246.
10. Saito, R. and Kataura, H. (2001) in *Carbon Nanotubes, Synthesis, Structure, Properties, and Applications* (eds M.S. Dresselhaus, G. Dresselhaus and P. Avouris), Springer-Verlag, Berlin, p. 213.
11. Wang, Q., Cao, F. and Chen, Q. (2005) Formation of carbon micro-sphere chains by defluorination of PTFE in a magnesium and supercritical carbon dioxide system. *Green Chem.*, **7**, 733–736.
12. Ayral, R.M., Rouessac, F. and Massoni, N. (2009) Combustion synthesis of silicon carbide assisted by a magnesium plus polytetrafluoroethylene mixture. *Mater. Res. Bull.*, **44**, 2134–2138.
13. Huczko, A., Lange, H., Chojecki, G., Cudziło, S., Zhu, Y.Q., Kroto, H.W. and Walton, D.R.M. (2003) Synthesis of novel nanostructures by metal-polytetrafluoroethylene thermolysis. *J. Phys. Chem. B*, **107**, 2519–2524.
14. Cudziło, S., Gachet, S., Huczko, A., Monthioux, M. and Trzciński, W.A. (2003) Synthesis of ceramic and carbon nanostructures by self-sustaining combustion of mixtures of halogenated hydrocarbons with reducers. VI. Seminar New Trends in Research of Energetic Materials, Pardubice, Czech Republic, April 22–24, p. 69.
15. Cudziło, S. (2005) Formation of carbon based nanostructures by combustion of reductant-halocarbons mixtures. 2nd Workshop on Pyrotechnic Combustion Mechanisms, Karlsruhe, Germany, June 27.

16. Cudziło, S., Huczko, A., Lange, H., Panas, A.J. and Trzcinski, W.A. (2003) Self-propagating synthesis of ceramics in B/PTFE and Si/PTFE compositions. EUROPYRO, St. Malo, France, June 2003, p. 547.

17. Huczko, A., Lange, H., Bystrzejewski, M., Rummeli, M.H., Gemming, T. and Cudziło, S. (2005) Studies on spontaneous formation of 1D nanocrystals of silicon carbide. *Cryst. Res. Technol.*, **40**, 334–339.

18. Huczko, A., Bystrzejewski, M., Lange, H., Fabianowski, A., Cudziło, S., Panas, A. and Szala, M. (2005) Combustion synthesis as a novel method for production of 1-D SiC nanostructures. *J. Phys. Chem. B*, **109**, 16244–16251.

19. Huczko, A., Lange, H., Bystrzejewski, M., Rotkowska, A., Cudziło, S., Szala, M. and Wee, A.T.S. (2006) Quasi one-dimensional ceramic nanostructures spontaneously formed by combustion synthesis. *Phys. Status Solidi*, **243**, 3297–3300.

20. Trcziński, W.A., Cudziło, S., Szala, M. and Gut, Z. (2007) Investigation of the combustion of calcium silicide/polytetrafluoroethylene mixtures. *Arch. Combust.*, **27**, 69–79.

21. Huczko, A., Osica, M., Rutkowska, A., Bystrzejewski, M., Lange, H. and Cudziło, S. (2007) A self assembly SHS approach to form silicon carbide nanofibres. *J. Phys. Condens. Matter*, **19**, 395022, (10pp).

22. Busiakiewicz, A., Klusek, Z., Huczko, A., Kowalczyk, P.J., Dąbrowski, P., Kozłowski, W., Cudziło, S., Datta, P.K. and Olejniczak, W. (2008) Scanning tunneling microscopy investigations of silicon carbide nanowires. *Appl. Surf. Sci.*, **254**, 4268–4272.

23. Busiakiewicz, A., Huczko, A., Lange, H., Kowalczyk, P.J., Rogala, M., Kozłowski, W., Klusek, Z., Olejniczak, W., Polański, K., Cudziło, S. and Datta, P.K. (2008) Silicon carbide nanowires: chemical characterization and morphology investigations. *Phys. Status Solidi B*, **245**, 2094–2097.

24. Cudziło, S., Szala, M., Huczko, A. and Bystrzejewski, M. (2006) Self-sustaining reductive defluorination of (CF)n and characterization of the reaction products. IX Seminar New Trends in Research of Energetic Materials, Pardubice, Czech Republic, April 19–21, p. 544.

25. Cudziło, S., Szala, M., Huczko, A. and Bystrzejewski, M. (2007) Combustion reactions of poly(carbon monofluoride), $(CF)_n$, with different reductants and characterization of the products. *Propellants Explos. Pyrotech.*, **32**, 149–154.

26. Yang, K., Yang, Y., Lin, Z.-M., Li, J.-T. and Du, J.S. (2007) Mechanical-activation-assisted combustion synthesis of SiC powders with polytetrafluoroethylene as promoter. *Mater. Res. Bull.*, **42**, 1625–1632.

27. Liu, Y.-Q., Zhang, L.-F., Yan, Q.-Z., Mao, X.-D., Feng, Q. and Ge, C.-C. (2009) Preparation of β-SiC by combustion synthesis in a large-scale reactor. *Int. J. Miner. Metall. Mater.*, **16**, 322–326.

28. Cudziło, S., Bystrzejewski, M., Lange, H. and Huczko, A. (2005) Spontaneous formation of carbon-based nanostructures by thermolysis-induced carbonization of halocarbons. *Carbon*, **43**, 1778–1782.

16
Vapour-Deposited Materials

For the first time, in 1990, Allford and Place reported about the feasibility of producing pyrolants from vapour phase-deposited materials. In addition to thermite-type systems such as Ti/Pb$_3$O$_4$, they also reported about the preparation of composite metal/fluorocarbon films produced by vapour phase deposition of Mg on polytetrafluoroethylene (PTFE) tape [1, 2]. The product obtained this way is a silvery tough but flexible tape. A schematic setup of the apparatus is given in Figure 16.1.

The tape, which is typically between 40 and 70 μm thick, obtains coatings on both sides, which range between 2 and 20 μm. As seen from the apparatus, a number of process parameters can be adjusted. These include the feed velocity of Mg, which influences the concentration and temperature of the Mg vapour, the coil speed, the angle of the depositing arrays θ and the pressure level in the apparatus, all of which influence the thickness and grain quality of the deposited Mg. Figure 16.2a depicts the cross section of a typical material obtained with $\theta = 45°$ at a pressure of ~5 × 10^{-2} Pa. The top view of the material is shown in Figure 16.2b. This material is often referred to as '*Firesheet*'. A material produced with lower Mg feed rate, lower pressure $p = 4 \times 10^{-3}$ Pa and different array angle $\theta = 15°$ yields an Mg deposit with a much finer grain structure as shown in Figure 16.2c. The latter material is often referred to as '*Pyrosheet*' [3]. Figure 16.2 shows the electron micrograph of a typical sample produced, after Ref. [2].

The properties of the materials are compared in Table 16.1.

In an independent work, the combustion behaviour of PTFE film Mg coated with either evaporation or sputtering technique has been compared [4]. The Mg: PTFE ratio was stoichiometric with PTFE substrate of 36 μm thickness and two 14 μm layers Mg each on both sides. The evaporation technique produces a granular structure similar to that shown in Figure 16.2b. The sputtering technique, however, yielded fine granules with a structure like in Figure 16.2c. The linear burn rates for laterally ignited samples varied between 444 and 583 mm s^{-1} for the sputtered material and between 297 and 304 mm s^{-1} for the evaporated material. Both Firesheet and Pyrosheet showed significant ageing on exposure to air. Thus these materials, despite their performance, were not further considered. However, a license agreement with then Imperial Chemical Industries (ICI) led to the development of Enerfoil®, a product made in principally the same way as Firesheet but additionally protected by a thin layer of aluminium (Figure 16.3) [5]. Enerfoil®

Metal-Fluorocarbon Based Energetic Materials, First Edition. Ernst-Christian Koch.
© 2012 Wiley-VCH Verlag GmbH & Co. KGaA. Published 2012 by Wiley-VCH Verlag GmbH & Co. KGaA.

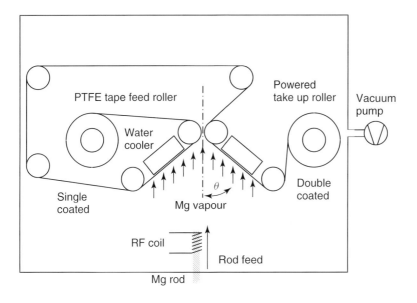

Figure 16.1 Schematic setup of double coating machine. (After Ref. [2].)

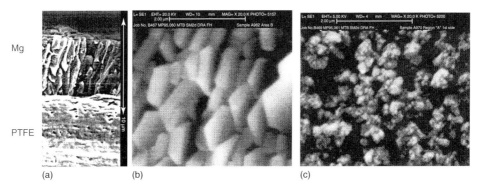

Figure 16.2 (a–c) Electron micrograph of vapour phase-deposited Mg on PTFE. (After Refs. [2, 3]. Reproduced with kind permission of the International Pyrotechnics Society USA Inc.)

Table 16.1 Material properties of Firesheet versus Pyrosheet.

Material	$d_{Mg}(\mu m)$	t_i (s)	$T_i(°C)$	u (cm s^{-1})	Q (kJ g^{-1})	σ_{488} nm (%)	E_{laser} (mJ))	UTS (MPa)
Firesheet	16.0	39.9	559	12	–	62.0	>2000	25
Pyrosheet	17.4	32.4	529	25	6760	12.7	4.5	–
Pyrosheet	6.5	32.2	537	29	3695			

was further modified to have a porous PTFE substrate that would exhibit a greater surface area [6]. In addition, texturing and partial punctuating of the surface would improve flame spreading [7]. Final modifications of Enerfoil® included coatings from common gas-generating materials such as sodium azide or nitrocellulose [8].

Being an anisotropic material, Enerfoil® shows a combustion rate strongly dependent on the respective geometrical configuration.

When stacked as shown in Figure 16.4a, it burns with a vertical propagation velocity of the order of $0-4-4$ cm s^{-1}. A similar stack ignited laterally (b) will burn at a speed varying between 2 and 100 cm s^{-1}, depending largely on the gap between the films. Finally, Enerfoil® as any other pyrolant material will show directed burning under confined conditions (c) and undergo fast deflagration in the order of 1000 m s^{-1}.

The particular flame-spreading behaviour of Enerfoil® has been investigated by Yeh *et al.* [10]. They found that for two aligned films, the maximum flame spread velocity would be obtained with a gap distance of $\sim$100 μm (Figure 16.5). Surprisingly, the flame-spread velocity drops with increasing pressure (Figure 16.6).

The influence of both pressure and type of atmosphere on flame-spreading rate in a two-film experiment as a function of gap distance is displayed in Figure 16.5.

The ignition performance of Enerfoil® in air bag igniters is discussed in Ref. [11].

The hazard data for 'Enerfoil®,' are given in Table 16.2. From this, it becomes obvious that these materials are actually less sensitive than the commonly produced granular Magnesium/Teflon/Viton (MTV) materials (see chapter 19).

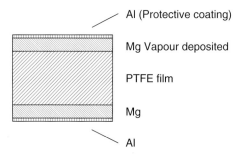

Al (Protective coating)

Mg Vapour deposited

PTFE film

Mg

Al

Figure 16.3 Principal structure of Enerfoil® film. (After Ref. [9].)

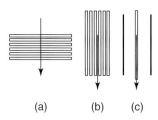

(a) (b) (c)

Figure 16.4 (a–c) Combustion modes of Enerfoil® in stacks and in tubes.

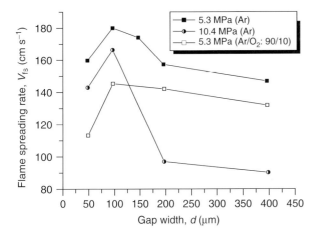

Figure 16.5 Flame spreading velocity as a function of gap width in different atmospheres [10].

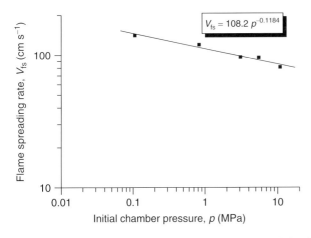

Figure 16.6 Flame spreading velocity as a function of initial chamber pressure [10].

Table 16.2 Material safety data on Enerfoil®.

Test	Value
BAM impact (J)	16
BAM friction (N)	> 360
ESD (J)	20
Thermal onset in DSC (°C)	> 450
Time to ignition (s)	30

Abbreviation: BAM, Bundesanstalt für Materialprüfung.

An intended application, although never commercially realized, was in fire transfer lines referred to as '*Flashtrack*' [12]. In a particular example, three 45 μm thick PTFE tapes with a coarse 16 μm thick Mg coating on both sides would be cut into 2 mm wide strips and were confined in a 4 mm inner diameter Viton® tube. After heat shrinking of the tube, it was read for use. The loading density was 1 g pyrolant per meter. When the tube is ignited, the pyrolant deflagrates with a velocity of the order of 100 m s^{-1} but leaves the tube intact.

It must be noted that Enerfoil™, even though superior in performance to other common igniter materials, did not become a commercial success. This is because the Airbag manufacturing industry would just stay with the old igniters (S.K. Chan, personal communication). A similar story recently happened with the igniters based on nanoporous silicon [13]. It is quite often experienced that new technologies merely serve to put producers of common technology under pressure to reduce their prices. It is questionable if such behaviour of the buyers in companies pays back on the long term.

References

1. Allford, F.G. and Place, M.S. (1990) The development and production of pyrotechnic systems from vapour phase – part one: thermite type materials. 15th International Pyrotechnics Seminar, Boulder, CO, July 9–14, p. 1.

2. Allford, F.G. and Place, M.S. (1990) The development and production of pyrotechnic systems from vapour phase – part two: the production of bulk reactive material. 15th International Pyrotechnics Seminar, Boulder, CO, July 9–14, p. 17.

3. Powell, G., Place, M., Leach, C., James, R. and Hall, J. (1997) Developments in vapour phase deposited pyrotechnic materials. 28th International Annual ICT Conference, Karlsruhe, Germany, June 24–27, p. 63.

4. Guoxin, A.L. and Yunliang, L. (2005) in *Producing Mg Film by Magnetron Sputtering and Analysis, Theory and Practice of Energetic Materials*, vol. **VI** (eds Z. Wang and P.L.S. Huang), Science Publisher, Beijing, p. 73.

5. Graham, S.J., Leiper, G.A. and Bishop, C.A. (1993) Pyrotechnic sheet material. EP Application 0,584,921 A2, United Kingdom.

6. Graham, S.J., Leiper, G.A. and Chan, S.K. (1993) Pyrotechnic sheet material. EP Application 0,584,922 A2, United Kingdom.

7. Chan, S.K., Graham, S.J., Kirby, I.J. and Leiper, G.A. (1994) Pyrotechnic material. EP Specification 0,645,354 B1, United Kingdom.

8. Chan, S.K., Graham, S.J. and Leiper, G.A. (1995) Pyrotechnic sheet material. EP Application 0,710,637 A1, United Kingdom.

9. Chan, S.K. (1997) Characterization of enerfoil®™ pyrotechnic film. Terminal Effects and Energetic Materials Advisory Committee (TEEMAC), Shrivenham, UK, November 4, p. 39.

10. Yeh, C.L., Mench, M.M. and Kuo, K.K. (1997) An investigation on flame spreading process of thin film Mg/PTFE/Mg pyrotechnics. *Combust. Sci. Technol.*, **126**, 271–289.

11. Chan, S.K., Graham, S.J. and Leiper, G.A. (1995) Enerfoil™ ignition film as an ignition charge. 26th International Annual Conference of ICT, Karlsruhe, Germany, July 4–7, p. 58.

12. Allford, F.G. (1993) Pyrotechnic material. US Patent 5,253,584, (The Secretary of State for defence in Her Britannic

Majesty's Government of the United Kingdom of Great Britain and Northern Ireland).

13. Koch, E.-C. and Clement, D.T. (2007) Silicon – an old fuel with new perspectives. *Propellants Explos. Pyrotech.*, **32**, 205–212.

17
Ageing

Pyrolant compositions based on finely divided Mg particles and fluorocarbons are susceptible to chemical degradation. This degradation also termed *ageing* is known to have at least two main causes: reaction of magnesium with moisture and an inherent reactivity between Mg and its basic compounds with fluoroelastomer binders prone to eliminate HF, such as Viton® A or polyvinylidene fluoride (PVDF).

Magnesium is reactive with water and generates hydrogen according to the following equation:

$$Mg + 2\,H_2O \longrightarrow Mg(OH)_2 + H_2 \uparrow \tag{17.1}$$

This assumption is verified by measurements on MTV samples aged in a humid atmosphere that show mass increase by up to 190% of the initial mass, which equals complete conversion of Mg to $Mg(OH)_2$ (Figure 17.1). Moreover, chemical analysis, both X-ray diffraction (XRD) and FTIR confirm the presence of $Mg(OH)_2$.

The increase in volume shown in Figure 17.1 is caused by the separation of a new filamentous phase made of $Mg(OH)_2$ (Figure 17.2). This can lead to serious degradation of flares made from MTV, as is shown in Figure 17.3 [2].

The heat evolution of a particular MTV (65/30/5) exposed to hot and humid storage has been investigated in Ref. [1]. It was found that on exposure, the measurable heat production initially drops a little, but after a while, the heat production increases linearly with time (Figure 17.4).

The heat production eventually slows down to finally decrease again linearly as seen in the overall plot in Figure 17.5.

Based on the simple triangular shape of the heat production curve (Figure 17.5), a model has been derived that is schematically depicted in Figure 17.6 [1].

It has been found that the temperature influences the time t_1 and the slope of the curve α_1, which is depicted in Figure 17.7.

For 80 °C and 20% relative humidity (RH), the following parameters have been derived to describe the degradation of MTV.

$$\alpha_1 = 8.0 \cdot 10^{-16} \cdot e^{(0.2197 \cdot T)} \cdot RH^2 (h^{-2}), \xi_{t2} = 0.15$$

Metal-Fluorocarbon Based Energetic Materials, First Edition. Ernst-Christian Koch.
© 2012 Wiley-VCH Verlag GmbH & Co. KGaA. Published 2012 by Wiley-VCH Verlag GmbH & Co. KGaA.

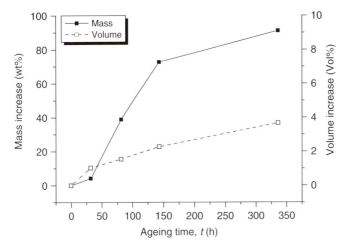

Figure 17.1 Increase in mass and volume of MTV (65/30/5) pellets upon exposure to 80 °C and 80% RH (~240 g H$_2$O m^{-3}). (After Ref. [1].)

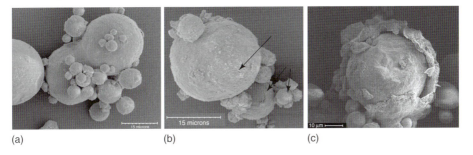

(a) (b) (c)

Figure 17.2 Electronmicrographs of Mg coated with 1–2 wt% Viton before (a) and after ageing at 80 °C 50% RH for 12 weeks (b) and 19 weeks (c). The arrows indicate cracks in the VitonTM coating on the Mg particle. (From Refs. [3, 4] Reproduced with kind permission of Dr. N. Davies.)

Figure 17.3 Aged M 206 flares with serious expansion of flare pellet [2]. (Reproduced with kind permission of IPSUSA.)

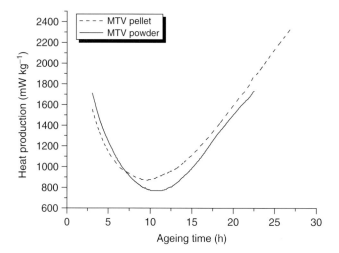

Figure 17.4 Heat production of MTV (65/30/5) at 80 °C and 20% RH (15 g m^{-3}).

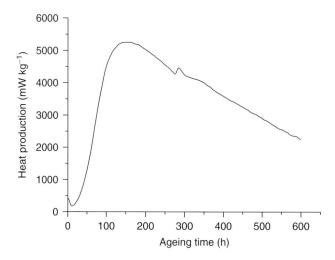

Figure 17.5 Heat production of MTV (65/30/5) at 80 °C and 20% RH (15 g m^{-3}).

with

$$\alpha_2 = -\alpha_1 \cdot \frac{\xi_{t2}}{1 - \xi_{t2}} (h^{-2})$$

$$t_2 = t_1 + \sqrt{\frac{\xi_{t2}}{0.5 \cdot \alpha_1}} (h)$$

$$t_e = t_2 + (t_2 - t_1) \cdot \frac{1 - \xi_{t2}}{\xi_{t2}} (h)$$

The main uncertainty of the model is the parameter ξ_{t_2}, which is probably dependent on the surface area and initial grade of oxidation of the Mg particles. At 80 °C and

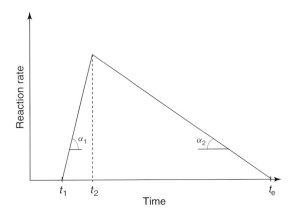

Figure 17.6 Schematic representation of the modelled reaction rate. In this graph, the following abbreviations apply. t_1 [h] = time at which the heat production starts to increase, t_2 [h] = time at which the heat production starts to decrease and t_e [h] = time at which all Mg in the MTV is converted. α_1 [h^{-2}] = slope of the curve of the reaction rate between t_1 and t_2. α_2 [h^{-2}] = slope of the curve at the reaction rate between t_2 and t_e.

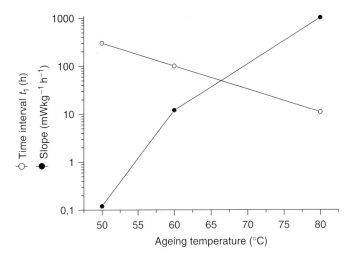

Figure 17.7 Influence of temperature on start time t_1 and slope of heat production.

20% RH, it equals 0.15. In this investigation, the Mg used was of an irregular type, which was scraped or rasped.

Griffiths *et al.* have investigated ageing of a variety of MTV formulations at 70 °C and 66% RH [5]. They studied compositions containing 5 wt% Tecnoflon® and variable amounts of both Mg and polytetrafluoroethylene (PTFE). The heat flow increases with the Mg content as shown in Figure 17.8. MTV (55/40/5) showed an increasing heat flow for 32 days, when it eventually reached a maximum of 2026 mW kg^{-1}. Thermal analysis of the composition showed the temperature of

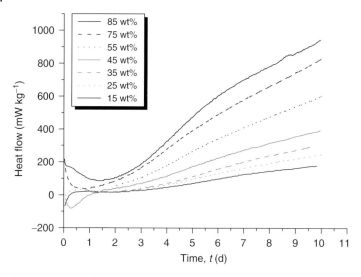

Figure 17.8 Heat flow curves for different MTV compositions, 200 mg sample mass each at 70 °C and 66% RH.

onset of ignition to rise from 550 to 623 °C for the aged composition. Griffiths *et al.* also noted that a sample of Mg coated with 5 wt% Tecnoflon would produce more heat than a sample coated with just 2.5 wt% Tecnoflon. Thus they reasoned a particular cause behind this.

Davies and Koch have emphasized on the potential incompatibility of Mg with binders prone to eliminate HF [6, 7]. Davies found that Viton-type binders accelerate corrosion of both Mg and alloys with aluminium under warm and humid storage [3, 4]. The degradation of MTV with different binder levels at various levels of humidity and temperature has been studied by Koc *et al.* [8].

Hancox has reported about the degradation in dynamic performance of both pressed and extruded MTV flares [9]. Figure 17.9 depicts radiant intensity of both control and aged MTV stored six months under B2 conditions at 350 kts in specific operational bandpass range. B2 storage is basically a sine wave pattern of reoccurring high and low temperatures (30–60 °C) and high and low relative humidities (20–65%), where high temperature coincides with low RH and the phase duration is about 12 h.

Degradation models for terminal growth of M206 MTV flare pellets in operational cartridges have been developed by Bixon *et al.* [2, 10].

In order to counter degradation of magnesium, extensive studies have been conducted with organic titanates, microcrystalline wax and polymer coatings [11]. Apart from ineffective titanates and wax, it was found that copolymers of ethylene/vinyl acetate – trade name (Elvax®) – coat Mg effectively and significantly reduce the hydrogen production under humid storage. Elvax-coated Mg used in MTV formulations yields comparable performance as standard binders using either Hycar® and Viton binder. It was also revealed that Elvax binder-coated Mg would lend the final grain a higher mechanical stability. Table 17.1 compares burn rate, spectral

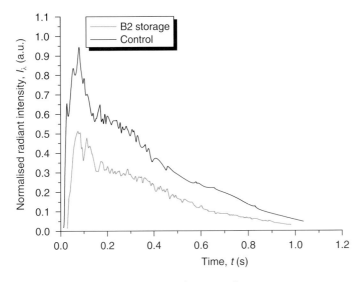

Figure 17.9 Dynamic radiometric performance of MTV at 350 kts for control, and aged sample stored six months under B2 climatic conditions. (After Ref. [9].)

Table 17.1 Burn rate, spectral efficiency and tensile strength of various Mg/PTFE formulations.

Composition	Burn rate (mm s^{-1})	Spectral efficiency (%)	Tensile strength (MPa)
Mg/PTFE/9 wt% Viton®	2.286	100	3.95
Mg/PTFE/5 wt% Hycar®	2.413	116	3.68
Mg/PTFE/5 wt% Elvax®	3.632	118	7.74

After Ref. [11].

efficiency and tensile strength for two baseline formulations and one advantageous formulation using Elvax binder.

Similar improvements have been disclosed in a patent [12]. Another attempt to protect Mg consisted in coating it with an amorphous aluminium phosphate layer commercially traded as Cerablak® [13]. Although the hydrogen production could be reduced, it was not totally satisfactory as it adversely affected the energy density of the final flare materials, affected ignitability and reduced burn rates of the formulations using non-irregular-scraped powders [14].

Current trends in protection of MTV against degradation comprise the investigation of siliane-based coatings [15] and the use of advanced Mg alloys less prone to undergo oxidation in hot and humid storage and less affected by HF-eliminating binders (J. Callaway, personal communication).

References

1. van Driel, C., Leenders, J. and Meulenbrugge, J. (1995) Ageing of MTV. 26th International Annual ICT Conference, July 4–7, Karlsruhe, Germany, p. V–31.

2. Bixon, E., Broad, R., DeSalvio, J., Gagliardi, F., Nagori, A., Poret, J. and Zimmer, A. (2008) Accelerated aging of the M206 countermeasure IR flare. 35th International Pyrotechnics Seminar, July 13–18, Fort Collins, CO, p. 449.

3. Callaway, J. and Davies, N. (2008) Ageing of magnesium coated with Viton. 35th International Pyrotechnics Seminar, July 13–18, Fort Collins, CO, p. 39.

4. Davies, N., Smith, P. and Callaway, J. (2009) Coated magnesium powder for pyrotechnic decoy flares for the protection of aircraft. Nano-Scale Energetic Materials: Fabrication, Characterization and Molecular Modeling, Strasbourg, France, June 8–12.

5. Claridge, R.P., Griffiths, T.T., Charsley, E.L., Goodall, S.J. and Rooney, J.J. (2006) Degradation studies on Infrared countermeasure compositions. 33rd International Pyrotechnics Seminar, Fort Collins, CO, July 16–21, p. 653.

6. Davies, N. (1999) Binders for pyrotechnic compositions. 26th International Pyrotechnics Seminar, Nanjing, China, October 1–4, p. 98.

7. Koch, E.-C. (2001) Wege zu Unempfindlicher Pyrotechnischer Munition, WIWEB Explosivstoff Forum, Unempfindliche Sprengstoffe, Swisttal-Heimerzheim, Germany, April 5.

8. Koc, S., Erogul, F. and Tinaztepe, H.T. (2009) Accelerated aging study for MTV igniter charges. 45th AIAA/ASME/SAE/ASEE Joint Propulsion Conference & Exhibit, Denver, CO, August 2–5, AIAA 2009-5276.

9. Hancox, R.J. (1994) Degradation studies on MTV flares. TTCP WTP-4 Workshop on Degradation of Pyrotechnic Fuels, Denver, CO, Paper Q.

10. Bixon, E., Broad, R., DeSalvio, J., Gagliardi, F., Nagori, A., Poret, J. and Zimmer, A. (2009) Estimation of the shelf life of the M206 flare. 36th International Pyrotechnics Seminar, Rotterdam, Netherlands, August 23–28, p. 87.

11. Taylor, F.R. and Jackson, D.E. (1987) Organic Coatings to Improve the Storageability and Safety of Pyrotechnic Compositions. Technical Report ARAED-TR-87022, Picatinny Arsenal, NJ.

12. Chen, G., Broad, R., Valentine, R.W. and Mannix, G.S. (2001) Magnesium-fueled pyrotechnic compositions and processes based on Elvax-Cyclohexane coating technology. US Patent 6,174,391, USA.

13. Gudgel, T.J., Chapman, F., Sambasivan, S., Gillard, T., Wilharm, C. and Douda, B. (2008) Inorganic barrier coating for the protection of magnesium powder against humidity-based aging. 35th International Pyrotechnics Seminar, Fort Collins, CO, July 13–18, p. 25.

14. Wilharm, C. (2008) Combustion performance of coated magnesium. 35th International Pyrotechnics Seminar, Fort Collins, CO, July 13–18, p. 31.

15. Fotea, C., Morgan, A., Smith, P. and Callaway, J. (2008) Surface modification of magnesium particulates with silanes presented as vapour: inhibition of atmospheric corrosion. J. Adhes., 84, 389–400.

18
Manufacture

18.1
Introduction

The salient properties of pyrotechnic compositions – performance and safety – very much depend on the processing of the ingredients. That is, pretreatment of the raw materials, mixing, binding, drying or curing, loading and pressing, machining, extrusion or casting all have a strong influence on the properties of the final pyrolant grain. Unless otherwise stated, the procedures discussed in the following sections apply for Magnesium/Teflon™/Viton™ (MTV) or modifications with different binders.

A typical flow chart for an MTV decoy flare production sequence is depicted in Figure 18.1. The production process starts with weighing out the appropriate amounts of chemicals for a batch, including the solvents. After mixing, the major amount of the composition is dried and consolidated to planks or final pellet geometry, whereas a small amount remains as pasty material to serve for priming the pellets after machining in necessary holes and grooves to facilitate spread of ignition. After drying, the primed grains, the pellets, are wrapped in self-adhesive aluminium foil and assembled with safety devices and felt pads in the cartridge that already bears the electrical squib. The cartridge is sealed with elastomer, packed and ready for transport to customer.

18.2
Treatment of Metal Powder

As discussed in the preceding chapter magnesium powder in pyrolants may deteriorate because of the combined action of oxygen and moisture trapped in the composition or diffusion of moisture into ammunition casings. In other cases, certain components may even prove incompatible with Mg. To prevent this from happening, either other fuels are substituted for Mg, such as Mg_4Al_3 alloy, or Mg powder is treated to obtain a protective coating that prevents it from degradation in a typical store's lifetime scale. State-of-the-art coating agents comprise ethylene-vinyl acetate copolymers such as Elvax®-40W and Elvax®-150W [1]. These are typically

Metal-Fluorocarbon Based Energetic Materials, First Edition. Ernst-Christian Koch.
© 2012 Wiley-VCH Verlag GmbH & Co. KGaA. Published 2012 by Wiley-VCH Verlag GmbH & Co. KGaA.

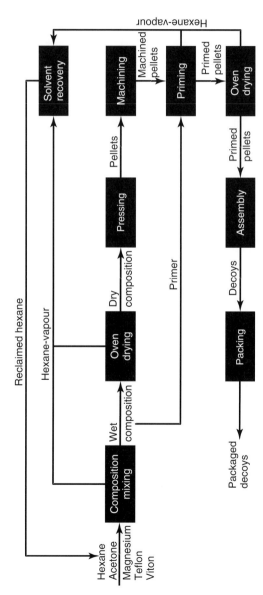

Figure 18.1 Flow chart of MTV flare production line.

applied by dispensing Mg powder in cyclohexane solutions of the corresponding polymer and successive vapourization of the solvent. The coated powder – having 5 wt% coating agent – is a free flowing material.

Even though inacceptable from an environmental and occupational health point of view, dichromate coatings have been applied extensively in the past to protect Mg effectively against corrosion [2].

18.3
Mixing

18.3.1
Shock Gel Process

The most common process to manufacture MTV is the so-called shock gel process also called 'coacervation' process. According to Douda, it had been first used in 1959 [3]. It is an adaptation of the well-known coagulation process of nitrocellulose binder practiced since the 1920s to vinylidenefluoride-hexafluoropropene-copolymer (70 : 30 mol%) binder (marketed as either Viton™ or Fluorel™) [4].

In the shock gel process, the fluoroelastomer binder is dissolved in any low-boiling ketone-type solvent such as acetone or methylethylketone and mixed subsequently with both magnesium and Teflon. After mixing, the polymer is precipitated out of solution by adding a non-solvent such as a low-boiling, non-polar hydrocarbon (n-hexane, n-heptane) or supercritical CO_2. In an effort to reduce Volatile Organic Carbon (VOC) emissions hexane is reclaimed from the exhaust lines of both mixing and drying operations as is shown in Figure 18.1.

18.3.1.1 Procedure A
The binder is dissolved in acetone to give a viscous solution with 8–20 wt%. Therefore, Viton™ or Fluorel™ slabs are chopped down to pieces ~1 × 1 × 1 cm and stirred with acetone overnight in a Cowles-type blender at ambient temperature. Dry magnesium and polytetrafluoroethylene (PTFE) powder are then added to the solution to yield a slurry. The slurry volume can be increased by adding a little n-hexane. However, the hexane : acetone ratio at this point must not exceed 1 : 5 as premature precipitation of Viton will occur [5]. After homogenization, a twofold volume of n-hexane is then added to the blender, causing immediate precipitation of the Viton® to coacervate both Mg and PTFE particles. A typical composition manufactured by this process is given below:

Composition according to MIL-Spec. MIL-I-82736A(OS) [15]

52.5–54.5 wt% magnesium
29.5–32.0 wt% PTFE
15.0–16.5 wt% Viton A or Fluorel 2125™

The composition is then washed several times with *n*-hexane or any other non-polar solvent to extract all the acetone. The composition is obtained as grey coarse grains. The material is transferred either manually in stainless steel trays (Figure 18.2) and dried in an oven as depicted in Figure 18.3 or transported in pneumatic feeding lines under inert gas as depicted in Figure 18.4.

In the shock gel process, for a 45 kg batch per kg MTV, ~0.5 l acetone and ~3 l hexane are necessary [7].

Figure 18.5 shows MTV granules obtained by conventional mixing after drying.

Figure 18.2 Manual transfer of MTV composition 'crumb' in a vacuum oven at Space Ordnance Systems, Canyon, CA, USA in 1980s. Safety note: standard protective clothing has changed since. Operations require grounding of the operator and Nomex-type overall with balaclava and additional faceshield.

Figure 18.3 Vacuum oven with hot water supply (see back) and vacuum line (see top part).

Figure 18.4 Instrumented blender with automatic feed lines, vacuum system and weighing (front lower part) of finalized composition.

Figure 18.5 MTV granules prepared by conventional mixing in conductive rubber beaker.

The dried granules are transferred either manually or automatically to the pressing section.

18.3.1.2 Procedure B

According to Diewald and Bickel [8], magnesium powder and PTFE powder and an equal mass of acetone are mixed and worked up in a colloid mill [9] filled with acetone. The homogenized dispersion is then pressure filtered at 50 kPa to leave a solvent-moist Mg/PTFE filter cake. The filter cake is then transferred to a planetary mixer. To this is then added a solution of the necessary amount of Viton binder in acetone. After mixing and forced evaporation of a part of the solvent, the mixer is switched off and about 15 wt% of the dry weight composition of n-heptane are added. After a few minutes of agitation, the composition attains a sand-like texture. The solvent is evaporated by blowing with air. Now the composition is granulated and dried in a vacuum oven. Typical compositions produced according to this process are shown below

Composition 1	Composition 2
58 wt% magnesium, <50 μm	58 wt% magnesium, <28 μm
40 wt% polytetrafluoroethylene	38 wt% polytetrafluoroethylene
2 wt% fluoroelastomer	4 wt% fluoroelastomer

The use of very fine Mg powder in composition 2 requires a higher percentage of fluoroelastomer binder to assure complete coating.

18.3.2
Conventional Mixing

Certain binders such as polyacrylates may not be processed by the shock gel technique. Thus the binder is dissolved in acetone, and successively, both Mg and PTFE powder are added to the mix. Blender equipped with a heating jacket as the one shown in Figure 18.7 is then used to drive off the solvent. The putty-like material obtained is spread on stainless steel trays and dried in way similar to the one described above. However, this material may form larger agglomerates that have to be either separated and/or comminuted in a granulator before further use. Material obtained after deagglomeration is depicted in Figure 18.5.

18.3.3
Experimental Super Shock Gel Process

An experimental shock gel process using supercritical CO_2 was explored in the late 1990s by Nauflett and Farncomb in an effort to reduce solvent use, concomitant emissions of VOC and danger of fire [10].

In this variant, both PTFE and Mg are mixed with the solution of Viton in acetone in a special vessel. After homogenization, the premix is stored in the same vessel. The premix is then fed through a line into a pressure vessel. Supercritical CO_2 is now added to the premix. At first, the $scCO_2$ will dissolve in the acetone. As the solubility limit is exceeded, the Viton will precipitate out of the combined acetone $scCO_2$ solution. The CO_2 overpressure will then be released via another valve. More supercritical CO_2 is fed into the vessel to allow all acetone to be removed from the MTV crumb. This is tracked by a hydrocarbon detector (Figure 18.7). The vaporized emissions are separated in the cyclone separator to give pure CO_2 and acetone. As soon as no acetone is detected in the MTV crumb, it can be considered acetone free and the $scCO_2$ supply line is shut (Figure 18.8).

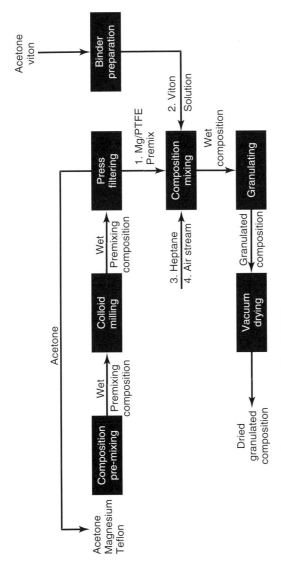

Figure 18.6 MTV production process. (After Ref. [8].)

Figure 18.7 Planetary blender with water-heated jacket.

Table 18.1 Process parameters for experimental MTV production with sc-CO_2.

Example	Batch size (g)	Acetone (weight percentage)	Acetone in CO_2 (weight percentage)	Yield (%)	Fluorel in MTV (weight percentage)	Fluorel in acetone (weight percentage)
1	200	41	60	96	12	>0.1
2	100	40	21	83	14–16	>0.1
3	100	40	41	92	16	>0.1

Viton A and Fluorel FC-2175 because of their difference in chemical nature display different solvation behaviours in scCO_2. Figure 18.9 depicts the cloud point temperature for FC-2175 dissolved in pure CO_2. Admixtures of acetone lower the necessary pressure significantly. In contrast, Viton A does not dissolve up to temperatures of 180 °C.

Table 18.1 shows the influence of acetone/scCO_2 ratio on yield and composition.

Even though this process does overcome the use of highly flammable and neurotoxic *n*-hexane, it did not prove cost efficient and was never realized at large scale (B.E. Douda, personal communication).

Typical composition according to Ref. [10]

56.0 wt% magnesium type II
32.1 wt% Teflon 7C
11.9 wt% Fluorel

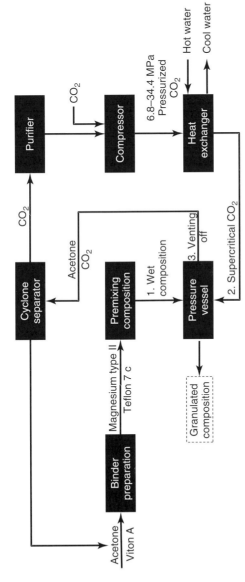

Figure 18.8 Supercritical shock gel process. (Adapted from Ref. [10].)

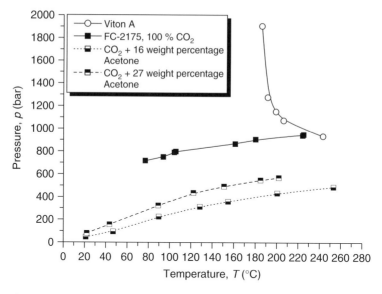

Figure 18.9 Cloud point curve for Viton A and FC-2175. (After Ref. [11].)

18.3.4
Experimental Dry Mixing Technique

Farnell has proposed to dry mix MTV to overcome the problems associated with highly flammable solvents acetone and hexane [12]. For this purpose, ground Viton was obtained as premix with Teflon from Dupont (Viton actually behaves like a highly viscous liquid; thus ground particles coalesce after a short period of storage). The actual mixing process is carried out in a combined screw feeder and mixer that opens feeds into a mold or extruder to give the final shape. Table 18.3 compares

Table 18.2 Composition tested for mechanical properties and performance [12].

Sample	Magnesium (weight percentage)	Teflon (weight percentage)	Viton (weight percentage)	Teflon/Viton (45/55) premix (weight percentage)
1 (wet)	62	30	8	–
2 (wet)	60	35	5	–
3 (wet)	55	40	5	–
4 (dry)	62	20.2	–	17.8
5 (dry)	60	28.9	–	11.1
6 (dry)	55	33.9	–	11.1
7 (dry)	60	40	–	–

Table 18.3 Stress, strain, toughness for wet and dry mixed MTVs [12].

Sample	Compressive strength (kPa × 10³)	Strength at maximum stress (cm⁻¹)	Toughness (kPa m⁻¹)
1 (wet)	30.3	0.042	868.7
2 (wet)	21.4	0.036	682.5
3 (wet)	29.6	0.038	744.6
4 (dry)	40.0	0.025	696.4
5 (dry)	38.6	0.023	613.6
6 (dry)	37.2	0.023	661.9
7 (dry)	21.4	0.010	124.1

Table 18.4 Combustion and radiative properties.

Sample	Burn rate (mm s⁻¹)	E_{lw} (J g⁻¹ sr⁻¹)	E_{sw} (J g⁻¹ sr⁻¹)
2 (wet)	2.95	132	288
5 (dry)	3.38	117	246
MTH[a]	3.51	174	—

[a]65% Mg, 30% PTFE, 5% Hycar 4051.

mechanical properties of a variety of both moist and dry blended grains as described in Table 18.2. Whereas the compressive strength of dry mixed grain is superior to wet blended material, on average about 40% higher, the strength at maximum stress is significantly lower with dry mixed composition. The toughness of dry blended material is a little lower than with wet mixed material. Sample 7 is a dry blended binary mix without any Viton binder. It has similar compressive strength but fails to reach both toughness and strength at maximum stress levels for both dry and wet mixed materials.

The dry mixed compositions burn faster than the wet mixed compositions because of significant higher grain porosity (Table 18.4). This also effects a lower spectral efficiency because of shorter residence time of Mg particles in the hot flame and thus less complete burn out. Dry mixing was never realized on a larger scale.

Composition 5(dry) has a reported resistance of 4×10^{10} Ohm and a conductivity of 1.5×10^{-3} S cm⁻¹ [12]. Another experimental dry mixing technique was used to prepare an Mg/PTFE-based flash mixture [13].

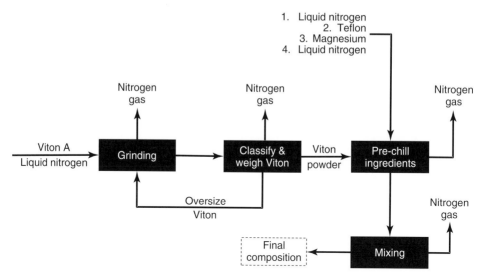

Figure 18.10 Flow diagram of cryogenic MTV process [14].

18.3.5
Experimental Cryo-N₂ Process

A cryogenic mixing process similar to that proposed by Farnell *et al.* was developed by Lateulere and coworkers [14]. They aimed at blending Mg, PTFE and Viton in liquid nitrogen. The principle flow diagram is depicted in Figure 18.10.

In this process, Viton is most advantageously grinded with a hammer mill below 150 μm particle size to obtain sufficient tensile strength after pressing or ram extruding the grain. However, even with particle sizes as low as 100 μm, the tensile strength of the final grain did reach only ~83% of the value obtained with shock gelled material. To overcome the electrostatic charge built up, the composition is modified with <1 wt% acetylene black. Apart from mechanical stability, the cryogenic mixed material after pressing successfully meets the requirements in MIL-PRF-82736A (OS). The liquid nitrogen mixing technique was never implemented for large-scale production of flares (B.E. Douda, personal communication).

18.3.6
Extrusion

18.3.6.1 Twin Screw Extrusion
Continuous extrusion of various energetic materials has been proposed in the 1980s [15–17]. Müller has proposed processing MTV via twin screw extrusion (TSE) for the first time in 1992 [18]. At about the same time, a TSE process for MTH using Hycar® was developed in the United States [7, 19] but was halted. In the late

1990s, TSE of MTV was picked up again [20] and transferred into production at Crane [21].

MTV/MTH Both MTV and MTH can be processed via TSE with high reproducibility. The clear advantage of TSE is the absence of dead spots in the homogenization process and the low solvent level necessary to run the process.

In the Mueller process, which is based on a 58 mm Werner and Pfleiderer extruder, Mg or Mg_4Al_3 and PTFE are wetted with ethanol before feeding into the system [18]. Downstream from the solid feed port, Viton™ dissolved in Acetone is injected. The materials are then transported to the kneading section where they are thoroughly mixed. A vacuum vent port at the end of the extrusion section is used to remove the solvent; a final extrusion die yields the desired grain geometry. Alternatively, the moist grain can be extruded and cut and dried batchwise before compacting to pellets. With one particular TSE model, throughputs of 50–80 kg/h are achieved with an instantaneous overall load of 2–3 kg. Grains obtained with this process could be furnished to yield burn rates between 0.3 and 50 mm s^{-1}.

Typical stoichiometries tested are given below.

Composition 3	Composition 4
52–56 wt% magnesium	52–56 wt% magnesium-aluminium
33–38 wt% polytetrafluoroethylene	33–38 wt% polytetrafluoroethylene
8–12 wt% Viton	8–12 wt% Viton

Representative in-process profiles for the extruder are depicted in Figure 18.11.

A MTH production-process developed in the United States, although using the same Werner and Pfleiderer TSE, used a different approach and consisted of a solid feedstream of Mg and a slurry feedstream of PTFE, Hycar and acetone. It used a weight ratio of 11.5 : 88.5 Hycar : acetone for the second feedstream, which allowed stirring, pumping and pouring of the slurry [7].

As a Viton/acetone solution at the same solvent level was too viscous to be transported, adaptation of that process to MTV was achieved by feeding all solids and acetone separately. It was found that a solvent level of 16.5–19 wt% acetone is satisfactory both to solvate Viton and disperse the composite. The viscosity η of an MTV with 19 wt% acetone feeding stream at 27 °C was found to relate with shear rate, γ, (s^{-1}) in the following manner:

$$\eta = 3.30 \times 10^4 \times \gamma^{-0.597} (\text{Pa·s})$$

The solvent level in the composition leaving the extruder is ~8 wt%. The overall solvent amount per kilogram MTV needed is only 0.225 l kg^{-1}, which is only ~6% the amount needed (3.5 l/kg) for the classical shock gel process [7].

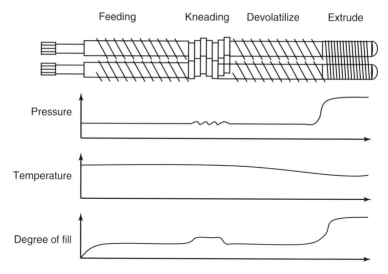

Figure 18.11 Typical *p*, *T*, *V* profile for Werner–Pfleiderer TSE. (After Ref. [18].)

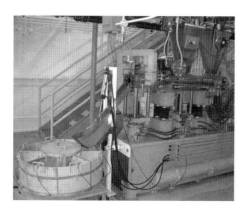

Figure 18.12 MTV TSE process according to Ref. [22].

MTV produced via TSE (Figure 18.12) shows the same performance and sensitivity as classical shock gel material. A typical cruciform grain for use in 1 × 1 × 8 M 206 flare is shown in Figure 18.13.

MTTP-TSE Process To completely overcome worries with VOC emissions and reduce the danger of fire in production, Nielson has invented magnesium/fluorocarbon formulations based on thermoplastic polymers (TPs) called MTTP [23–25]. Initially polystyrene was considered as material of choice but in subsequent was substituted by polyvinylchloride–polyvinylacetate copolymer (PVC-*co*-PVAc), which shows lower softening temperature. The softening agent

Figure 18.13 Cruciform $1 \times 1 \times 8$ in. flare grain obtained from TSE composition after ram extrusion process. (From Ref. [20].)

is dimethylphthalate (DMP). Table 18.5 displays some physical data of MTTP based on Mg/PTFE/PVC-*co*-PVAc/DMP. The process is depicted in Figure 18.14. The compositions required a significant higher amount of Mg to reach the same performance level as MTV-based compositions.

Whereas the sensitivity properties of the MTTP grain are similar to that of the standard MTH grain, the mechanical properties are inferior as is depicted in Table 18.5.

Despite the advantageous elimination of VOC emission and improved process safety, MTTP was never realized for large-scale production (B. Douda, personal communication).

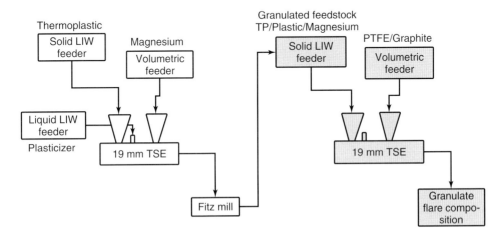

Figure 18.14 Processing diagram for experimental MTTP production.

Table 18.5 Mechanical properties of consolidated compositions.

Composition	Consolidation pressure (MPa)	Density	Crush force (kN)
Standard MTH	82.7	1.787	1.003
MTTP	82.7	1.691	0.230
MTTP	103.4	1.680	0.228

After Ref. [24].

18.4
Pressing

Once the composition is dried and granulated, it is either pressed or ram-extruded into its final shape. In a particular process, ~1400 g MTV crumb from the shock gel process or from TSE is ram-extruded to give planks of about 1 m in length. These planks are reformed to the desired grain geometry and then separated with a saw into seven final segments of MJU-32B type pellets [26]. The pellets are then machined at one end to obtain grooves and a hole to hold the igniter.

Extrusion imparts longitudinal orientation of PTFE fibres and assists to obtain greater mechanical stability to a grain [27]. However, extrusion is not available to all manufacturers and cannot account for complex 3D shapes. These require either pressing and/or milling.

Schematic pressing of a 218 flare pellet is depicted in Figure 18.15. It comprises the following steps: **1** filling the mold, **2** pressing the pellet, **3** and **4** expulsion of the pellet.

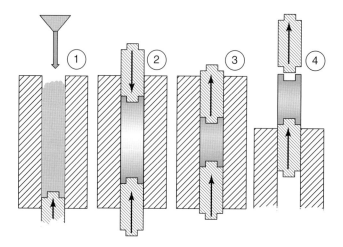

Figure 18.15 2 × 1 × 8 in. flare pellet pressing – schematic sketch.

Figure 18.16 Pressing 2 × 1 × 8 in. flare pellet with longitudinal grooves at Space Ordnance Systems, Canyon, CA, USA in 1980s. Safety note: standard protective clothing has changed since. Operations require grounding of the operator and Nomex-type overall with balaclava and additional faceshield.

A twin double-side press for 2 × 1 × 8 in. flare pellets is depicted in Figure 18.16. It shows an operator inspecting a pellet with pressed longitudinal grooves on the smallest side as shown in Figure 18.17.

Pressing occurs with adjusted pressure in a pre-determined way. A typical 1 × 1 × 8 in. grain is depicted in Figure 18.18. The lateral surface grooves are either remotely milled or pressed with a second tool as depicted in Figure 18.19 for the RR-119 pellet press process.

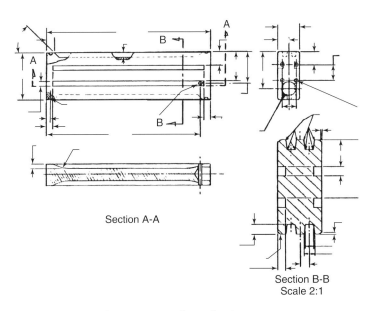

Section A-A

Section B-B
Scale 2:1

Figure 18.17 Typical 2 × 1 × 8 in. flare pellet design.

Figure 18.18 Typical 1 × 1 × 8 in. flare pellet and cross section of grain (1) with applied intermediate composition (2).

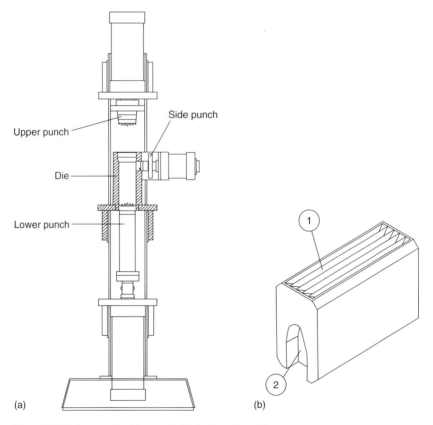

(a)   (b)

Figure 18.19 Press with side punch (a) for 3 × 2 × 5 in. RR119 flare pellet (b) reproduced from Ref. [28] with kind permission by Elsevier Publishers.

Figure 18.19b shows a decoy pellet with longitudinal grooves (1) and a base hole (2) to contain the safe and arm device.

In general, if the grooves are not pressed they can be milled remotely under water or under inert gas layer.

18.5
Cutting

Apart from its use in decoy flares, MTV is also used in propellant igniters, and therefore, MTV crumb obtained from shock gel process is extruded in small round or cruciform strands of 2.3–2.4 mm diameter. The extrusion is carried out at 121 °C, 4 MPa for round and 8.3 MPa for cruciform strands. Finally, the strand is cut with guillotine type cutter or band saw into length of 3.8–5.8 mm [29, 30].

18.6
Priming

The grain in its final geometry is applied with both intermediate and first fire to facilitate ignition transfer and ignition of the pellet. Figure 18.20 depicts a cross section of a 1 × 1 in. flare grain with applied intermediate fire based on MTV slurry. Figure 18.21 depicts ram-extruded cruciform 1 × 1 in. pellet with both first and intermediate fire applied [24].

18.7
Miscellaneous

Radiative polymerization of monomeric binder chlorotrifluoroethylene, C_2ClF_3, has been investigated for the MK 48 Decoy flare but was never realized in large scale [31]. Brown and Conkling have proposed to use liquid perfluorinated

Figure 18.20 Typical 1 × 1 × 8 in. flare pellet cross section of grain (1) with applied intermediate composition (2).

Figure 18.21 1 × 1 × 8 in. extruded Fe$_2$O$_3$-modified MTTP grain with applied first fire (black) and intermediate fire (grey). (Taken from Ref. [24].)

polyethers as oxidizers in pyrolants that would exclude use of flammable solvents in processing. It is claimed that aqueous dispersions of these fluorocarbons can be mixed with metal powders, which upon air or oven drying would yield gelly-like mixtures [32].

18.8
Accidents and Process Safety

The high impact and electrostatic sensitivity of Magnesium/Teflon®-based compositions have caused a number of accidents in the manufacturing industry. Table 18.6 gives an overview of accidents involving MTV-type pyrotechnic compositions (mainly MTV) in the United States in the time frame: 1968–2011.

Although this listing is not complete, it emphasizes on both mixing and pressing steps to be the most accident prone. We will discuss exemplary accidents for both steps in the following sections.

18.8.1
Mixing

Dillehay has described the 1993 incident listed above at Longhorn Army Ammunition Plant. A fire occurred during hexane washing of shock gel procedure of MTV in a Cowles Mixer [33]. The geometry of the Cowls dissolver is depicted in Figure 18.22.

The mixing procedure followed is given below:

Step	Description
1.	3 min blending commercially available fluoroelastomer/acetone solution with PTFE.
2.	Adding magnesium from remote vibrational feeder yielding a total of 56 kg MTV.
3.	10 min blending.
4.	Shock precipitation with 121 l n-hexane while the Cowles blades rotate to keep all the particles in contact with the non-solvent.
5.	Shutting off the blades, allow solids to settle. Supernatant fluid (acetone/n-hexane) is siphoned from the bowl with a pump.
6.	Start the blades of the dissolver again, and add additional n-hexane to wash the mix.
7.	Stop the blades, allow solids to settle and remove solvent.
8.	Repeat steps 6 and 7.
9.	Stop the blade and raise the mixing head to allow for removal of the bowl to a safety bay for dumping and mix recovery.

Table 18.6 Accidents with MTV-type pyrotechnic compositions [28, 30, 33–36].

Date	Location	Operation	Fatality	Injury
16 August 1968	NSWC Crane	Flare assembly	3	1
18 December 1968	NSWC Crane	Extruder	1	1
13 November 1975	Celesco industry	Mixing	0	0
9 December 1975	Celesco industry	Mixing	0	0
7 January 1980	Longhorn AAP	Mixing	1	1
26 July 1983	Longhorn AAP	Pressing	0	2
3 December 1983	Longhorn AAP	Pressing	0	1
19 December 1983	Tracor MBA	Assembly	2	19
20 January 1984	Longhorn AAP	Pressing	0	0
18. July 1985	Bermite Division	Mixing	0	0
8 August 1985	Tracor MBA	Emptying tumbler	3	5
28 August 1985	NSWC Indian Head	Cutting strands	0	2
20 October 1985	Longhorn AAP	Mixing	0	0
19 May 1986	Dela Trek	Mixing	1	2
11 August 1986	NSWC Indian Head	Cutting strands	0	0
28 August 1986	Kilgore Corp.	Drying	2	1
29 September 1987	Kilgore Corp.	Extruder	0	0
26 July 1989	Longhorn AAP	Mixing	0	0
9 November 1989	Tracor Aerospace	Mixing/drying	0	0
November 1993	Longhorn AAP	Mixing	0	0
1989–2006[a]	20 incidents	Not-specified	8	19
14 September 2010	Kilgore Corp.	MJU 7assembly	1 [46]	6
13 April 2011	Kilgore Corp.	Mixing	0	3

[a]Then the last incident occurred in 2003.

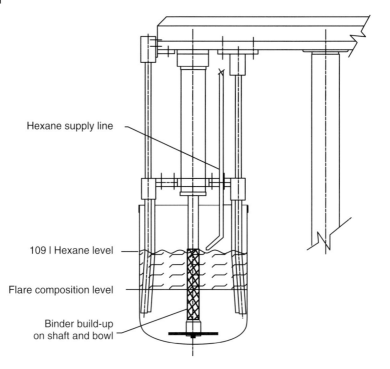

Hexane supply line

109 l Hexane level

Flare composition level

Binder build-up
on shaft and bowl

Figure 18.22 Cut view of Cowles dissolver. (From Ref. [33].)

Another operation carried out in the plant is cross-mixing half of a batch (28 kg) with a new batch, giving a total of 84 kg MTV.

In the reported incident, a filled bowl caught fire after a half batch was added at step 4 and hexane was added for a second wash (step 8).

A fault tree was developed and included the following

- Human initiated error
- ESD

    – Agitator
    – Distance between shaft and hexane
    – Flow of hexane
    – Failure of antistatic material
    – Failure of grounding

- Foreign matter
- Metal-to-metal contact
- Friction between compacted material and blades or agitator
- Mechanical failure

    – Fuel hopper
    – Blender
    – Hexane pump

– Moving cart for bowl
– Vapour removal system

• Environment.

The exact cause of the fire could not be determined. However, it was deduced that either

• electrostatic ignition of hexane vapours followed by ignition of composition was one likely cause of the event or
• the build-up of consolidated material under the mixing blades that would undergo frictional heating would be the likely cause.

In an earlier incident at the same facility, a Cowles Mixer also caught fire, and it was found that the net potential difference between the Hexane supply line and the mixer shaft could be as high as 1800 V, thus giving rise to ESD that could ignite the hexane vapours above the vessel [34].

18.8.2
Pressing

Diercks has reported about a series of pressing-related incidents with pressing RR-119 flare pellets [35]. The composition is given below:

RR-119 flare composition [3]

---

62 wt% magnesium
38 wt% Teflon
2 wt% polyester binder

It was observed that pellets (see Figure 18.19 for press and pellet) resting on the lower punch after expulsion of the press would ignite upon manual removal. Pellets were probed for electrostatic voltages. It was found that the voltage of the pellets resting on the lower punch after expulsion would range between 1200 and 3600 V. This voltage would dissipate within 30 s down to 200–1400 V. It was observed that a decrease of surface charge without any measures generally occurred on a level of several minutes.

As a countermeasure, ionized air nozzles were installed such that the pellets would be projected by the ionized air stream. This proved effective in removing surface charges within a few seconds [35].

Hearn and Singh investigated the electrostatic surface potentials on a consolidation press manufacturing of 55 mm MTV decoy flare pellets [37]. A typical voltage-trace for filling consolidation and removal of the pellet as is depicted in Figure 18.23.

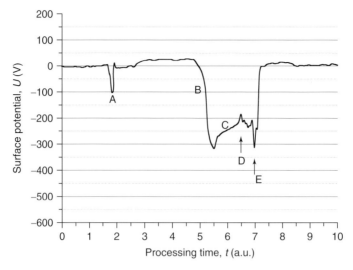

Figure 18.23 Typical surface potential trace for filling, consolidation and pellet removal step in press [37]. (A) depicts filling MTV composition in the die. At (B), the consolidate pellets emerges from the die. The surface charge decay is seen at (C). Both (D) and (E) indicate grasping and lifting of the pellet by robot claws. It was also observed that at one particular press, the grasping and lifting action would coincide with spark discharges.

18.8.3
Process Analysis

Hubble *et al.* have examined the likeliness of ESD in various operation steps of MTV/MTH flare production from mixing composition to final loading of the wrapped flare grain in the cartride case [38]. Table 18.7 gives an overview on static electricity build-up in the various steps of the production process.

It was found that the magnitude of charge build-up depends on

- the amount of handling on grain
- if gloves are worn on handling

    – the type of gloves worn

- whether the relative humidity level <40% RH – 60% RH or >60% RH
- the air ionization.

18.8.4
Personal Protection Equipment (PPE)

The mixing of pyrotechnic compositions is a dangerous process, and as such, is carried out remotely [39]. However, charging chemicals, discharging composition and handling bare flare pellets for priming and/or assembly may require direct

Table 18.7 Results of static electricity build-up survey.

Process step	State evaluated	Process	Material handling
Mixing	Composition	Not investigated	–
Air drying	Composition	No	No
Oven drying	Composition	No	500–1000 V when dry
Extrusion	Plank	No	Cutting: 500 V
Normalization	Plank	No	Up to 20 000 V
Plank cutting	Grain	Yes, but depends on technology used	Up to 2000 V
Machining	Grain	Unknown, difficult to measure	Up to 1500 V
Conveyor belt	Grain	Yes	150–200 V
Ignition slurry application	Grain	No if designed to minimize grain handling and exposure to solvent vapours	3000–7000 V
Oven slurry drying	Grain	Not investigated	–
Connect igniter to grain	Grain	No	100–7000 V
Inserting igniter pellet	Grain	No	2000–3000 V
Aluminium foil	Grain	No	2000–3000 V
Flare case loading	Grain	No	–

manipulation by workers. Hence employees have to wear suitable personal protection equipment (PPE) to be safeguarded in the event of an accidental ignition of the composition. Moreover, it should also be considered that manual handling must not alter the safety of the process, for example by causing electrostatic discharges.

PPE in all operations involving MTV should include the following (N. Brune, personal communication):

- Conductive safety shoes.
- Fire-protective and ESD-safe overall or smock made from suitable multilayer, quilted inner fire-resistant liners and aluminized coated fabrics such as aluminized NOMEX® [40–43] or aluminized PBI®.
- Fire-protective long-sleeve shirt and long underpants or underwear made from suitable fabric such a NOMEX® [42, 43].
- Fire protection for hands made of goat skin leather gloves over ESD-safe Nomex® or PBI® liners.
- Fire-protective and ESD-safe balaclava made from aluminized NOMEX® or PBI® with impact and IR-reflective protection safety shield.
- Grounding (anti-ESD) wristband.

Figure 18.24 Demolished MJU 7 assembly box after MTV blast at Kligore Toone/TN 14 September 2010 [44].

Analysis of past accidents has revealed that the people involved often lack experience because they are either insufficiently trained on the procedures and/or not advised about the particular hazards of the operations.

In other cases, experienced personnel may have become sloppy as a result of the eventless daily routine. Very often, it was then tried to 'improve' the production procedure by accumulating more explosive material than allowed on the workbench in order to increase the daily up output [44]. These actions may double-cross available safety measures and result in fire and blast effects. Figure 18.24 shows a press photograph taken from the demolished MJU 7 assembly box at Kligore/Toone facility [45].

To advertise, both workers have to receive in-depth, regular and continuous training. Education is necessary to maintain the awareness of workers involved in the manufacture of MTV and decoy flares. Demonstrating the effect of blast and fires with flare material, for example on the facility's own testing ground, will compliment any verbal instruction by its own force.

References

1. Chen, G., Broad, R., Valentine, R.W. and Mannix, G.S. (2001) Magnesium-fueled pyrotechnic compositions and processes based on elvax-cyclohexane coating technology. US Patent 6,174,391, USA.

2. Ellern, H. (1968) *Military and Civilian Pyrotechnics*, Chemical Publishing Company, New York, p. 303.

3. Douda, B.E. (2009) Genesis of Infrared Decoy Flares, Naval Surface Warfare Center, Crane, IN.

4. Fedoroff, B.T. and Sheffield, O.E. (eds) (1966) *The Encyclopedia of Explosives and Related Items*, vol. 3, p. C 352.

5. Potter, L.J. (1988) Magnesium teflon viton igniter sensitivity investigation results. 13th International Pyrotechnics

Seminar, Grand Juction, CO, July 11–15, pp. 639–659.

6. N.N. (1988) Ignition Pellets, Magnesium-Fluorocarbon, MIL-PRF-82736A(OS), Naval Sea Systems Command, Indian Head, USA, 16 November 1988, *http://www.every-spec.com/MIL-SPECS/MIL+SPECS+%28MIL-M%29/MIL-PRF-82736A_14750/* (accessed 1 September 2011).

7. Stevenson, B. Campbell, J. and Rose, M. (2001) Continuous Processing of MTV Based IR Decoy Compositions. 28th International Pyrotechnics Seminar, Adelaide, Australia, November 4–9, pp. 649–660.

8. Diewald, G. and Bickel, H. (1977) Pyrotechnic compositions, Oerlikon Werkzeugmaschinenfabrik. GB Patent 1,600,576, Switzerland.

9. Teipel, U. (2005) *Energetic Materials, Particle Processing and Characterization*, Wiley-VCH Verlag GmbH, Weinheim, p. 36.

10. Nauflett, G.W., Farncomb, R.E. and Chordia, L. (1998) Preparation of Magnesium-Fluoropolymer pyrotechnic material. US Patent 5,886,293, USA.

11. Cocchiaro, J.E., DiNoia, T.P., McHugh, M.A. and Morris, J.B. (1997) *Supercritical Carbon Dioxide Based Processing of PEP Binder Polymers*, CPIA Publication No. 647, Vol. I, CPIA, March 1997, pp. 257–270.

12. Taylor, F.R. and Farnell, P.L. (1998) New solventless process for safer manufacturing of pyrotechnic flares. CPIAC Meeting, pp. 949–963.

13. Aikman, L.M. Shook, T.E. Lehr, R.M. Robinson, E. and McIntyre, F. (1987) Improved mixing, granulation and drying of highly energetic pyromixtures. *Propellants Explos. Pyrotech.*, **12**, 17–25.

14. Lateulere, M. (1991) Cryogenic processing of Magnesium-Teflon-Viton (MTV) pyrotechnics. Joint International Symposium on Energetic Materials Technology, New Orleans, October 5–7, pp. 282–290.

15. Müller, D. and Schubert, H. (1984) Method for producing plastically bonded propulsion powders and explosives. US Patent 4,608,210, Germany.

16. Müller, D. (1984) Kontinuierliche Herstellung von Treibladungspulvern. 25th International Annual ICT Conference, Karlsruhe, Germany, June 27–29, p. 97.

17. Stewart, J.E. (1984) Considerations for using twin-screw extruders for manufacturing propellants & explosives. 25th International Annual ICT Conference, Karlsruhe, Germany, June 27–29, p. 119.

18. Müller, D. (1992) Continuous processing of MTV decoy flares. International Pyrotechnics Seminar, Breckenridge, CO, July 13–17, pp. 635–648.

19. Dillehay, D.R., Turner, D.W., Wingfield, H.L. and Blackwell, J.A. (1996) Energetic materials processing technique, US Patent 5,565,150, USA.

20. Sotsky, L. and Jasinkiewicz, K. (2002) Twin screw Mixing/Extrusion of M206 Infrared (IR) decoy flare composition. 33rd International Annual Conference of ICT, p. V - 35.

21. Backer, S.L., Campbell, J.R., Stevenson, B.D. and Limberger, T.R. (2004) Hazard assessment testing of in-process infrared decoy composition. Crane Division, Naval Surface Warfare Center, Crane, IN.

22. Rose, M. (2001) Twin Screw Extruder Production Engineering Study, ATK Systems, *http://www.pica.army.mil/jocg/files/2001/04%20-%20Continuous%20Processing%20of%20MTV%20Based%20IR%20Decoy%20Composition.pdf.* (accessed 1 September 2011).

23. Nielson, D.B. and Lester, D.M. (2002) Extrudable black body decoy compositions. US Patent 6,432,231, USA.

24. Campbell, C. (2005) Twin Screw Extruder Production of MTTP Decoy Flares. Report No. SERDP WP-1240, Thiokol Propulsion, Brigham, UT.

25. Ray, M.A., Ashcroft, B.N., Blau, R.J., Campbell, C.J., Rose, M.T., Chen, G. and Campbell, J. (2006) An example of Magnesium-Teflon©-Thermoplastic (MTTP), a viable blackbody emitting composition that has been produced via a solvent free twin-screw extrusion process. 33rd International Pyrotechnics Seminar, Fort Collins, CO, July 16–21, p. 297.

26. Hubble, B., Stevenson, B. and Benstin, M. (2004) Experimental evaluation of the static electricity characteristics associated with processing MTV. 31st International Pyrotechnics Seminar, Fort Collins, CO, July 11–16, p. 211.

27. Renner, R.H., Farncomb, R.E., Nauflett, G.W. and Deiter, S.J. (1995) Characterization of MTV made by the shock-gel process. International Symposium on Energetic Materials Technology, Phoenix, Arizona, September 24–27, 1995.

28. Diercks, B.V. (1986) Electrostatic sensitivity of consolidated Magnesium-Teflon compositions. *J. Hazard. Mater.*, **13**, 3–15.

29. Hong, G.T. (1998) Recovery of Pyrotechnic Ingredients Using Supercritical Fluids, Naval Surface Warfare Center, Crane, IN.

30. Potter, L.J. (1988) Magnesium teflon, viton igniter sensitivity investigation results. 13th International Pyrotechnics Seminar, Grand Junction, CO, p. 639.

31. Parrish, C.F., Short, J.E. and Biggs, W.T. (1973) Radiation-Polymerization Binder for MK 48 Decoy Flares. RDTR No. 232, Naval Ammunition Depot, Crane, IN.

32. Brown, J.S. and Conkling, J.A. (1997) Energetic compositions containing no volatile solvents. US Patent 5,627,339, USA.

33. Dillehay, D.R. and Leander, R.C. (1994) Cowles-dissolver fire involving IR flare mix. Proceedings of the 26th DoD Explosives Safety Seminar, Miami, FL, August 16–18.

34. Johnson, J.A. (1990) Special study of safety in pyrotechnics manufacturing. 24th DOD Explosives Safety Seminar, St. Louis, MO, August 28–30, p. 2081.

35. Diercks, B.V. Hawley, J.E. and Naron, M.L. (1986) Muller mixer fire lessons learned. 22nd DoD Explosives Safety Seminar, Anaheim, CA, August 26–28, p. 1967.

36. Stevenson, B. and Campbell, J. (2006) Twin screw extruder processing of MTV based IR decoy compositions.

Compositions 14th Continuous Mixer and Extruder Users Group Meeting, Picatinny Arsenal, NJ, October 11–12, *http://www.pica.army.mil/JOCG/archives/2006.aspx.*

37. Hearn, G.L. and Singh, S. (1987) Hazards from electrostatics in the manufacture of infra-red decoy flares. *Inst. Phys. Conf. Ser.*, **85** (Section 3), 229–234.

38. Hubble, B. Stevenson, B. and Benstin, M. (2006) Experimental evaluation of the static electricity characteristics associated with processing MTV. 31st International Pyrotechnics Seminar, Fort Collins, CO, 16–21 July, pp. 211–232.

39. Berufsgenossenschaft Chemie (2001) Berufsgenossenschaftliche Vorschrift für Sicherheit und Gesundheit bei der Arbeit. BGV B 5, Explosivstoffe – Allgemeine Vorschrift vom 1 April 1995 in der Fassung vom 1 April 2001.

40. *http://www2.dupont.com/Personal_Protection/en_US/products/Nomex/nomexind/nomex_industrial_faq.html#7QD.* (accessed 1 September 2011).

41. *http://www.hainsworth.co.uk/downloads/110225_TI-TECHNOLOGY_Final7.pdf.* (accessed 1 September 2011).

42. Burn, S.L. (2006) Assessment of personal protective clothing for protection against fire hazards. 33rd International Pyrotechnics Seminar, Fort Collins, CO, July 16–21, p. 493.

43. Burn, S.L. (2007) Personal protective equipment. 5th International Workshop on Pyrotechnic Combustion Mechanisms, Beaune, France, October 6.

44. Hunter, N.B. (2011) Report: Load Fed Kilgore Blast, *http://www.jacksonsun.com/fdcp/?unique=1302897465921.*

45. Chemring Press release Report: Incident at Kilgore Flares, *http://www.chemring.co.uk/media/press-releases/2010/2010-09-15.aspx.* (accessed 1 September 2011).

46. *http://www.myfoxmemphis.com/dpp/news/tennessee/kilgore-employee-dies-from-explosion-injuries-mfo-20110929.* (accessed 30 September 2011).

19
Sensitivity

19.1
Introduction

The sensitivity of an energetic material towards accidental stimuli such as fast heating, electrostatic discharge, friction, impact and shock determines its safety characteristic and, hence, decides whether a certain material may be actually used and whether it can be safely transported. Corresponding test methods are described in Test Series 3 of the UN 'Orange Book' [1a]. These tests determine the sensitiveness to impact, friction (and impacted friction), thermal stability and response of the substance to fire.

In addition to safe handling and transport of energetic materials, requirements have evolved that call for energetic materials that are less vulnerable to certain accidental stimuli and give a more benign response as opposed to explosion and detonation. Corresponding tests for energetic materials are described in Test Series 5 [1b] and 7 [1c]. Substances passing Test Series 5 are classified as hazard division 1.5, and those passing Test Series 7 a–f (substances tests) are considered extremely insensitive detonating substances (EIDS). In the military area, NATO AC326 Munition Safety Panel has issued STANAG 4439 via Standardization Agency (NSA) [2]; STANAG 4439 describes tests to probe the vulnerability and response of stores containing energetic materials. Materials passing UN Test Series 5 and/or 7 (substance tests) are referred to as being '*insensitive*'.

In contrast, substances that pass Test Series 3 are provisionally accepted into Class 1 which does not imply that these substances are insensitive by default.

In addition to sensitivity, chemical deterioration (ageing) of an energetic material upon climatic storage and/or when brought into contact with another (energetic) material may severely restrict its use in certain military applications. Therefore, tests have been devised at both national [3] and NATO levels [4].

Metal-Fluorocarbon Based Energetic Materials, First Edition. Ernst-Christian Koch.
© 2012 Wiley-VCH Verlag GmbH & Co. KGaA. Published 2012 by Wiley-VCH Verlag GmbH & Co. KGaA.

19.2
Impact Sensitivity

19.2.1
MTV

The impact sensitivity of granular Magnesium/Teflon/Viton (MTV) (grain size <1.5 mm) scatters between that of dry crystalline RDX and Hexal 70/30, which translates to impact energies with the BAM testing machine between 4 and 10 J (very sensitive–sensitive). However, larger spherical pieces (exceeding 2–3 mm diameter) may undergo no reaction even at 50 J impact energy. The influence of magnesium particle size on impact energy is shown in Figure 19.1. For 49 μm Mg, the minimum impact energy of 4 J is obtained, whereas at lower and higher Mg concentrations, the impact energy is above 5 J. Coarser Mg (78 μm) yields slightly less impact-sensitive formulations. The influence of polytetrafluoroethylene (PTFE) particle size on impact energy of Mg/PTFE compositions with Bureau of Energy (BoE) impact tester [5] is listed in Table 19.1.

19.2.2
Titanium/PTFE/Viton and Zirconium/PTFE/Viton

Pyrolants from micro-metric Ti (22 μm) and Zr (10 μm) and 5 μm PTFE with 12 wt% Viton are less sensitive to impact than MTV. TiTV with 40 wt% Ti is nearly impact insensitive with an impact energy of 47 J. However, at higher metal content, TiTV shows impact energies of 13 J (60 wt%) and 14 J (80 wt%) respectively. ZrTV shows similar impact energies 13 J (60 wt% Zr) and 11 J (80 wt% Zr) [9].

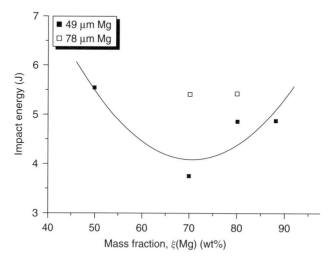

Figure 19.1 BAM impact energy of MTV as a function of stoichiometry and Mg particle size [6].

Table 19.1 BoE impact energy of various metal/PTFE compositions [7, 8].

Components	FW 231	FW 306	M 22
Magnesium, 44–74 µm	54	54	75
PTFE, 149 µm	46	–	–
PTFE, 250 µm	–	46	–
PTFE, 275 ± 100 µm	–	–	10
Nitrocellulose	2.6	2.6	–
Fluoroelastomer binder	–	–	15
Impact energy (J)	3.4	9.2	3.4

19.2.3
Metal–Fluorocarbon Solvents

A number of combinations of reactive metal powders, granules and shavings with fluorocarbon solvents such as Freon^{TM} MF and Freon^{TM} TF are impact sensitive. Among the materials tested, combinations of barium with fluorocarbons are the most sensitive ones. This is surprising in view of the low reactivity observed for barium in Grignard reactions [10]. Table 19.2 lists the 50% impact energy required to initiate the corresponding combination. The Jet Propulsion Laboratory impact sensitivity tester was used to determine the sensitivity. Comp B was tested for reference and yielded an impact energy of 4.6 J. The maximum impact energy obtained with this apparatus was limited to 28 J. For comparison, combinations with tetrachloromethane are also given.

19.2.4
Viton as Binder in Mg/NaNO$_3$

The ignition behaviour of bare Mg/NaNO$_3$ (50/50) and compositions with 5% binders each (Elvax 210$^®$, Lithographic varnish and Viton$^®$) has been tested at high strain rate (Figure 19.2) [11]. Pellets of 85% TMD were subjected to impact in the Hopkinson bar apparatus. It turns out that organic binders actually sensitize Mg/NaNO$_3$ against initiation by impact. This is evident from the critical velocity that is on an average 1.5–2 m s^{-1} higher with pure Mg/NaNO$_3$. Among the binders tested, Viton appears to have the strongest sensitizing effect (Figure 19.3). It has been speculated about viscous heating and/or localization of energy in shear bands as the cause of this.

19.3
Friction and Shear Sensitivity

Because of the low friction coefficient of PTFE/PTFE ($\mu = 0.04$), PTFE-based materials usually are not very prone to ignition by friction. Thus, there is no

Table 19.2 Fifty percent of impact energy of metal/fluorocarbon combinations (J).

Solvent	Al (13 μm)	Mg (600–700 μm)	Ti (44 μm)	Ba[a]	Li[a]	Be[c]	BeH$_2$[c]	Al[b]	Mg[b]	B (< 5 μm)
Freon MF®, CCl$_3$F	F, 28	0	0	E, 11.5 F, 8.6	E, 28 F, 22.4	0	0	0	0	0
Freon TF®, C$_2$Cl$_3$F$_3$	F, 28	0	F, 28	F, 5.7 E, 2.3	E 11.5	0	0	0	0	0
CCl$_4$	E, 28	F, 28	0	E, 8.6 F, 5.7	E, 10.3	0	0	0	0	0

[a] Shavings.
[b] Filings.
[c] Fine powder.
F, flashes; E, explosion [10].

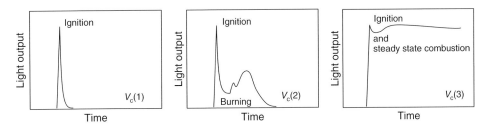

Figure 19.2 Qualitative light output for Mg/NaNO$_3$ composition at increasing critical velocities $v_c(1) < v_c(2) < v_c(3)$.

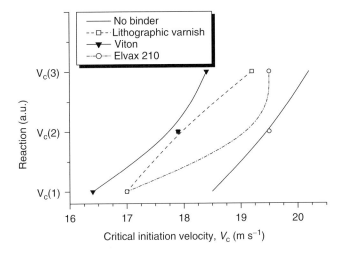

Figure 19.3 Variation of critical velocity as a function of binder v_c.

reaction observed with MTV up to 360 N friction force with the BAM apparatus [5, 12] or any other of the tested material [7, 8].

19.3.1
Metal/Fluorocarbon

A number of commercial polychlorotrifluoroethylene products such as Fluorolube® and Kel-F® (oils, grease and powders) when used for machining aluminium yield explosive reactions under high shear loads. This kind of reactivity is rarely observed with pure fluorocarbons. Thus, it is assumed that the high reactivity stems from the chlorine connected to the polymer that is more easily abstracted than fluorine. In addition, thermal reaction between Al and Kel-F oil has been observed at temperatures as low as 218 °C, whereas a fully fluorinated grease does not react until about 648 °C [13]. Tribochemical-induced reaction between fluorocarbons and certain metals has been confirmed by Electron Spectroscopy for Chemical Analysis (ESCA) investigations on PTFE samples filled with both aluminium and zinc [14, 15].

19.4
Thermal Sensitivity

Thermal sensitivity is an important criterion with pyrolants as it determines both safety and functionality in terms of reliability of ignition and propagation. Preliminary indicators for the thermal stability are any means of thermal measurements such as Differential Scanning Calorimetry (DSC) (see Chapter 5), Differential Thermal Analysis (DTA) and Thermogravimetry (TG) or Differential Thermogravimetry (DTG). More precisely, the Heat Flow Calorimetry (HFC) measurements at a fixed temperature are able to probe the intrinsic stability of an energetic material [16].

To assess the overall thermal sensitivity of an energetic material in Germany, a number of tests related to the ignition towards standard ignition sources such as cerium–iron alloy sparks, bickford fuse flame jet, gas flames, red hot steel bar and match flame are conducted. In addition, the response of the material towards ignition under various levels of enclosure is tested, the so-called Koenen test [17].

19.4.1
MTV

The thermal tests (except for Koenen test) for three metal–fluorocarbon pyrolants are listed in Table 19.3.

The thermal sensitivity of MTV and Magnesium/Poly(carbon monofluoride)/ Viton[TM] MPV is significant, and the substances are classified easily ignitable. Ignition temperatures for MTV formulations are listed Table 19.4.

Farnell et al. investigated the ignitability of pyrotechnic compositions by solvent vapour flames [18]. In addition, they also looked at the ignition delay of compositions moistened with the solvent. The ignition delays of mixtures of 5 g FW − 306[a] with 4 cm^3 acetone and alcohol each are 83 and 174 s, respectively.

Table 19.3 Thermal sensitivity of three MTV/MPV formulations.

Sample	DSC first reaction °C	HFC at 89 °C (J g^{-1})	Cerium–iron spark	Bickford fuze flame jet	Gas flame	Red hot steel bar, 5 mm	Burning match	Wood's metal bath	Red hot steel bowl
MTV (53/42/5)	–	12	No ignition	Ignition	Ignition	No ignition	Ignition	N.r. at 360 °C	Instant ignition
MTV (57/33/10)^a	263	–	No ignition	Ignition	No ignition	Ignition	No ignition	N.r. at 360 °C	Instant ignition
MPV (45/50/5)	347	17	–	–	–	–	–	–	–

^{a}0.1 wt% graphite.

Table 19.4 Autoignition temperatures of various metal/PTFE compositions [7, 8].

Components	FW 231	FW 306	FW 306[a]	M 22	SR 886
Magnesium, 44–74 μm	54	54	43[a]	75	55 ± 5
PTFE, 149 μm	46	–	36[a]	–	–
PTFE, 250 μm	–	46	–	–	–
PTFE, 275 ± 100 μm	–	–	–	10	40 ± 5
Nitrocellulose	2.6	2.6	21[a]	–	–
Fluoroelastomer binder	–	–	–	15	5
Autoignition temperature (°C)	510	510	600	469	550

[a]Data on FW 306 from Ref. [18], which gives different stoichiometries and no details on particle size distribution.

19.5
ESD Sensitivity

Electrostatic discharges have been made responsible for many accidents in the manufacture of magnesium/Teflon®-based compositions [19–21] (see also Chapter 18). Hubble et al. recently reviewed the complete manufacturing process of cylindrical 36 mm countermeasure flares and identified a number of possible operation steps prone to yield a large-surface electrostatic potential [22] (Table 18.7).

Investigations of aluminized loaded solid propellants have established a typical ignition mechanism [23–25]. On exposure such material to an electric field, a breakdown path is created at first starting from the surface between single particles and finally creating a total conductive duct through the bulk of the material. A similar process can be assumed for Magnesium/Teflon-based energetic materials.

Kent and Rat have analysed the processes occurring upon exposure of a metalized energetic material to an applied direct current [23]. When a dc voltage is applied across a sample at voltages lower than the critical voltage, U_c, the current versus voltage relation at a given temperature is linear in shape (Figures 19.4 and 19.5). For

Table 19.5 Compositions[a] of samples tested for dielectric breakdown.

	SR 886B[b]	SR 886E	SR 886F
Mg (wt%)	55	55	55
PTFE (wt%)	40	40	40
Viton A (wt%)	5	5	5
Burn rate (mm s^{-1})	2.2–12.0	12.0	2.2

[a]Compositions differ in the type of PTFE used (Fluon G3 and Fluon L169).
[b]Blend of SR 886E and SR 886F.
From Refs. [26, 27].

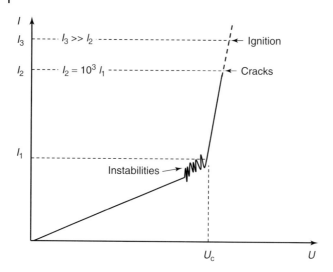

Figure 19.4 Evolution of conduction current, I_c, versus applied potential, U, according to [23].

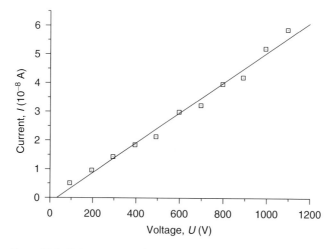

Figure 19.5 Voltage/current characteristics of a 1 mm thick MTV sample. (After Ref. [26].)

voltages $U \sim U_c$, the sample current becomes instable and fluctuates (Figures 19.4 and 19.6). This is indicative of some kind of breakdown process in the sample. In this voltage regime, the sample resistance still temporarily recovers. If the field is maintained for longer duration or voltage significantly increased, the samples totally loose resistance and breakdown occurs (Figure 19.7). If the breakdown occurs at lower current I_2 after longer exposure, ignition occurs not very often (1 in 50 trials); however, if the current is that sufficient level, I_3 breakdown will be accompanied by ignition. At I_3, micro-explosions also take place that shatter the

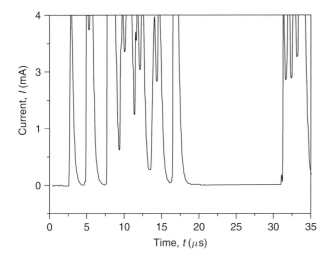

Figure 19.6 Current instabilities at U = 580 V for a 1 mm thick MTV sample.

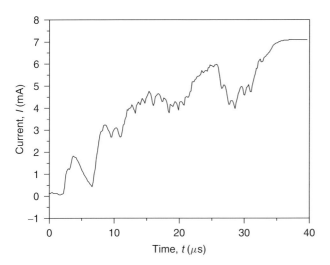

Figure 19.7 Steady current increase at U = 1000 V for 1 mm thick MTV sample.

sample sometimes without igniting it as shown for a pellet in Figure 19.8 [26]. These samples also show distinct cavities that form along the breakdown path (Figure 19.9).

The electrical field strength at breakdown for a pellets decreases with increasing thickness (Figure 19.10). The breakdown voltage, however, remains essentially constant.

Influence of sample composition and PTFE particle size on conductivity and breakdown conditions is listed in Table 19.6. The addition of conductive additives

(a)   (b)

Figure 19.8 Cratered MTV pellet (b) with ejected piece (a) and due to micro-explosion after ESD. (From Ref. [26] reproduced with kind permission by the International Pyrotechnics Society.)

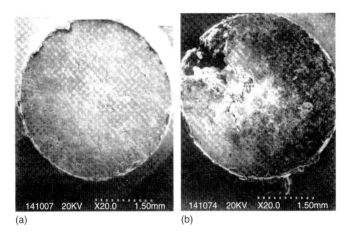

(a)   (b)

Figure 19.9 MTV pellet before (a) and after (b) dielectric breakdown [28]. Reproduced with kind permission by N. Forichon-Chaumet and the Association Francaise de Pyrotechnie.

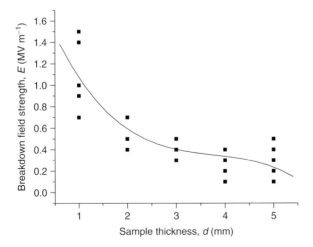

Figure 19.10 Breakdown field strength of MTV as a function of sample thickness.

Table 19.6 Conductivity, σ, and dielectric breakdown conditions of various MTV samples.

Composition	Thickness (mm)	U_c (kV)	E_c (MV m^{-1})	Number samples	σ_{before} (Ω^{-1} m^{-1})	σ_{after} (Ω^{-1} m^{-1})
SR 886E	1	3	3	10	1.18×10^{-10}	3.25×10^{-3}
SR 886E + 1%Cu	1	0.8	0.8	6	1.44×10^{-10}	1.41×10^{-2}
SR 886E + 7%~Cu	2	0.2	0.1	8	1.86×10^{-7}	1.38×10^{-2}
SR 886E + 7.5% Cu	2	0.07	0.035	7	6.65×10^{-7}	1.1×10^{-2}
SR 886F	2	1.2	0.6	10	1.63×10^{-10}	1.24×10^{-2}
PTFE coated with Viton A	1	2.5	2.5	8	1.37×10^{-10}	8.04×10^{-4}
PTFE tape	75 µm	1.2	16	6	7×10^{-12}	2.37×10^{-4}

further facilitates breakdown as is seen for SR 886F modified with increasing Cu content.

Analysis of the samples that experienced breakdown shows the formation of conductive ducts. These ducts in the cases where ignition takes place undergo Joule heating and create hot spots that effect sample ignition.

The ignition energies for a series of samples of different composition and thickness are shown in Figure 19.11. For three series of MTV compositions, SR 886 B, F and E, which differ only in the type and ratio of PTFE particles used, the minimum ignition energy rises with sample thickness. The electrical setup is shown in Figure 19.12.

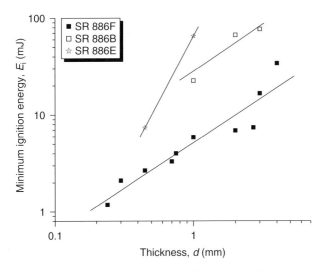

Figure 19.11 Minimum ignition energy for circular pellets with both 7 and 5 mm diameter as a function of thickness for three compositions types SR 886B, F and E [29].

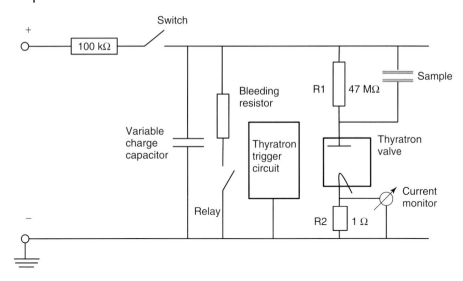

Figure 19.12 Schematic electrical circuit for determining minimum ignition energies. (After Ref. [29].)

The dielectric breakdown voltage of MTV igniter compositions as a function of modifiers added has been studied in [28, 30].

The modification of MTV with nano-scale aluminium (Alex®: 100–200 nm diameter) and nano-laminate powder (ALNC) (a polymer–aluminium composite obtained from Sigma Lab) has been studied by Wilharm [31, 32]. Two standard fuel-rich (54–55 wt% Mg) control formulations were investigated for reference. Although Alex is a sensitive material, incorporation in a Viton A-based system actually desensitizes the resulting Alex/MTV composite. Hence, the no-fire energies climb up to 1 J for the substitution of one-third of magnesium by Alex. The last entry in Table 19.7, however, shows an increased sensitivity for MTV modified with Alex and a high amount of Viton binder. The authors failed to deliver a reasonable explanation for this sensitization [32].

19.6
Insensitive Munitions Testing

19.6.1
Introduction

An accidentally ignited pyrotechnic illumination flare on board USS Oriskany (Figure 19.13) caused a catastrophic fire that subsequently killed 44 men and injuring 156. Two helicopters and four aircraft were severely damaged [33].

As a consequence of this accident and others that followed, US Navy initiated the development of explosives that would be less vulnerable to unplanned stimuli

Table 19.7 Ignition sensitivities of MTV composition containing 16 wt% Viton A.

Comments	Mg (wt%)	PTFE (wt%)	ALNC (wt%)	Alex wt(%)	Viton wt(%)	ESD no-fire (mJ)
	54–55	??[a]	0	0	??	10
	16	??	16	0	??	100
	0	??	0	32	??	10
	27–27.5	??	0	27–27.5	??	80
	16	??	0	16	??	30
	??	??	0	0	16	80
	5/6	??	0	1/6	16	200
	2/3	–	0	1/3	16	200
	5/6	–	0	1/6	16	1250
Pre-coated Alex	2/3	–	0	1/3	16	1013
Pre-coated Alex	2/3	–	0	1/3	18	5

[a]indicates unavailable data.

Figure 19.13 USS Oriskany burning on 26 October 1966. (After Ref. [34].)

such as fuel fire and bullet impact (BI) than explosives used at that time. The US Navy also accounts for the first insensitive munitions standard, DOD-STD-2105 (NAVY) [35], issued in 1982 that describes tests such as slow cookoff (SC) and fast cookoff (FC) and BI. Nowadays, NATO STANAG 4439 [2] and associated AOP 39 [36] define the international standard in insensitive munitions testing.

Contrary to high explosives that have been designed to detonate, explode and deflagrate, pyrotechnic compositions have been designed to burn fiercely or in a slow manner. Hence, it is perhaps that catastrophic reactions of pyrotechnic ammunition are not expected to occur upon stimuli such as fast heating or BI. In this context, the term catastrophic reaction refers to the munition response types I (detonation), II (partial detonation), III (explosion) and IV (deflagration). The definition of the reaction types is given in the grey box [36]. The threats defined in STANAG 4439 define a series of threats likely to be encountered for a munition during its life cycle. These threats are listed in Table 19.8.

Table 19.8 Threat, definition and minimum pass requirement [7, 8].

Threat acronym	Pass-requirement	Definition	Scenario
Fast Cookoff FC	No response more severe than type V (burning)	Average temperature between 550 and 850 °C until all munitions reactions completed. About 550 °C reached within 30 s from ignition	Magazine/store fire or aircraft/vehicle fuel fire
Slow Cookoff SC	No response more severe than type V (burning)	Between 1 and $30°\mathrm{C\,h^{-1}}$ heating rate from ambient temperature	Fire in an adjacent magazine, store or vehicle
BI	No response more severe than type V (burning)	From one to three 12.7 mm (armour piercing) round velocity between 400 and $850\,\mathrm{m\,s^{-1}}$	Small arms attack
Fragment impact (FI)	No response more severe than type V (burning)	Steel fragment from 15 g with velocity up to $2600\,\mathrm{m\,s^{-1}}$ and 65 g with velocity up to $2200\,\mathrm{m\,s^{-1}}$.	Fragmenting munitions attack
Shaped charge jet impact (SCJ)	No response more severe than type III (explosion)	Shaped charge calibre up to 85 mm	Shaped charge weapon attack
Sympathetic reaction SR	No propagation of reaction more severe than type III explosion	Detonation of donor in appropriate configuration	Most severe reaction of same ammunition in magazine, store aircraft or vehicle.

Response Types as Taken from AOP-39

Detonation (Type I)

This is the most violent type of munition reaction where the energetic material is consumed in a supersonic decomposition.

Primary evidence of a type I reaction is the observation or measurement of a shock wave with the magnitude and timescale of a purposely detonated calibration test or calculated value and the rapid plastic deformation of the metal casing contacting the energetic material with extensive high shear rate fragmentation.

Secondary evidence may include the perforation, fragmentation and/or plastic deformation of a witness plate and ground craters of a size corresponding to the amount of energetic material in the munition.

Partial Detonation (Type II)

This is the second most violent type of munition reaction where some of the energetic material is consumed in a supersonic decomposition.

Primary evidence of a type II reaction is the observation or measurement of a shock wave with magnitude less than that of a purposely detonated calibration test or calculated value and the rapid plastic deformation of some, but not all, of the metal casing contacting the energetic material with extensive high shear rate fragmentation.

Secondary evidence may include scattered burned or unburned energetic material; the perforation, fragmentation and/or plastic deformation of a witness plate and ground craters.

Explosion (Type III)

This is the third most violent type of munition reaction with sub-sonic decomposition of energetic material and extensive fragmentation.

Primary evidence of a type III reaction is the rapid combustion of some or all of the energetic material once the munition reaction starts and the extensive fracture of metal casings with no evidence of high shear deformation resulting in larger and fewer fragments than observed from the purposely detonated calibration tests.

Secondary evidence may include significant long-distance scattering of burning or unburned energetic material, witness plate damage, the observation or measurement of overpressure throughout the test arena with a peak magnitude significantly less than and significantly longer duration than that of a purposely detonated calibration test and ground craters.

Deflagration (Type IV)

This is the fourth most violent type of munition reaction with ignition and burning of confined energetic materials, which leads to a less violent pressure release.

Primary evidence of a type IV reaction is the combustion of some or all of the energetic material and the rupture of casings resulting in a few large pieces that might include enclosures and attachments. At least one piece (e.g. casing, packaging or energetic material) travels (or would have been capable of travelling) beyond 15 m and with an energy level greater than 20 J based on the distance versus mass relationships in Figure B-1 (*not reproduced here*). A reaction is also classified as type IV if there is no primary evidence of a more severe reaction and there is evidence of thrust capable of propelling the munition beyond 15 m.

Secondary evidence may include a longer reaction time than would be expected in a type III reaction; significant scattered burning or unburned energetic material, generally beyond 15 m and some evidence of pressure in the test arena that may vary in time or space.

Burn (Type V)

This is the fifth most violent type of munition reaction where the energetic material ignites and burns non-propulsively.

Primary evidence of a type V reaction is the low pressure burn of some or all of the energetic material. The casing may rupture resulting in a few large pieces that might include enclosures and attachments. No piece (e.g. casing, packaging or energetic material) travels (or would have been capable of travelling) beyond 15 m and with an energy level greater than 20 J based on the distance versus mass relationships shown in Figure B-1 (*not reproduced here*). There is no evidence of thrust capable of propelling the munition beyond 15 m. A small amount of burning or unburned energetic material relative to the total amount in the munition may be scattered, generally within 15 m but no more than 30 m.

Secondary evidence may include some evidence of insignificant pressure in the test arena and for a rocket motor a significantly longer reaction time than if initiated in its design mode.

No Reaction (Type VI)

This is the least violent type of munition response where any reaction is self-extinguished immediately upon the removal of the external stimulus.

Primary evidence of a type VI reaction is no reaction of the energetic material without a continued external stimulus; the recovery of all or most of the energetic material with no indication of sustained combustion and no fragmentation of the casing or packaging greater than from a comparable inert test item.

Secondary evidence–none.

19.6.2
Cookoff

Cookoff tests consider the response of an energetic material under various rates of heating and/or confinement. These tests aim at determining the response of a store filled with a certain energetic material exposed to an adjacent fire (slow heating) or to a direct fire (fast heating).

Experiments with MTV (61.1/33.9/5) obtained with the shock-gel process indicate that fast heating under heavy confinement (Figure 19.14) yields a benign reaction with full consumption of the material (Table 19.9).

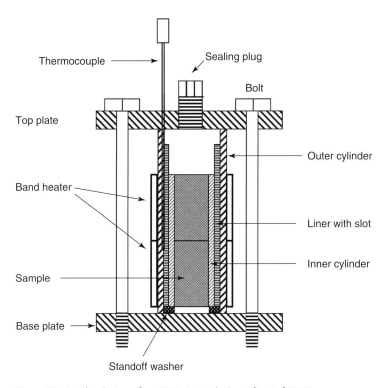

Figure 19.14 The design of an SSCB test vehicle. (After Ref. [37].)

Table 19.9 Response of MTV to slow and fast cookoff [38].

Heating rate	Temperature (°C)		Time (s)	Reaction
	Bomb	Pyrotechnic		
Slow (6.667 K min^{-1})	>531	513	>2610	No reaction[a]
Fast (37.5 K min^{-1})	527	471	744	Mild burning.

[a]The MTV cylinders had physically swollen by 1.2 mm.

SC of MTV payload in the RARDE SBCT test vehicle as shown in Figure 19.15 at a heating rate of 28.125 K min^{-1} effects mild burning with venting of the reaction products.

These results contrast a restricted report in which deflagration and partial detonation was observed when subjecting a real store to FC conditions (90 K min^{-1}) occurred [39]. Thus, it is obvious that the confinement may play a crucial role in determining the degree of response of an energetic material as well [40].

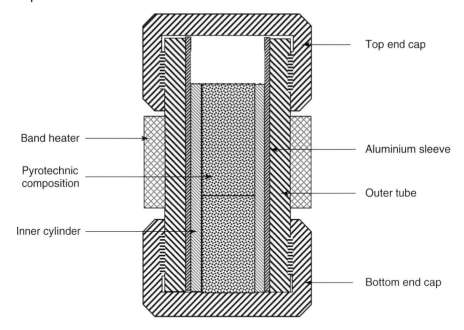

Band heater

Pyrotechnic composition

Inner cylinder

Top end cap

Aluminium sleeve

Outer tube

Bottom end cap

Figure 19.15 RARDE SBCT test vehicle.

Unspecified cookoff tests have been conducted with both packaged 218 Mk 2 and MJU 10B flares that resulted in a fire with an overall irradiation level below second burn threshold at 5 m distance [41].

19.6.3
Bullet Impact

Small arms attack on decoy flares in either logistic package or loaded in a dispenser is a very likely scenario for helicopters in urban theatres. Thus, it is of considerable relevance to test response of MTV against this type of threat.

Tests on 218 Mk 2 Decoy Flares (255 g MTV) packaged in M480 ammunition box (Figure 19.16) with 12.7 mm (armour piercing) AP round at 850 m s^{-1} yielded a deflagration event (not evaluated according to AOP 39) of the stores and ejection of flares. Six out of 18 ejected flares did not react [42]. In another test series, it was found that the projectile velocity correlates with the reaction type. Thus, velocities < 750 m s^{-1} are likely to produce a deflagration, whereas at higher velocities, an explosion (type III) with significant pressure impulse (Figure 19.17) is found to occur [41]. Furthermore, high-velocity impact can lead to projection of burning flares over a wide area (15 m radius), thus causing further hazard to nearby personnel and equipment.

For the same event, the irradiation or heat dose (J m^2) was measured at 5 m distance to the packaged store. These reactions result in second-degree burns to unprotected skin very rapidly (Figure 19.18).

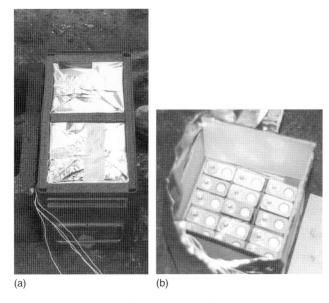

(a)   (b)

Figure 19.16 Setup of M480 ammunition filled with
2 × 218 flare boxes (b) containing 15 pieces each.

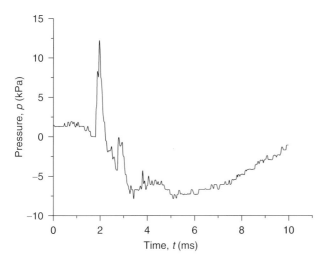

Figure 19.17 Pressure impulse at 10 m distance to 218 Mk2
store at 850 m s^{-1} impact velocity.

The correlation between impact velocity and heat dose for 218 Mk 2 flares is
listed in Table 19.10. The very high heat dose developed at low velocity was due to
the combustion of all 150 flares.

The explosion flame plume responsible for the strong radiation is shown in
Figure 19.19.

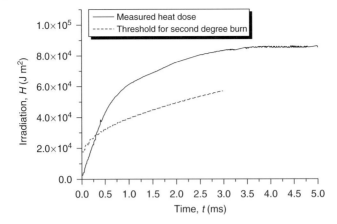

Figure 19.18 Irradiation at 5 m distance to 218 Mk 2 store at 850 m s^{-1} impact velocity.

Table 19.10 Response of 218 Mk 2 MTV flares to bullet impact [41, 43].

Configuration	Bullet velocity (m s^{-1})	Net mass of MTV (kg)	Blast overpressure at 10 m (kPa)	Irradiation at 5 m (kJ m^2)	Duration) (s)	Irradiation exceeded second-degree burn threshold	IM response
5 × A480boxes with 30 flares each	840	7.8	n.m.	110	4	Yes	III/IV[a]
Same	804	7.8	11	85	5	Yes	III/IV[a]
Same	459	7.8	1.8	350	9	Yes	IV/V[b]

[a] Live flares projected over 10 m radius.
[b] No flares projected.

Figure 19.19 Visual response of 218 Mk 2 store to bullet impact at 850 m s^{-1} velocity.

Table 19.11 Response of MJU 10B MTV flares to bullet impact [41, 43].

Configuration	Bullet velocity (m s^{-1})	Net mass of MTV (kg)	IM response
1 × H83 box with four flares each	808	4	Two flares penetrated by bullet burned. All flares ejected from box.
Same	721	4	Two flares burned, although only one was hit by bullet. Two others remained in box unburned.
Same	590	4	On flare ignited others remained in box without igniting.
Same with 40 mm plywood mitigation	569	4	Tumbling bullets struck two flares that ignited. Others thrown out without ignition.

The results obtained with 4 MJU 10B flares packed in H83 ammunition boxes are listed in Table 19.11. The reaction appears less violent as probed with the above 218 Mk 2 flares.

19.6.4
Sympathetic Reaction

Sympathetic reaction testing is described in STANAG 4396. A test with MTV containing stores 218 Mk 2 and MJU 10B has been reported to exclusively yield type V reaction. The general test setup for 218 Mk 2 flares is shown in Figure 19.20. The results are listed in Table 19.12.

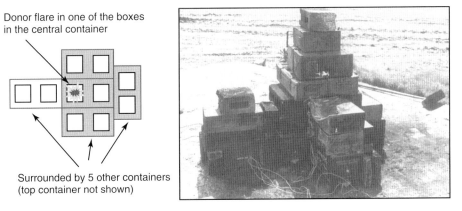

Donor flare in one of the boxes in the central container

Surrounded by 5 other containers (top container not shown)

Figure 19.20 Test setup for sympathetic reaction with 218 Mk 2 flares.

Table 19.12 Response of 218 Mk 2 and MJU/10B MTV
flares to sympathetic reaction [41, 43].

Configuration	Net mass of MTV (kg)	Blast over- 10 m/kPa	Irradiation at 5 m/kJ m²	Duration/ s	Irradiation exceeded second-degree burn threshold	IM response
1 × A480 boxes with 24 × 218 Mk 2 flares	6.2	n.m.	80	5	Yes	IV[a]
1 x A480 boxes with 27 × 218 Mk 2 flares + 2 MJU 10B flares	9	1.8	162	6	Yes	III/IV[a]

[a]Two large reactions occurred in the test. The first was due to the donor and sympathetic reaction of 14 acceptor flares in the same box. The second reaction that occurred 2 min, 52 s later was due to fast cookoff of 15 flares in the other box in the same container.

19.6.5
IM Signature Summary

MTV stores are sensitive to BI. Depending on the impact velocity, the reaction of the store may vary between explosion (type III) at high velocities and deflagration (type IV) at low velocities (Table 19.13).

The radiant intensity produced by the attacked flares is probably greater at low velocity as less scattering of the payload occurs, and a greater fireball is generated. The radiant intensity may cause more than second-degree burns to persons exposed in ≤5 m distance to an attacked store.

Table 19.13 IM signature of MTV containing stores.

Store	SCO[a]	FCO	BI	SR
218 Mk 2	V	V	III–IV	IV
MJU 10/B	V	V	IV	IV–V

[a]Estimated from the reaction reported in the SSCB test.

19.7
Hazards Posed by Loose In-Process MTV Crumb and TNT Equivalent

The accumulation of bulk amounts of the MTV composition in production poses a significant hazard. The reaction of loose MTV in a number of different configurations with respect to the amount of composition, type of container and type of ignition has been investigated [44–47].

Jenkins et al. investigated an MTV tracer composition

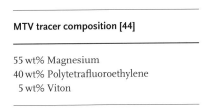

MTV tracer composition [44]

55 wt% Magnesium
40 wt% Polytetrafluoroethylene
 5 wt% Viton

1 kg MTV was initiated in a paper-mache pot (13 cm height, 13 cm diameter) by

- electric match head
- L2A1 detonator
- L2A1 detonator plus tetryl booster pellet.

Although both electric match head and single detonator inserted in the bulk composition would yield only a mild explosion, the use of the detonator with a booster yields a 'severe explosion', that is, according to [44]: air blast that ruptured foil gauges. The equivalent mass of TNT per unit mass of MTV is 0.23.

Backer et al. have carried out large-scale testing with both 5 and 10 kg MTV crumb of a similar composition with different PTFE/Viton proportions [45]. Tests showed that at solvent levels equal and greater than 10 wt% acetone both 5 and 10 kg MTV ignited with an electric quick match on top of the bulk in a 30 cm diameter, 30 cm height contained deliver a combustion reaction with a mass consumption rate inversely proportional to the amount of the solvent. At 5 wt% acetone, a sharp report is obtained. At 0 wt% acetone, according to the pressure probe, ignition initially yields a deflagration that after a few micro-seconds shocks up to a detonation as is seen by the steep pressure rise in Figure 19.21. However, transferring 10 kg dry MTV in a paper bag yields a deflagration reaction only with no steep pressure pulse.

Figure 19.22 shows the mass consumption rate as a function of the solvent level. The change in regime below 10 wt% acetone can be clearly seen.

As a consequence of the testing, the circular array of collector stainless steel cans (Figure 19.23a) after the central twin screw extrusion (TSE) mixing unit was substituted by a rack-holding Velostat® bags to avoid confinement (Figure 19.23b).

TNT equivalents were calculated with two established models CONWEP and DDESB Blast Effects Computer V. 5.0. The confinement heavily determines the equivalency. In the paper bag configuration, the TNT equivalent is as low as

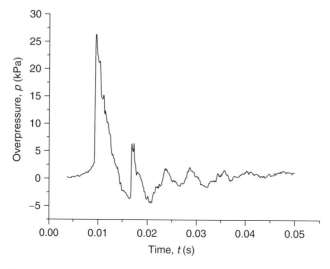

Figure 19.21 Time pressure curve for 10 kg MTV reaction in 30 × 30 cm² stainless steel bucket at as 9 m distance.

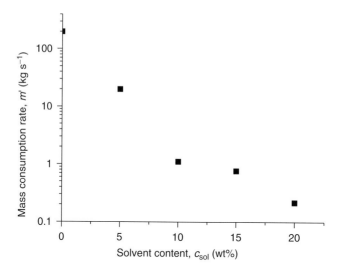

Figure 19.22 MTV mass consumption rate as a function of solvent level [45].

1.1–1.7% TNT (CONWEP) between 6 and 12 m distance, whereas semi-confined MTV in the stainless steel cans yield 20–33% TNT. A peculiarity of MTV detonation is the increasing equivalency with increasing distance, an effect observed with many non-ideal explosive showing high-temperature after-burn reactions.

Davies et al. conducted heat flux, heat dose and overpressure measurements on three different types of MTV on spherical charges weighing between 200 g and 36 kg [47].

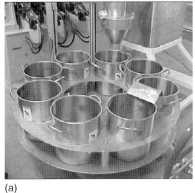

(a) (b)

Figure 19.23 (a) Stainless steel collecting cans yielding 25–30% TNT equivalent and (b) Velostat collecting bags yielding 1–2% TNT equivalent [45]. Reproduced with kind permission by International Pyrotechnics Society.

References

1. (a) (2009) *Recommendations on the Transport of Dangerous Goods, Manual of Tests and Criteria*, 5th revised edn, United Nations, Geneva, Switzerland, pp. 69–121, (b) pp. 129–141, (c) pp. 157–175.
2. NATO Standardization Agency (2010) Policy for Introduction and Assessment of Insensitive Munitions (IM), STANAG 4439 JAIS (EDITION 3), Brussels, Belgium.
3. Lissel, E. (1997) Untersuchungen pyrotechnischer Sätze im BICT. CCG Seminar Grundlagen der Pyrotechnik, Weil am Rhein, Deutschland, Mai 12–15.
4. NATO Standardization Agency (2005) Chemical Compatibility of Ammunition with Explosives (Non-Nuclear Applications), STANAG 4147 (EDITION 3), Brussels, Belgium.
5. Suceska, M. (1995) *Test Methods for Explosives*, Springer, Berlin.
6. Kuwahara, T., Matuso, S. and Shinozaki, N. (1997) Combustion and sensitivity characteristics of Mg/TF pyrolants. *Propellants Explos. Pyrotech.*, **22**, 198–202.
7. Aikman, L.M., Shook, T.E., Lehr, R.M., Robinson, E. and McIntyre, F. (1987) Improved mixing, granulation and drying of highly energetic pyromixtures. *Propellants Explos. Pyrotech.*, **12**, 17–25.
8. McIntyre, F.L. (1980) *A Compilation of Hazard and Test Data for Pyrotechnic Compositions*, Computer Sciences Corporation, NSTL Station, Mississippi, pp. 268–269.
9. Kuwahara, T. (1998) Combustion and sensitivity of Metal/Teflon (Mg/Ti/Teflon) pyrolant. *Kayaku Gaikkaishi*, **59**, 18–23.
10. Westerhausen, M. (2001) 100 Jahre nach Grignard: Wo steht die metallorganische Chemie der schweren Erdalkalimetalle heute? *Angew. Chem.*, **113**, 3063–3065.
11. de Yong, L., Gray, R. and Redman, L. (1991) Assessment of the behaviour of pyrotechnic compositions at high strain rate. 16th International Pyrotechnics Seminar, Jönköpping, Sweden, June 24–28, p. 229.
12. Lee, P.R. (1998) in *Explosive Effects and Applications* (eds J.A. Zukas and W.P. Walters), Springer, New York, pp. 259.
13. Armed Services Explosives Safety Board (1967) Additional Information Potential Incident. Report No. 38, Fluorolube–Aluminum Detonation Point, Washington, DC.
14. Gong, D., Zhang, B., Xue, Q. and Wang, H. (1990) Investigation of adhesion wear of filled polytetrafluoroethylene by ESCA, AES and XRD. *Wear*, **57**, 25–39.

15. Gong, D., Xue, Q. and Wang, H. (1991) ESCA study on tribochemical characteristics of filled PTFE. *Wear*, **148**, 161–169.

16. Bohn, M.A. (2008) *Heat Flow Calorimetry on Energetic Materials*, Fraunhofer IRB Verlag, Germany.

17. Lissel, E. (1998) Qualifikation pyrotechnischer Sätze für Munition der Bundeswehr. CCG Seminar Grundlagen der Pyrotechnik, Weil am Rhein, Deutschland, Mai 25–29.

18. Farnell, P.L. and Beardell, A.J. (1975) Ignitability of Energetic Compositions in the Presence of Flammable Liquids. Technical Report No. 4724, Picatinny Arsenal, Dover, NJ.

19. Havron, H.C. (1984) Electrostatic sensitivity of Magnesium-Teflon compositions. 21st DOD Explosive Safety Seminar, Houston, TX, August 28–30, pp. 495–500.

20. Diercks, V.B. (1986) Electrostatic sensitivity of consolidated Magnesium-Teflon compositions. *J. Hazard. Mater.*, **13**, 3–15.

21. Hearn, G.L. and Singh, S. (1987) Hazards from electrostatics in the manufacture of infra-red decoy flares. *Inst. Phys. Conf. Ser.*, **85** (Section 3), 229–234.

22. Hubble, B. Stevenson, B. and Benstin, M. (2004) Experimental evaluation of the static electricity characteristics associated with processing MTV. 31st International Pyrotechnics Seminar, Fort Collins, CO, July 11–16, p. 211.

23. Kent, R. and Rat, R. (1985) Static electricity phenomena in the manufacture and handling of solid propellants. *J. Electrostat.*, **17**, 299–312.

24. Gyure, M.F. and Beale, P.D. (1989) Dielectric breakdown of a random array of conducting cylinders. *Phys. Rev.*, **40 B**, 9533–9540.

25. Gallier, S. (2010) Numerical modelling of dielectric breakdown in solid propellant microstructures. *Int. J. Mult. Comp. Eng.*, **8**, 523–533.

26. Haq, I.U. and Chaudhri, M.M. (1989) Dielectric breakdown and ignition of MTV compositions. 14th International Pyrotechnics Seminar, Jersey, Channel Islands, September 18–22, p. 135.

27. Hall, A.R. (1980) Some Combustion Characteristics of a Magnesium/Fluorocarbon Pyrotechnic Composition. SR 886b. Technical Report 179, Propellants, Explosives and Rocket Motor Establishment, Westcott, Aylesbury, Bucks, declassified 09 March 2006.

28. Forichon-Chaumet, N. and Espagnacq, A. (1995) Étude du comportement diélectrique de la composition MTV. Europyro 95, 6e Congrès International de Pyrotechnie, Tours, France, Juin 5–9, p. 557.

29. Chaudhri, M.M., Al-Ramadhan, F.A. and Haq, I.U. (1993) *Dielectric Breakdown and its Influence on Ignition*, Cavendish Laboratory University of Cambridge.

30. Forichon-Chaumet, N. and Espagnacq, A. (1996) Étude du comportement diélectrique des materiaux energetiques. Journees d'etudes sur la sensibilite des composants 95, 6e Congrès International de Pyrotechnie, Tours, France, Juin 5–9, p. 557.

31. Poehlein, S.K., Shortridge, R.G. and Wilharm, C.K. (2001) A comparative study of ultrafine aluminium in pyrotechnic formulations. 28th International Pyrotechnics Seminar, Adelaide, Australia, November 4–9, p. 597.

32. Wilharm C.K. and Shortridge, R.G. (2002) Effect of binder loading and application om mixture homogeneity, ignition sensitivities, and burn times for pyrotechnic compositions with Alex®. 29th International Pyrotechnics Seminar, Westminster Colorado, July 14–19, p. 213.

33. *http://www.baconlinks.com/USS_Oriskany/ LifeMag/Fire34.htm* (accessed 1 September 2011).

34. *http://en.wikipedia.org/wiki/File:USS_ Oriskany_(CV-34)_on_fire,_26 October_1966.jpg* (accessed 1 September 2011).

35. Department of Defense (1982) Military Standard Hazard Assessment Tests for Navy Non-Nuclear Ordnance. DOD-STD-2105, (NAVY), Washington, DC.

36. NATO Standardization Agency (2010) Allied ordnance publication. 39 Guidance on the assessment and development of Insensitive Munitions (IM), Brussels, Belgium.

37. Parker, R. (1989). Establishment of a Super Small-Scale Cookoff Bomb (SSCB) Test Facility at MRL, MRL-TR-89-9, Maribyrnong, Australia.

38. DeYong L. and Redman, L. (1991) Response of pyrotechnic compositions fo fast and slow cookoff. 17th International Pyrotechnics Seminar, p. 169.

39. Seubert, J.G. and Wildridge, D.K. (1988) Insensitive Munitions Advanced Development Program. NSWC/CR/RDTR-4155, Crane Naval Weapons Support Center, Indiana.

40. DeYong L. and Redman, L.D. (1992) Cookoff Behaviour of Pyrotechnics. Report No. MRL-TR-91-44, Materials Research Laboratory, Australia.

41. Elphick, R. (2002) Meeting insensitive munitions requirements for pyrotechnics in the UK. Thesis. Cranfield University, UK.

42. Barnes, P., Shepperson, C., Cardell, A. and Queay, J. (1994) The vulnerability of pyrotechnic stores. 20th International Pyrotechnics Seminar, Colorado Springs, July 25–29, p. 71.

43. Wallace, I.G., Elphick, R. and Barnes, P. (2002) Insensitive munitions considerations for pyrotechnic devices. 30th DOD Safety Seminar, Atlanta, August 13–15.

44. Jenkins, J.M., Nicholls, R.F. and Peer, M. (1980) Explosive power of pyrotechnic compositions. 19th DOD Explosive Safety Seminar, Los Angeles, CA, September 9–11, p. 77.

45. Backer, S.L., Campbell, J.R., Stevenson, B.D. and Limberger, T.R. (2004) Hazard assessment testing of in-process infrared decoy composition. 31st DOD Explosive Safety Seminar, San Antonio, TX, August 24–26.

46. Backer, S.L., Campbell, J.R., Stevenson, B.D. and Limberger, T.R. (2004) Hazard assessment testing. 33rd International Pyrotechnics Seminar, Fort Collins, CO, July 16–21, p. 541.

47. Davies, N., Williams, M. and Dunne, L. (2011) Blast and heat flux characterisation of MTV compositions. EUROPYRO 2011, Reims, France, May 16–19, p. P8.

20
Toxic Combustion Products

Metal–fluorocarbon reactions yield fluorides and carbonaceous products as the main products, which can be potentially harmful to both humans and the environment. Thus, in the following sections, the formation and release of these materials is discussed.

20.1
MTV Flare Composition

The thermochemical modelling of pyrotechnic compositions has been used as an approach to assess the quality and amount of potential combustion products [1]. However, thermochemical equilibrium calculations fail to give an exact answer to the actual species involved in dynamic processes. Thus it is necessary to take into account reaction kinetics in order to obtain a realistic insight into Magnesium/Teflon/Viton (MTV) combustion [2]. Three typical MTV formulations are given in Table 20.1.

The initial anaerobic combustion products of these formulations at their corresponding temperature are given in Table 20.2. HF is seen to be a minor ingredient.

Its content does vary with either Viton or polytetrafluoroethylene (PTFE) content as is explained with its role as the sole hydrogen source. From highly resolved FTIR spectra of MTV combustion flames of flares having formulations similar to MTV-3, distinct HF emission bands are observed, confirming the predicted occurrence Weiser [3] (see also Figure 9.33).

Figure 20.1 depicts the calculated HF mole fraction for formulations using either 5 or 16 wt% level Viton.

With increasing amount of air mixed with the primary combustion products, the HF concentration decreases as is depicted for the three flare formulations in Figure 20.2. The investigation of HF formation on combustion of a payload similar to MTV-3 had been carried out in Ref. [4]. The combustion had been carried out in a combustion chamber of $3.6\,m^3$ volume lined with chlorinated polyvinylchloride (PVC). Fragments of either 12 and 25 g mass were burned and yielded between 1.41 and 3.88 mg HF per gram of MTV [4].

Metal-Fluorocarbon Based Energetic Materials, First Edition. Ernst-Christian Koch.
© 2012 Wiley-VCH Verlag GmbH & Co. KGaA. Published 2012 by Wiley-VCH Verlag GmbH & Co. KGaA.

Table 20.1 Compositions investigated in Ref. [2].

Component	Unit	MTV-1	MTV-2, SR 886	MTV-3
Magnesium	wt%	54	55	61
Polytetrafluoroethylene	wt%	30	40	34
Viton A	wt%	16	5	5
Application	–	US flare	UK flare	US igniter

Table 20.2 Initial state of combustion products for MTV/air reaction in Ref. [2].

Flame temperature (K)	MTV-1 2707 moles	MTV-2 2847 moles	MTV-3 2476 moles
$Mg_{(g)}$	0.3463	0.3436	0.4598
MgF	0.1354	0.1731	0.1143
$MgF_{2(g)}$	0.1147	0.1520	9.276×10^{-2}
$MgF_{2(l)}$	2.358×10^{-3}	$1,625 \times 10^{-2}$	5.551×10^{-2}
$MgF_{2(s)}$	2.343×10^{-2}	1.252×10^{-3}	8.081×10^{-3}
$C_{(s)}$	0.2910	0.2839	0.2389
HF	2.036×10^{-2}	2.748×10^{-2}	2.344×10^{-2}

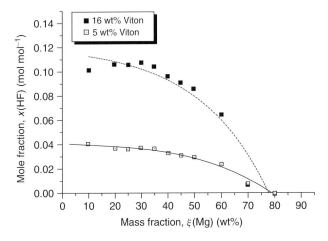

Figure 20.1 Calculated HF concentration as a function of Mg and Viton content. (After Ref. [2].)

The operational use of MTV composition in decoy flares occurs in tactical altitudes between 500 and 10 000 m. Thus, any combustion product released into the atmosphere will undergo sufficient dilution with air before reaching the ground.

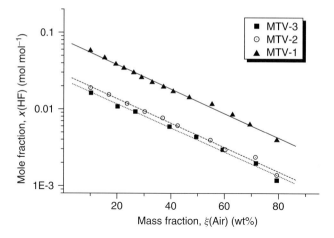

Figure 20.2 Development of HF concentration with increasing admixtures of air for three different MTV formulations. (After Ref. [2].)

However, testing MTV flares in a test tunnel or incineration of MTV composition in a facility with a low smoke stack may cause problems. The open burning of bulk MTV as a measure of waste treatment in flare manufacturing facilities has been the subject of an investigation [5]. Open burning is different from both flare testing and chamber incineration in that much greater masses are involved and delayed afterburn reactions take place. Figure 20.3 depicts a general scenario of a so-called open burn/open detonation (OBOD). The dispersion of OBOD inventory is mathematically described in Refs. [6, 7]. In a particular example, MTV has

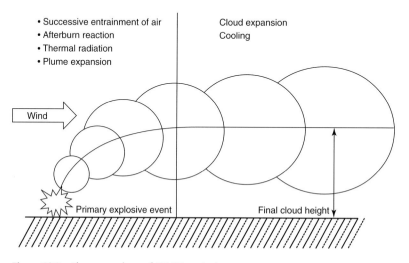

Figure 20.3 Phenomenology of OBOD emission.

been burned with diesel fuel, the composition and ratio of which is given in Table 20.3.

For waste burned at 55% RH, and 8 m s^{-1} wind speed, Figure 20.4 depicts the concentration profile along the downwind axis as a function of time. Early in the burn process, air entrainment into the plume causes NO and NO$_2$ formation and thus decrease of CO level. In addition, the high temperatures predicted are responsible for strong transient MgF formation, thus also affecting the HF level, which jumps right after undergoing a temperature increase.

With respect to MTV combustion, hydrogen fluoride emissions pose the greatest toxicological concern [8]. In addition, mass burning may also yield soot particles that are often contaminated with low-molecular-weight aromatic compounds, which can be either carcinogenic or irritant in nature. However, investigation of MTTP combustion residues despite the use of PVC binder gave no indications for PCDD or PCDF formation [9].

Table 20.3 Composition of MTV flare waste incineration with fuel.

Component	Amount (kg)
Magnesium	32.5
Polytetrafluoroethylene	18.9
Viton A	2.7
#2 diesel fuel	34.5

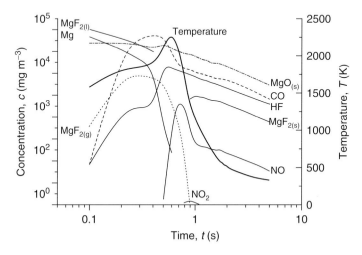

Figure 20.4 Predicted evolution of plume temperature and specie profiles downwind an open pit deflagration of MTV flare waste mixed with diesel fuel. (After Ref. [2].)

20.2
Obscurant Formulations

Fluorocarbon binders such as polyvinylidene fluoride (PVDF) have been proposed as oxidizers in obscurant formulations [10]. For a formulation similar as the one shown below, the content of polyaromatic compounds in the soot has been determined. Six hundred grams of the payload were burned, and the generated aerosol was collected from the combustion chamber on a filter with a volume flow rate of 2 $m^3 h^{-1}$.

IR-opaque obscurant [10]
20 wt% magnesium
27 wt% polyvinylidene fluoride
5 wt% chlorinated paraffin
48 wt% anthraquinone

Three PAH were detected, which according to the German MAW catalogue are considered carcinogenic A2: benzo[a]anthracene, benzo[a]pyrene and benzo[k] fluoranthene, that is, these substances have shown to be carcinogenic in animal testing and could pose similar risk to human (Table 20.4).

However, taking into account the singularity of the event firing a salvo of smoke rounds for vehicle protection and being exposed to the aerosol for a maximum of 140 s, a person incorporates 3.35 ng benzo[a]pyren, which equals the amount of daily uptake in an urban atmosphere.

Besides of PAH exposition, particle size of aerosol will determine whether it is transported into the alveoli. The above obscurant yields particles with 90% having

Table 20.4   Concentration of PAHs detected in IR obscurant aerosol.

Component	Concentration	Unit
Fluorene	0.95	$ng\,m^{-3}$
Phenanthrene	21.64	$\mu g\,m^{-3}$
Anthracene	27.84	$\mu g\,m^{-3}$
Fluoranthene	5.71	$\mu g\,m^{-3}$
Pyrene	2.74	$\mu g\,m^{-3}$
Benz[a]anthracene	50	$ng\,m^{-3}$
Chrysen	n.d.	–
Benzo[a]pyrene	29.05	$ng\,m^{-3}$
Benzo[b]fluoranthene	n.d.	–
Benzo[k]fluoranthene	28.55	$ng\,m^{-3}$
Benzo[g,h,i]perylene	n.d.	–
Indeno[1,3,3-c,d]pyrene	n.d.	–

Abbreviation: n.d., (not detected)

a particle diameter between 5 and 90 μm. Thus these particles remain in the upper respiratory tract and will be cleared naturally.

20.3
Fluorine Compounds

20.3.1
Hydrogen Fluoride

As hydrogen fluoride is lighter than air unlike other combustion products such as HCl or CO, it will rise in the air. Thus, even semi-dynamic flare firings will not lead to significant HF concentrations, provided the burn out of the payload is accomplished well above the ground level.

However, in enclosed space such as radiometric test tunnels, incomplete ventilation may result in concentration of HF. The effects of HF are well described in the literature [8] and are based on the necrosis of the living tissue by precipitation of calcium known, which can lead to systemic hypocalcaemia, which can have fatal effects.

The 5 min and 60 min LC_{50} values for HF in comparison to HCl are given in Tables 20.5 and 20.6 [11].

20.3.2
Aluminium Fluoride

AlF_3 is a possible reaction product from decoy flare compositions using MgAl alloy. Although the fluoroaluminate $Mg_2[AlF_5]$ is a likely main combustion product, there

Table 20.5 HF lethality compared to HCl in 5 min test [11].

Gas	Target animal	5 min LC_{50} (ppm)
HF	Rats	18 200
HCl	Rats	40 989
HF	Mice	6247
HCl	Mice	13 745

Table 20.6 HF lethality compared to HCl in 60 min test [11].

Gas	Target animal	60 min LC_{50} (ppm)
HF	Rats	1395
HCl	Rats	3124
HF	Mice	342
HCl	Mice	1108

Table 20.7 Safety data of HF, MgF$_2$, AlF$_3$.

	Unit	HF	MgF$_2$	AlF$_3$
CAS-No	–	7664-39-3	7783-40-6	7784-18-1
Density	kg m^3	0.901 (at 0.1 MPa)	3130	3197
State	–	Gaseous	Solid	Solid
Mp	°C	−83	1263	–
Hazard group	–	–	n.a.	–
Bp	°C	19	2262	1275
MAK	mg m^{-3}	0.83	1	1
R sentences	–	26/27/28-35	20, 22	22−36/37/38
S sentences	–	1/2/7/9/26/36/37/45	22−25	26
H sentences	–	330-310-300-314	–	302-315-319-335
P sentences	–	260-264-280-284-301 + 310-302 + 350	–	261-305+351+338
Solubility in water at 20 °C	G l^{-1}	Miscible with water	4.6×10^{-4}	5.6

is barely any information available on it. The biological effects of fluoroaluminate are numerous and are discussed in Ref. [12].

20.3.3
Magnesium Fluoride

Owing to its low solubility product, $pK_L = 10.3$ ($L = 7.32 \times 10^{-6}$ mol l^{-1}) [13], fluoride ion concentration is well below any physiological active level. Thus, effects from MgF$_2$ can be strictly considered to be due to its particulate behaviour.

Table 20.7 gives the summary of the safety data of the fluorides.

References

1. Howe, P.D., Dobson, S., Malcolm, H.M. and Griffiths, T.T. (2008) Environmental assessment of pyrotechnic combustion products. 35th International Pyrotechnics Seminar, July 13–18, Fort Collins, CO, p. 9.
2. Christo, F.C. (1999) Thermochemistry and Kinetics Models for Magnesium/Teflon/Viton Pyrotechnic Compositions, Weapons Systems Division Aeronautical and Maritime Research Laboratory, Melbourne, Australia.
3. Weiser, V. (2005) FTIR Spectroscopy.
4. Machatschek, A. (2003) Ermittlung der Fluorwasserstoff-Emissionen beim Abbrand der Wirkmasse eines Scheinziels Wehrwissenschaftliches Institut für Werk-, Explosiv- und Betriebsstoffe, Swisttal-Heimerzheim, Deutschland.
5. Beach, A.B., Crews, R.C., Harrison, D.D., Masonjones, M.C., Moussa, N.A. and Zhang, X.J. (1998) Atmospheric dispersion of reacting chemicals from energetic materials, 4th Conference on Life Cycles of Energetic Materials, March 29–April 1, Fullerton, CA, p. 332.

6. Bjorklund, J.R., Bowers, J.F., Dodd, G.C. and White, J.M. (1998) Open Burn/Open Detonation Dispersion Model (OBODM) Users Guide, Volume I. User's Instructions, West Desert Test Center, US Army Dugway Proving Ground, Dugway, UT.

7. Bjorklund, J.R., Bowers, J.F., Dodd, G.C. and White, J.M. (1998) Open Burn/Open Detonation Dispersion Model (OBODM) Users Guide, Volume II. Technical Description, West Desert Test Center, US Army Dugway Proving Ground, Dugway, UT.

8. Bertolini, J.C. (1992) Hydrofluoric acid – a review of toxicity. *J. Emerg. Med.*, **10**, 163–168.

9. Campbell, C. (2005) Twin screw Extruder Production of MTTP Decoy Flares, SERDP WP-1240, Thiokol Propulsion, Brigham, Utah.

10. Schneider, J. and Büsel, H. (1995) Composition generating an IR-opaque smoke, US Patent 5,389,308, Germany.

11. Speitel, L.C. (1995) Toxicity Assessment of Combustion Gases and Development of a Survival Model. Airport and Aircraft Safety Research and Development Division FAA Technical Center Atlantic City International Airport, NJ.

12. Strunecka, A., Patocka, J., Blaylock, R.L. and Chinoy, N.J. (2007) Fluoride interactions: from molecules to disease. *Curr. Signal Transduct. Ther.*, **2**, 190–213.

13. Lide, D.R. (ed.) (2005) *CRC Handbook of Chemistry and Physics*, CRC Press, Boca Raton, FL, pp. 8–131.

21
Outlook

Current trends in the development of energetic materials aim at improving both performance and safety by engineering materials on a nanometric scale. In this respect, the results obtained in the 1990s with vapour phase-deposited materials deem to be revisited. With the development of Enerfoil® thin layer Mg/PTFE (polytetrafluoroethylene) material, it was impressively demonstrated that it is possible to manufacture materials that are comparingly powerful and still more safe to handle than the classical Magnesium/Teflon/Viton (MTV) [1]. It was already emphasized that the missing commercial success then was primarily due to cost-driven decisions and not based on any failed performance or safety criteria. In view of the technological advances made with vapour phase deposition in the meantime, it is worth reconsidering this technology for good.

The use of nanometric Mg or Al in conventional blending processes, although a popular token, is not advised in view of the inherent sensitivity of these materials [2]. However, it is worth exploring nanometric and mesoporous fuels that are typically not very reactive towards perfluorocarbons, such as boron and silicon [3, 4]. In this context, grafting of fluoropolymers with either boron or silicon may come into play [5]. Aluminium particles bearing fluorocarboncarboxylate chains $(C_{13}F_{27}CO_2)^-$ (WARP) have been prepared but display very high sensitivity towards ESD and have shown accidental ignition on agitation in test tubes [6].

Composites of nickel and PTFE have been prepared by electroplating from solutions containing nickel salts and dispersed PTFE [7]. The same principle can be envisaged to prepare composites from more electropositive metals in non-aqueous solvents.

Potential oxidizers based on highly strained fluorocarbon compounds have been investigated by the author earlier [8, 9]. From this, the fluorofullerene $C_{60}F_{48}$ should be the most interesting compound with respect to both energy content and reactivity. The author has predicted stratified carbon compounds with alternating NF_2-substitutents as possible oxidizers in pyrolants [10].

Meller has investigated perfluoroalkylated borazine compounds [11–15]. Meller pointed out that because of the high energy content of some these compounds, they can undergo explosive decomposition (A. Meller, personal communication). Thus these materials are interesting starting molecules for both homogenous

Metal-Fluorocarbon Based Energetic Materials, First Edition. Ernst-Christian Koch.
© 2012 Wiley-VCH Verlag GmbH & Co. KGaA. Published 2012 by Wiley-VCH Verlag GmbH & Co. KGaA.

non-classical high-explosive materials and advanced metal/fluorocarbon based energetic materials.

High nitrogen compounds with perfluoroalkyl-substituents have been shown to be superior oxidizers in Mg-based pyrolants [16–18]. Most recently, Schaller demonstrated the applicability of fluorinated ionic liquids as possible oxidizers in Mg-based flare formulations [19]. Shreeve has synthesized ionic liquids with pentafluorosulfanyl moieties [20, 21] that are interesting oxidizers for liquid metal/fluorocarbon formulations.

The author proposed organometallic polymers constituted from electropositive metals and fluorinated carbon chains as homogeneous pyrolants [22].

Just before submission of this book manuscript, the author demonstrated the feasibility of using ytterbium, thulium, samarium and europium in flare formulations together with oxygeneous and fluorocarbon-based oxidizers [23]. Right before the editorial deadline Son *et al.* could show that energetic materials made from aluminium and piezolectric fluoropolymers based on polyvinylidene fluoride show impact sensitivity that can be influenced in magnitude by an electric current. This is an important finding as it could lead to say primers which are sensitive to stabbing when under current and insensitive when disconnected [24].

In summary, the above discussion shows that there are plenty of possibilities to further improve metal/fluorocarbon-based energetic materials. The author assumes that the reader will come up with new concepts to further develop this interesting field of energetic materials.

References

1. Chan, S.K. (1997) Characterization of enerfoil™ pyrotechnic film, terminal effects and energetic materials advisory committee TEEMAC. Open Workshop on Ignition, Shrivenham, UK, 4 November, p. 39.

2. Koch, E.-C. and Schaffner, D. (2009) Sensitivity of Nanoscale Energetic Materials II. Fuel, Pyrolants and Propellants, L-162, NATO Munitions Safety Information Analysis Center (MSIAC), Brussels, Belgium.

3. Koch, E.-C. (2006) Pyrotechnischer Satz, Verfahren zu dessen Herstellung und seine Verwendung, DE Patent 10 2005/003,579 B4, Germany.

4. Mason, A.B., Son, S.F., Cho, K.Y., Yetter, R.A. and Asay, B.W. (2009) Combustion performance of porous silicon-based energetic composites. 45th AIAA/ASME/SAE/ASEE Joint Propulsion Conference & Exhibit, Denver, CO, August 2–5, AIAA 2009-5081.

5. Dubois, C., Brousseau, P., Roy, C. and Lafleur, P. (2004) In-situ polymer grafting on ultrafine aluminum powder. 35th International Annual Conference of ICT, June 29–July 2, Karlsruhe, Germany, p. V–12.

6. Jouet, J.R. and Schuman, A.J. (2006) MIC/Al incidents at Indian head 1998–2005. MIC Safety Meeting, February 13, Los Alamos National Laboratory, USA.

7. Chong, Y.-B. (1995) in *Fluoride-Carbon Materials, Chemistry, Physics and Applications* (ed. T. Nakajima), Marcel Dekker, New York, p. 381.

8. Koch, E.-C. (2003) Theoretical considerations on the performance of various fluorocarbons as oxidizers in pyrolant systems. 34th International Annual Conference of ICT, June 24–27, Karlsruhe, Germany, p. V40.

9. Koch, E.-C. (2004) Metal/fluorocarbon pyrolants: V. Theoretical evaluation of the combustion performance of

metal/fluorocarbon pyrolants based on strained fluorocarbons. *Propellants Explos. Pyrotech.*, **29**, 9–18.

10. Koch, E.-C. (2005) Pyrotechnic composition containing poly-carbon-difluoroamine as oxidising agent, used in infrared decoy devices such as flares for defence against heat-seeking anti-aircraft missiles. DE Application 102004/018,861, Germany.

11. Meller, A., Schlegel, R. and Gutmann, V. (1964) B-Tetra fluoropropoxy- und unsymmetrische propoxy-borazinderivate. *Monatsh. Chem.*, **95**, 1564–1572.

12. Meller, A., Wechsberg, M. and Gutmann, V. (1965) Fluoroalkyl- und Fluorarylborazinderivate, 1. *Mitt.*, *Monatsh. Chem.*, **96**, 388–395.

13. Meller, A., Gutmann, V. and Wechsberg, M. (1965) 1,3,5-Tris(Pentafluorophenyl)-2,4,6-trichloroborazine. *Inorg. Nucl. Chem. Lett.*, **1**, 79–81.

14. Meller, A., Wechsberg, M. and Gutmann, V. (1966) N-Pentafluorphenylborazinderivate und einige neue perfluorarylsubstituierte Borazine. *Monatsh. Chem.*, **97**, 619–632.

15. Meller, A., Wechsberg, M. and Gutmann, V. (1966) Fluoralkyl- und Fluorarylborazinderivate. *Monatsh. Chem.*, **97**, 1163–1176.

16. Crawford, M.-J., Hahma, A., Klapötke, T.M., Radies, H. and Koch, E.-C. (2009) Synthesis and characterization of perfluorinated nitriles and the corresponding 5-perfluoroalkyltetrazolate salts, 19th Winter Fluorine Conference, St. Petersburg, FL, January 11–16.

17. Koch, E.-C., Hahma, A., Klapötke, T.M. and Radies, H. (2010) Metal-fluorocarbon pyrolants: XI. Radiometric performance of pyrolants based on magnesium, perfkluorinated

tetrazolates and viton A. *Propellants Explos. Pyrotech.*, **35**, 248–253.

18. Koch, E.-C., Klapötke, T.M., Radies, H., Lux, K. and Hahma, A. (2011) Metal-fluorocarbon pyrolants. XII: calcium salts of 5-perfluoroalkylated tetrazoles – synthesis, characterization and performance evaluation as oxidizers in ternary mixtures with magnesium and viton™. *Z. Naturforsch.*, **66b**, 378–386.

19. Schaller, U. (2011) Triazolium based energetic ionic liquids. 8th Workshop on Pyrotechnic Combustion, Reims, France, May 14.

20. Gao, H., Ye, C., Winter, R.W., Gard, G.L., Sitzmann, M.E. and Shreeve, J.M. (2006) Pentafluorosulfanyl (SF_5)-containing energetic salts. *Eur. J. Inorg. Chem.*, 3221–3226.

21. Garg, S. and Shreeve, J.M. (2011) Trifluoromethyl- or pentafluorosulfanyl-substituted poly-1,2,3-triazole compounds as dense stable energetic materials. *J. Mater. Chem.*, **21**, 4787–4795.

22. Koch, E.-C. (2010), Energy-producing material, US Patent 7,678,209, Germany.

23. Koch, E.-C., Weiser, V., Roth, E. and Kelzenberg, S. (2011) Consideration of 4f – metals as new flare fuels – thermochemical and combustion properties of europium, samarium, thulium and ytterbium metal and their pyrolants, 42nd International Annual ICT Conference, Karlsruhe, Germany, June 28–July 1, p. V–1.

24. Janesheski, R.S., Groven, L.J. and Son, S.F. (2011) Fluoropolymer and aluminum piezoelectric reactives, 17th Biennial International Conference of the APS Topical Group on Shock Compression of Condensed Matter, 26 June–1 July, Chicago, IL.

Index

Metal-Fluorocarbon Based Energetic Materials, First Edition. Ernst-Christian Koch.
© 2012 Wiley-VCH Verlag GmbH & Co. KGaA. Published 2012 by Wiley-VCH Verlag GmbH & Co. KGaA.